W0254596

RNDr. Miroslav Krbec, CSc.
Born 1950 in Pelhřimov. Research fellow in the Department of Methods of Mathematical Physics, Mathematical Institute of the Czechoslovak Academy of Sciences, Prague.
Fields of research: Theory of function spaces and integral operators linked with PDE's.

Prof. RNDr. Alois Kufner, DrSc.
Born 1934 in Plzeň. Corresponding member of the Czechoslovak Academy of Sciences. Research fellow of the Mathematical Institute Czechoslovak Academy of Sciences, and Professor of Mathematics of the Technical University, Plzeň.
Fields of research: Function spaces,PDE's.

RNDr. Bohumír Opic, CSc.
Born 1948 in Libunec. Research fellow in the Department of Methods of Mathematical Physics, Mathematical Institute of the Czechoslovak Academy of Sciences, Prague.
Fields of research: Function spaces and their applications to PDE's.

RNDr. Jiří Rákosník, CSc.
Born 1950 in Varnsdorf. Research fellow in the Department of Methods of Mathematical Physics, Mathematical Institute of the Czechoslovak Academy of Sciences, Prague.
Fields of research: Function spaces, integral operators and application to PDE's.

Nonlinear Analysis, Function Spaces and Applications, Vol. 4
<4, 1990, Roudnice nad Labem>:
Proceedings of the Spring School held in Roudnice nad Labem, 1990;
ed. by Miroslav Krbec; Alois Kufner; Bohumír Opic; Jiří Rákosník. -
1. Aufl. - Leipzig : B. G. Teubner Verlagsges., 1990. - 256 S.
(Teubner-Texte zur Mathematik ; 119)
NE: Krbec, Miroslav [Edit.] ; GT

ISBN 978-3-663-01273-3 ISBN 978-3-663-01272-6 (eBook)
DOI 10.1007/978-3-663-01272-6
ISSN 0138-502X

1. Auflage
VLN 294-375/65/90
Lektor: Dipl.-Math. Erika Roth

Gesamtherstellung: Typodruck Döbeln, Bereich Leisnig

TEUBNER-TEXTE zur Mathematik · Band 119

Nonlinear Analysis, Function Spaces and Applications Vol. 4

Proceedings of the Spring School held in Roudnice nad Labem 1990

Editors:
Miroslav Krbec, Alois Kufner, Bohumír Opic, Jiří Rákosník

This book represents a free continuation of the Proceedings published by the Teubner Publishing House in 1979, 1982 and 1986 (TEUBNER-TEXTE zur Mathematik, Vols. 19, 49 and 93). It contains lectures delivered at the above mentioned Spring School dealing with the following topics: Abstract theory of weighted function spaces, classical and weighted norm inequalities, properties of special function spaces, nonlinear scattering theory, theory of monotone operators with applications and elliptic boundary value problems.

Dieses Buch ist eine freie Fortsetzung der 1979, 1982 und 1986 im Teubner-Verlag erschienenen Sammelbände (Bände 19, 49 und 93 der TEUBNER-TEXTE zur Mathematik). Es enthält Vorträge, die an der oben erwähnten Frühlingsschule gehalten wurden und die folgenden Themenkreise betreffen: Abstrakte Theorie der gewichteten Funktionenräume, klassische und gewichtete Ungleichungen, Eigenschaften spezieller Funktionenräume, nichtlineare Streuungstheorie, Theorie monotoner Operatoren mit Anwendungen und elliptische Randwertaufgaben.

Ce livre représente une suite libre aux comptes rendus publiés chez Teubner en 1979, 1982 et en 1986 (voir TEUBNER-TEXTE zur Mathematik, vol. 19, 49 et 93). Il contient le texte des conférences traitant des sujets suivantes: la théorie abstraite des fonctions des espaces poids, les inégalités classiques et avec poids, les proprietés des espaces spéciaux des fonctions, la théorie nonlinéaire de dispersion, la théorie des opérateurs monotones avec applications et les problemes aux limites elliptiques.

Эта книга представляет собой вольное продолжение сборников изданных издательством Тейбнер в 1979, 1982 и 1986 гг. (TEUBNER-TEXTE zur Mathematik, тома 19, 49 и 93). Она содержит лекции прочитанные на выше упомянутой Весенней школе и посвященные следующим вопросам : Абстрактная теория весовых пространств функций, классические и весовые неравенства, свойства специальных пространств функций, нелинейная теория рассеяния, теория монотонных операторов с применениями и эллиптические краевые задачи.

PREFACE

These Proceedings contain the lectures presented at the Spring School

"Nonlinear Analysis, Function Spaces and Applications IV"

held in May 21 - 25, 1990, in Roudnice nad Labem near Prague and organized by the Mathematical Institute of the Czechoslovak Academy of Sciences.

The School has continued the preceding three held under the same name in Horní Bradlo, 1978, in Písek, 1982, and in Litomyšl, 1986, the Proceedings of which were published by the TEUBNER Publishing House, Leipzig, in 1979, 1982, and 1986 as volumes 19, 49 and 93, respectively, of the series "Teubner-Texte zur Mathematik".

Analogously as in the foregoing Spring Schools, the organizers concentrated on survey lectures presented by a smaller number of speakers. Moreover, a highly attended poster session was organized and there was enough possibility for discussions and exchange of ideas.

The Spring School took place on the bank of the Elbe river in the pleasant setting of the old Czech town *Roudnice nad Labem*; excursions to local memorabilities (the monastery, the picture gallery, the legendary hill *Říp*) formed an integral part of the School.

A list of participants can be found at the end of the volume, the total number being 84 mathematicians and 3 accompanying persons from the following countries: Argentina (1), Australia (2), Belgium (1), Canada (3), Czechoslovakia (25), Federal Republic of Germany (2), Finland (4), German Democratic Republic (9), Hungary (3), Japan (1), Poland (5), Spain (7), Sweden (6), United Kingdom (1), USA (2), USSR (13), Vietnam (1) and Yugoslavia (1).

Two other mathematicians from Prague participated in organization of the Spring School and preparation of the Proceedings, namely, Dr. P. Gurka, and Dr. L. Pick. Also our thanks go to Mrs. R. Pachtová for excellent typewriting and to Mrs. H. Hřebcová for her help in Prague during the School. The editors wish to thank to Mrs. E. Roth from the Teubner Publishing House for her efficient collaboration leading to a prompt publication of these Proceedings.

Prague, June 1990 The Editors

CONTENTS

FACTORIZATION OF OPERATORS AND WEIGHTED NORM INEQUALITIES

José García-Cuerva
Madrid, Spain

The purpose of these lectures is to show how the theory of factorization of operators developed by B. Maurey in the 1970's can be applied to obtain very interesting results about weighted norm inequalities. The idea to carry out this program is due to José Luis Rubio de Francia. He constructed a beautiful theory, which culminates in the extrapolation theorem. This theory is presented in chapter VI of our book [8] in the context of L^p spaces. Here we have chosen to work in a more general class of Banach function spaces, an approach that José Luis Rubio also adopted in some later works [20], [21], [22]. There are two reasons to do this. First of all, the presentation of the main results becomes much clearer, and besides there are very nice applications to Banach lattices to be discussed in Section 5. There are several approaches to extrapolation, giving rise to different results. We have chosen the original approach of José Luis Rubio de Francia, but we have completed the theory so that all the known results become part of it.

§ 1. Banach function spaces

Let $(\Sigma, d\sigma)$ be a complete σ-finite measure space. We shall denote by $\mathcal{M}$ the collection of all extended real-valued measurable functions on Σ and by $\mathcal{M}^+$ the subcollection of $\mathcal{M}$ consisting of those functions whose values lie in $[0,\infty]$.

Definition 1.1. A mapping $\rho : \mathcal{M}^+ \to [0,\infty]$ is called *a function norm* if, for all f, g, f_n $(n = 1,2,3,\dots)$ in $\mathcal{M}^+$, for all constants $a \geq 0$ and for all measurable subsets E of Σ, the following properties hold:

1) $\rho(f) = 0 \Leftrightarrow f = 0$ a.e.; $\rho(af) = a\rho(f)$ and $\rho(f+g) \leq \rho(f) + \rho(g)$

2) $0 \leq g \leq f$ a.e. $\Rightarrow \rho(g) \leq \rho(f)$

3) $0 \leq f_n \uparrow f$ a.e. $\Rightarrow \rho(f_n) \uparrow \rho(f)$

4) $|E| < \infty \Rightarrow \rho(\chi_E) < \infty$

5) $$|E| < \infty \rightarrow \int_E f \, d\sigma \le C_E \rho(t)$$

for some constant C_E, $0 < C_E < \infty$, depending on E and ρ but independent of f.

If ρ is a function norm, the collection $X = X(\rho)$ of all functions f in $\mathcal{M}$ for which $\rho(|f|) < \infty$ is called *a BANACH FUNCTION SPACE*. For each $f \in X$, we define

$$\|f\|_X = \rho(|f|) .$$

The following result is easy to establish (see [1]):

<u>Theorem 1.2.</u> *Let* ρ *be a function norm, and consider* $X = X(\rho)$. *Then* $(X, \| \|_X)$ *is a Banach space and the following properties hold for all* f, g, f_n $(n = 1, 2,\ldots)$ *in* $\mathcal{M}$ *and all measurable subsets* E *in* Σ

a) *(lattice property) If* $|g| \le |f|$ *a.e. and* $f \in X$, *then* $g \in X$ *and* $\|g\|_X \le \|f\|_X$

b) *(Fatou property) Suppose* $f_n \in X$, $f_n \ge 0$ $(n = 1,2,\ldots)$ *and* $f_n \uparrow f$ *a.e. If* $f \in X$, *then* $\|f_n\|_X \uparrow \|f\|_X$ *whereas if* $f \notin X$, *then* $\|f_n\|_X \uparrow \infty$.

c) *(Fatou's lemma) If* $f_n \in X$ $(n = 1,2,\ldots)$, $f_n \to f$ *a.e. and* $\liminf_{n\to\infty} \|f_n\|_X < \infty$, *then* $f \in X$ *and* $\|f\|_X \le \liminf_{n\to\infty} \|f_n\|_X$.

d) *Every simple function belongs to* X.

e) *To each set* E *of finite measure there corresponds a constant* C_E, $0 < C_E < \infty$, *such that*

$$\int_E |f| \, d\sigma \le C_E \|f\|_X \quad \textit{for all} \quad f \in X .$$

f) *If* $f_n \to f$ *in* X, *then* $f_n \to f$ *in measure on every set of finite measure; in particular, some subsequence of* f_n *converges to* f *a.e.*

In view of theorem 1.2, we shall use the names BANACH FUNCTION SPACE or BANACH LATTICE interchangeably. Here are some examples of Banach lattices:

1) The Lebesgue spaces $L^p = L^p(d\sigma)$, $1 \le p \le \infty$, and the weighted Lebesgue spaces $L^p(v) = L^p(v d\sigma)$, $1 \le p \le \infty$, given for a weight function $v \ge 0$ by:

$$L^p(v) = \{f \in L^0 : \|f\|_{L^p(v)} = \left(\int_\Sigma |f(\sigma)|^p \, v(\sigma) \, d\sigma \right)^{1/p} < \infty\}$$

where $L^0 = \{f \in \mathcal{M} : |f(\sigma)| < \infty \text{ a.e.}\}$.

2) The Lorentz spaces $L(p,q)$, $1 \le p,q \le \infty$, with the exception of $L(1,\infty)$ which is not even a normed space. These are defined by:

$$L(p,q) = \{f \in L^0 : \|f\|_{L(p,q)} = \left((q/p)\int_0^\infty \left(t^{1/p} f^*(t)\right)^q dt/t\right)^{1/q} < \infty\}$$

where $f^*(t) = \inf\{s > 0 : |\{|f| > s\}| \le t\}$ is the non-increasing rearrangement of f.

3) The Orlicz spaces $\Phi(L)$ where Φ is convex, strictly increasing in $[0,\infty)$ and $\Phi(0) = 0$,

$$\Phi(L) = \{f \in L^0 : \int_\Sigma \Phi(|f(\sigma)|/a)\, d\sigma < \infty \quad \text{for some} \quad a > 0\}$$

with $\|f\|_{\Phi(L)} = \inf\{a > 0 : \int_\Sigma \Phi(|f(\sigma)|/a)\, d\sigma \le 1\}$.

4) The mixed-norm spaces $L^{p_1,p_2}(\Sigma,d\sigma)$ if $\Sigma = \Sigma_1 \times \Sigma_2$ and $d\sigma = d\sigma_1 \otimes d\sigma_2$ $1 \le p_1,p_2 \le \infty$, defined by the condition:

$$\|f\|_{L^{p_1,p_2}} = \left(\int_{\Sigma_2}\left(\int_{\Sigma_1} |f(\sigma_1,\sigma_2)|^{p_1}\, d\sigma_1\right)^{p_2/p_1} d\sigma_2\right)^{1/p_2} < \infty .$$

Given a Banach lattice $X = X(\rho)$ of functions on $(\Sigma,d\sigma)$, its dual space X^* can not always be identified with a Banach lattice of functions on $(\Sigma,d\sigma)$. This leads us to consider the associate space, which we now define. First of all we consider the associate norm ρ' defined by

$$\rho'(g) = \sup\{\int_\Sigma f(\sigma)\, g(\sigma)\, d\sigma : f \in \mathcal{M}^+, \ \rho(f) \le 1\} .$$

It is easy to see that ρ' is also a function norm, and it makes sense to give the following

Definition 1.3. Given a Banach lattice $X = X(\rho)$, the Banach lattice $X(\rho')$ determined by ρ' will be called *the associate space* of X and it will be denoted by X'.

The main properties of the correspondence $X \to X'$ are collected in the following result, whose proof can be seen in [1]:

Theorem 1.4. a) $X' = \{g \in L^0 : fg \in L^1$ *for all* $f \in X\}$ *and*

$$\|g\|_{X'} = \sup\{\left|\int_\Sigma f\, g\, d\sigma\right| : f \in X, \ \|f\|_X \le 1\} .$$

b) *Every Banach lattice* X *coincides with its second associate space* X''. *In other words, a function* f *belongs to* X *if and only if it belongs to* X'', *and in that case* $\|f\|_X = \|f\|_{X''}$.

c) X' *is (canonically isometrically isomorphic to) a closed norming subspace of* X^*. *Norming means that*

$$\|f\|_X = \sup \left\{ \left| \int_\Sigma f\, g\, d\sigma \right| : g \in X', \ \|g\|_{X'} \le 1 \right\}$$

for all $f \in X$.

Also in [1] we find this nice characterization of the Banach lattices X for which $X' = X^*$.

<u>Theorem 1.5.</u> *The Banach space dual* X^* *is (canonically isometrically isomorphic to) the associate space* X' *if and only if every* $f \in X$ *satisfies the following property:*

(1.6) *for every* $\varepsilon > 0$ *there is* $\delta > 0$ *such that* $|E| < \delta$ *implies* $\|f\chi_E\|_X < \varepsilon$.

Property (1.6) is referred to by saying that the function f has absolutely continuous norm. When this happens for every $f \in X$, we say that X has absolutely continuous norm. Theorem 1.5 can be rephrased by saying: $X' = X^* \Leftrightarrow X$ has absolutely continuous norm. For example, all the L^p spaces have absolutely continuous norm except L^∞ , and we have: $(L^\infty)' = L^1 \subsetneq (L^\infty)^*$.

The absolute continuity of the norm gives us a version of the dominated convergence theorem.

<u>Proposition 1.7.</u> $f \in X$ *has absolutely continuous norm if and only if the following holds: whenever* f_n $(n = 1,2,\ldots)$ *and* g *are measurable functions satisfying* $|f_n| \le |f|$ *for all* n *and* $f_n \to g$ *a.e., then* $\|f_n - g\|_X \to 0$.

As a consequence of theorem 1.5, we have the following nice characterization of the reflexive Banach lattices:

<u>Theorem 1.8.</u> *A Banach lattice* X *is reflexive if and only if both* X *and its associate space* X' *have absolutely continuous norm.*

<u>Proof.</u> If X and X' have absolutely continuous norm, then successive applications of theorem 1.5 give: $X^{**} = (X^*)^* = (X')^* = (X')' = X'' = X$. Since all the identifications are the canonical ones, we conclude that X is reflexive. Suppose, conversely that X is reflexive . Recall (theorem 1.4 c)) that X' is a closed norming subspace of X^* . If X' were a proper subspace of X^* , by the Hahn-Banach theorem, there would exist a nonzero functional $\Lambda \in X^{**}$ such that $\Lambda(X') = 0$. The reflexivity of X allows us to represent Λ as

$$\Lambda(f) = \int_\Sigma f\, g\, d\sigma$$

for some $g \in X$ and all $f \in X'$.

But $\Lambda(f) = 0$ for all $f \in X'$. Since X' is norming, this implies $g = 0$ a.e. But then $\Lambda = 0$, which is a contradiction. Thus $X' = X^*$ and, according to theorem 1.5, X has absolutely continuous norm. From this, and the fact that X is reflexive, we get $(X')^* = (X^*)^* = X = X'' = (X')'$. Applying once more theorem 1.5, we get that X' also has absolutely continuous norm. □

For X a Banach lattice we shall use the notation

$$X_+ = \{x \in X : x(\sigma) > 0 \ \text{a.e.}\}.$$

Definition 1.9. For X a Banach lattice of measurable functions on $(\Sigma, d\sigma)$ and $a > 0$ we shall consider $X^a = \{y \in L^0 : |y| = x^a$ for some $x \in X\}$ and for $y \in X^a$ we shall define $\|y\|_{X^a} = \||y|^{1/a}\|_X^a$.

Proposition 1.10. *For* $0 < a < 1$, $X^a \cdot X^{1-a} = X$ *and* $\|x\cdot y\|_X \le \|x\|_{X^a} \|y\|_{X^{1-a}}$.

Proof. We may assume $\|x\|_{X^a} = 1 = \|y\|_{X^{1-a}}$. Then

$$|x\cdot y| = (|x|^{1/a})^a (|y|^{1/(1-a)})^{1-a} \le a|x|^{1/a} + (1-a)|y|^{1/(1-a)}$$

and, consequently,

$$\|x\cdot y\|_X \le a\||x|^{1/a}\|_X + (1-a)\||y|^{1/(1-a)}\|_X = 1$$

□

Proposition 1.11. *If* $0 < a \le 1$, $\|\ \|_{X^a}$ *is a norm and* X^a *is a Banach lattice. If* $a > 1$, $\|\ \|_{X^a}$ *is, in general, only a* $(1/a)$*-norm.*

Proof. If $a > 1$, we have

$$|x+y|^{1/a} \le |x|^{1/a} + |y|^{1/a} \quad \text{and this implies}$$

$$\|x+y\|_{X^a}^{1/a} = \||x+y|^{1/a}\|_X \le \||x|^{1/a}\|_X + \||y|^{1/a}\|_X = \|x\|_{X^a}^{1/a} + \|y\|_{X^a}^{1/a}.$$

If $0 < a < 1$, we have

$$\||x+y|^{1/a}\|_X = \||x+y|\ |x+y|^{(1/a)-1}\|_X \le$$

$$\le \||x|\ |x+y|^{(1-a)/a}\|_X + \||y|\ |x+y|^{(1-a)/a}\|_X.$$

Now we use proposition 1.10 to conclude that

$$\||x+y|^{1/a}\|_X \le (\|x\|_{X^a} + \|y\|_{X^a})\||x+y|^{1/a}\|_X^{1-a},$$

i.e. $$\|x+y\|_{X^a} \le \|x\|_{X^a} + \|y\|_{X^a}.$$

□

For some Banach lattices X, X^a is still a Banach lattice for some $a > 1$. For example if $X = L^p$, $p > 1$, then $X^a = L^{p/a}$ and this is a Banach lattice for $a \le p$ and not only for $a \le 1$. This fact characterizes the p-convex Banach lattices to be defined below.

Note that for $X = L^1$, proposition 1.10 is simply Hölder's inequality.

Definition 1.12. Let $1 \le p,q \le \infty$. The Banach lattice X is said to be

a) *p-convex* if

$$\left\| \left(\sum_{j=1}^{n} |x_j|^p \right)^{1/p} \right\|_X \le M \left(\sum_{j=1}^{n} \|x_j\|_X^p \right)^{1/p}$$

(or $\| \sup_{1\le j\le n} |x_j| \|_X \le M \sup_{1\le j\le n} \|x_j\|_X$ if $p = \infty$)

b) *q-concave* if

$$\left(\sum_{j=1}^{n} \|x_j\|_X^q \right)^{1/q} \le M \left\| \left(\sum_{j=1}^{n} |x_j|^q \right)^{1/q} \right\|_X$$

(or $\sup_{1\le j\le n} \|x_j\|_X \le M \| \sup_{1\le j\le n} |x_j| \|_X$ if $q = \infty$).

The main properties we shall need concerning these notions are collected in the following statement whose proof can be seen in [19].

Proposition 1.13.

a) *Every Banach lattice is* 1*-convex and* ∞*-concave.*

b) *If* X *is* p_0*-convex and* q_0*-concave, then it is also* p*-convex for every* $1 \le p \le p_0$ *and* q*-concave for every* $q_0 \le q \le \infty$.

c) *If* X *is* p*-convex and* q*-concave, an equivalent norm can be defined so that inequalities* a) *and* b) *in definition* 1.12 *hold with* $M = 1$.

d) X *is* p*-convex if and only if* X^p *is a Banach lattice, with* X *renormed according to* c).

e) X *is* p*-convex (resp.* q*-concave) if and only if* X' *is* p'*-concave (resp.* q'*-convex) where* p' *is the exponent conjugate to* p *given by* $\frac{1}{p} + \frac{1}{p'} = 1$.

Definition 1.14. a) We say that an operator $T : E \to Y$ from the vector space E to the Banach lattice Y , is *sublinear* if it satisfies the following two conditions:

1) $|T(af)| = |a| \, |Tf|$, a.e., $a \in \mathbb{R}$, $f \in E$

2) $|T(f + g)| \le |Tf| + |Tg|$ a.e., $f, g \in E$.

b) We say that $T : E \to Y$ is *linearizable* if there exists $S : E \to Y(B)$ linear such that $Tf(\sigma) = \|Sf(\sigma)\|_B$ for a.e. $\sigma \in \Sigma$, where B is a certain

Banach space and

$$Y(B) = \{y : \Sigma \to B \ \text{ s.t. } \ \|y(\cdot)\|_B \in Y\} .$$

c) If $T : X \to Y$ is sublinear and X and Y are both Banach lattices, we say that T is *positive* if $|f| \le g$ a.e. implies $|Tf| \le Tg$ a.e.

<u>Observation 1.15.</u> Note that a linearizable operator is sublinear. But it is also non-negative in the sense that $Tf \ge 0$ for every f. Every sublinear operator satisfies the condition

$$\big| |Tf| - |Tg| \big| \le |T(f - g)| \quad \text{a.e.}$$

However, if T is non-negative sublinear, this condition becomes

$$(1.16) \qquad |Tf - Tg| \le |T(f - g)| \quad \text{a.e.}$$

This condition also holds for linear operators. All the sublinear operators to be considered will be either linear or non-negative sublinear. This justifies the convention which we shall adopt, of calling T sublinear if and only if (1.16) holds for every f, g. Accordingly when we have $T : E \to B$ and B is simply a Banach space, we shall say that T is sublinear if and only if

$$(1.17) \qquad \|Tf - Tg\| \le \|T(f - g)\| .$$

Observe that, with this restricted meaning, a sublinear operator $T : A \to B$ between two normed spaces is continuous if and only if it is continuous at 0, and this happens if and only if T is bounded, in the sense that $\|Ta\| \le C\|a\|$ for every $a \in A$. When for a given sublinear operator T we say that T is bounded from X to Y and also from Z to W, we shall implicitly assume that $X \cap Z$ is dense in both X and Z. This implies the uniqueness of its extension, so that it is reasonable to consider it as the same operator.

§ 2. Factorization of operators

<u>Definition 2.1.</u> Let X be a Banach lattice of measurable functions on $(\Sigma, d\sigma)$. Let $T : X \to B$ be an operator sublinear and continuous into the Banach space B. We say that T *factors through* $L^p = L^p(\Sigma, d\sigma)$ if there exist a continuous operator $T_0 : L^p \to B$ and a function $g(\sigma) > 0$ such that the following diagram is commutative:

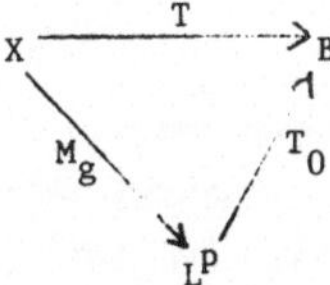

where M_g is the multiplication operator defined by $M_g(x)(\sigma) = x(\sigma)\cdot g(\sigma)$.

In order for M_g to map X into L^p we must have $|x\cdot g|^p \in L^1$ for every $x \in X$ or equivalently $|g|^p \in (X^p)'$. Thus g must belong to $((X^p)')^{1/p}$. If this is the case, we have $\|M_g(x)\|_{L^p} \le C\|x\|_X$ where $C = \|g\|_{((X^p)')^{1/p}}$.

Also since $T_0(fg) = T(f)$, the continuity of T_0 means

$$\|Tf\|_B^p \le C^p \int_\Sigma |f(\sigma)|^p g(\sigma)^p \, d\sigma$$

that is: $T : L^p(v) \to B$ with $v = g^p \in (X^p)'$. This is a Banach lattice if X is p-convex.

Definition 2.2. Given a p-convex Banach lattice X, $1 \le p < \infty$, we shall write $X_p = (X^p)'$.

We shall prove that if X is p-convex and has absolutely continuous norm, the factorization of $T : X \to B$ through L^p is equivalent to the fact that the vector extension $\tilde{T}$ defined by sending each sequence (x_j) of vectors in X to the sequence (Tx_j), is bounded from $X(\ell^p)$ to ℓ_B^p. This means that we have an inequality:

$$\left(\sum_j \|Tx_j\|_B^p \right)^{1/p} \le C \left\| \left(\sum_j |x_j|^p \right)^{1/p} \right\|_X .$$

Actually we shall formulate a slightly more general theorem valid for a family of perators.

Theorem 2.3. *Let $\mathcal{T}$ be a family of sublinear operators $T : X \to B$ where X is a p-convex Banach lattice, $1 \le p < \infty$, and B a Banach space. Then the sufficient condition for the inequality*

$$\left(\sum_j \|T_j x_j\|_B^p \right)^{1/p} \le C \left\| \left(\sum_j |x_j|^p \right)^{1/p} \right\|_X ; \quad T_j \in \mathcal{T}, \quad x_j \in X, \tag{2.4}$$

to hold is that there exists $v \in \tilde{X}_p$, $v > 0$, with $\|v\|_{\tilde{X}_p} \le 1$, such that:

$$\|Tx\|_B^p \le C^p \int_\Sigma |x(\sigma)|^p v(\sigma) \, d\sigma , \quad T \in \mathcal{T}, \quad x \in X .$$

If X has absolutely continuous norm, the condition is also necessary.

The proof of theorem 2.3 will depend on the following version of the mini-max lemma.

Lemma 2.5. *Let A and K be convex subsets of some real vector spaces, and suppose that K is endowed with a topology that makes K compact. Let $\Phi : A \times K \to \mathbb{R} \cup \{+\infty\}$ be a mapping such that:*

(i) $\Phi(\cdot,b)$ *is concave for each fixed* $b \in K$

(ii) $\Phi(a,\cdot)$ *is convex for each fixed* $a \in A$

(iii) $\Phi(a,\cdot)$ *is lower semicontinuous for each* $a \in A$.

Then $\min_{b\in K} \sup_{a\in A} \Phi(a,b) = \sup_{a\in A} \min_{b\in K} \Phi(a,b)$.

For the proof of this lemma we refer to [8].

Proof of Theorem 2.3. Suppose that v exists. Let us prove (2.4):

$$\sum_j \|T_j x_j\|_B^p \le C^p \int_\Sigma \sum_j |x_j(\sigma)|^p \, v(\sigma)\, d\sigma \le$$

$$\le C^p \Big\| \sum_j |x_j|^p \Big\|_{X^p} \|v\|_{(X^p)'} \le C^p \Big\| \Big(\sum_j |x_j|^p \Big)^{1/p} \Big\|_X^p .$$

Conversely, suppose that X has absolutely continuous norm and (2.4) holds. Let

$$A = \{ \sum_{j=1}^n |x_j|^p : x_j \in X \text{ and } \sum_{j=1}^n \|T_j x_j\|_B^p \le 1 \text{ for some } T_j \in \mathcal{T} \}$$

A is clearly a convex subset of X^p.

Let $K = \{z \in \tilde{X}_p : z \ge 0 , \|z\|_{\tilde{X}_p} \le 1\}$. Since X^p has absolutely continuous norm, $\tilde{X}_p = (X^p)' = (X^p)^*$, K is a convex subset of $\tilde{X}_p$ and if we consider the weak-* topology (the one given by X^p), K is also compact.
Consider now $\Phi : A \times K \to \mathbb{R}$ given by

$$\Phi(y,z) = - \int_\Sigma y(\sigma)\, z(\sigma)\, d\sigma .$$

Φ is actually bilinear and is continuous in z. Therefore, lemma 2.5 applies.
Note that

$$\min_{z\in K} \Phi(y,z) = - \sup_{z\in K} \int_\Sigma y(\sigma)\, z(\sigma)\, d\sigma = - \|y\|_{X^p} \le - \frac{1}{C^p} \quad \text{because of (2.4).}$$

Then $\min_{z\in K} \sup_{y\in A} \Phi(y,z) = \sup_{y\in A} \min_{z\in K} \Phi(y,z) \le - \frac{1}{C^p}$. This means that there exists $z \in K$ such that for every $y \in A$, $\Phi(y,z) \le - \frac{1}{C^p}$. If we take $y = |x|^p/\|Tx\|_B^p$, we get

$$- \int_\Sigma |x(\sigma)|^p \, z(\sigma)\, d\sigma \le - \frac{1}{C^p} \|Tx\|_B^p$$

which is what we wanted with $v = z$. □

Now we want to consider the dual situation of factoring an operator $T : B \to Y$ where B is a Banach space and Y is a Banach lattice of measurable functions on $(\Omega, d\omega)$.

Definition 2.6. $T : B \to Y$ sublinear and continuous factors through $L^p = L^p(\Omega, d\omega)$ if there exist a continuous operator $T_0 : B \to L^p$ and a function $g(\omega) > 0$ such that the following diagram is commutative:

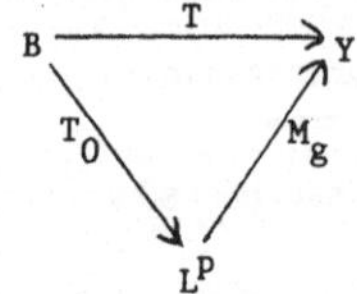

Proposition 2.7. *Assume* Y *is* p-*concave. Then* M_g *takes* L^p *into* Y *if and only if* $g \in (((Y')^{p'})')^{1/p'}$ *, which is a Banach lattice. In that case* M_g *is a continuous operator whose norm coincides with the norm of* g *in the above mentioned lattice.*

Proof. $fg \in Y \quad \forall f \in L^p \Longleftrightarrow hfg \in L^1 \quad \forall f \in L^p$, $\forall h \in Y' \Longleftrightarrow hg \in L^{p'}$ $\forall h \in Y' \Longleftrightarrow k|g|^{p'} \in L^1 \quad \forall k \in (Y')^{p'} \Longleftrightarrow |g| \in (((Y')^{p'})')^{1/p'}$. This is a lattice because Y' is p'-convex, which is equivalent to the fact that Y is p-concave. Also

$$\|M_g(f)\|_Y = \int_\Omega f\ g\ h\ d\omega \quad \text{for some } h \text{ with } \|h\|_{Y'} \le 1 .$$

Thus $\|M_g(f)\|_Y \le \|f\|_{L^p} \ \| |gh|^{p'} \|_{L^1}^{1/p'} \le \|g\| \ \|f\|_{L^p}$. □

Note that, since $T_0 f = (Tf)/g$, the continuity of T_0 means

$$\int_\Omega |Tf(\omega)|^p \ g(\omega)^{-p} \ d\omega \le C^p \|f\|_B^p$$

that is $T : B \to g \cdot L^p$ is continuous where $g \in (((Y')^{p'})')^{1/p}$. If $p < \infty$, $g \cdot L^p = L^p(u^{-1})$ where $u = g^p \in (((Y')^{p'})')^{p/p'}$.

Definition 2.8. Given Y a p-concave Banach lattice, $1 < p < \infty$, we shall write $\hat{Y}_p = (((Y')^{p'})')^{p/p'}$.

Under certain conditions, we shall prove that the factorization of $T : B \to Y$ through L^p is equivalent to the boundedness of the vector extension $\tilde{T}$ from ℓ_B^p to $Y(\ell^p)$. This means that we have the inequality

$$\left\| \left(\sum_j |Tf_j|^p \right)^{1/p} \right\|_Y \le C \left(\sum_j \|f_j\|_B^p \right)^{1/p} .$$

As in the previous case, we shall formulate a general theorem valid for a family of operators.

Theorem 2.9. *Let* $\mathcal{T}$ *be a family of sublinear operators* $T : B \to Y$ *where* B *is a Banach space and* Y *is a* p-*concave Banach lattice,* $1 < p < \infty$ *. Then a* *sufficient condition for the inequality*

$$(2.10)\qquad \left\| \left(\sum_j |T_j f_j|^p \right)^{1/p} \right\|_Y \le C \left(\sum_j \|f_j\|_B^p \right)^{1/p} \; ; \quad T_j \in \mathcal{T}, \quad f_j \in B,$$

to hold is that there exists $u \in \hat{Y}_p$, $u > 0$ *with* $\|u\|_{\hat{Y}_p} \le 1$, *such that*

$$\int_\Omega |Tf(\omega)|^p (u(\omega))^{-1} \, d\omega \le C^p \|f\|_B^p , \quad T \in \mathcal{T}, \quad f \in B .$$

If $(Y')^{p'}$ *is reflexive, the condition is also necessary.*

<u>Proof.</u> The key observation is that

$$(2.11)\qquad \|y\|_Y = \min \left\{ \left(\int_\Omega |y(\omega)|^p \, z(\omega)^{-p/p'} d\omega \right)^{1/p} : z \in ((Y')^{p'})', \ \|z\| \le 1 \right\}.$$

Indeed, if $y \in Y$, we have, for some $y' \in Y'$ with $\|y'\| = 1$:

$$\|y\|_Y = \int_\Omega yy' = \int_\Omega yz^{-1/p'} y' z^{1/p'} \le \left(\int_\Omega |y|^p \, z^{-p/p'} \right)^{1/p} \left(\int_\Omega |y'|^{p'} z \right)^{1/p'} \le$$

$$\le \left(\int_\Omega |y|^p \, z^{-p/p'} \right)^{1/p} \quad \text{for every} \quad z \in ((Y')^{p'})'_+ \quad \text{with} \quad \|z\| \le 1$$

and equality is achieved for some z .
Assuming u exists, let us prove (2.10):

$$\left\| \left(\sum_j |T_j f_j|^p \right)^{1/p} \right\|_Y^p \le \int_\Omega \sum_j |T_j f_j(\omega)|^p \, u(\omega)^{-1} \, d\omega \le C^p \sum_j \|f_j\|_B^p .$$

Conversely, assume that (2.10) holds and also that $(Y')^{p'}$ is reflexive. We shall apply again the mini-max lemma 2.5. In order to do that, we define

$$A = \left\{ \sum_{j=1}^n |T_j f_j|^p : \sum_{j=1}^n \|f_j\|_B^p \le 1 \right\} .$$

This is a convex set of Y^p . Let

$$K = \left\{ z \in ((Y')^{p'})' : z \ge 0 , \ \|z\| \le 1 \right\} .$$

This is a convex set and, since $((Y')^{p'})' = ((Y')^{p'})^*$ (theorem 1.8) it is also weak-* compact.

Define $\Phi : A \times K \to \mathbb{R} \cup \{+\infty\}$ by

$$\Phi(x,z) = \int_\Omega x(\omega) \, z(\omega)^{-p/p'} \, d\omega .$$

Φ is linear in x , and therefore concave. The convexity of Φ in z follows from the fact that the mapping $t \mapsto t^{-a}$, $a > 0$, is convex in $[0,\infty)$. Finally, in order to see that Φ is lower-semicontinuous in z , we need to see that, for every x and every a, the set $E = \{z \in K : \Phi(x,z) \le a\}$ is closed in the weak-*

topology. But this set is convex. Also, since $((Y')^{p'})^{**} = (Y')^{p'}$, the weak-* topology coincides with the weak topology. These two facts imply that we just need to see that E is closed in the norm topology of $((Y')^{p'})'$ (see, for example [23], theorem 3.12).

Now if $z_j \to z$ in the norm, there is a subsequence converging a.e.. Fatou's lemma (theorem 1.2 c)) can be applied to show that $z_j \in E$ implies $z \in E$. We are in a position to apply lemma 2.5 to Φ:

$$\min_{z \in K} \Phi(x,z) = \| |x|^{1/p} \|_Y^p \le C^p .$$

Thus

$$\min_{z \in K} \sup_{x \in A} \Phi(x,z) = \sup_{x \in A} \min_{z \in K} \Phi(x,z) \le C^p .$$

In other words: there exists $z \in K$ such that for every $x \in A$ $\Phi(x,z) \le C^p$. In particular, if we take $x = |Tf|^p / \|f\|_B^p$ for some $f \in B$ and some $T \in \mathcal{T}$, we get

$$\int_\Omega |Tf(\omega)|^p \, z(\omega)^{-p/p'} \, d\omega \le C^p \|f\|_B^p .$$

This is what we wanted with $u = z^{p/p'} \in \hat{Y}_p$ which satisfies $\|u\|_{\hat{Y}_p} \le 1$. □

There is a version of theorem 2.9 for $p = \infty$. In that case, note that $g \in Y$ and T maps B into $g \cdot L^\infty = \{f \in L^0 : \|f/g\|_\infty < \infty\}$.

Theorem 2.12. *Let $\mathcal{T}$ be a family of sublinear operators $T : B \to Y$. Then the inequality*

$$(2.13) \qquad \Big\| \sup_j |T_j f_j| \Big\|_Y \le C \sup_j \|f_j\|_B , \quad T_j \in \mathcal{T}, \quad f_j \in B ,$$

holds if and only if there exists $u \in Y_+$ with $\|u\|_Y \le 1$ such that

$$\|Tf\|_{u \cdot L^\infty} = \Big\| \frac{Tf}{u} \Big\|_\infty \le C \|f\|_B , \quad T \in \mathcal{T}, \quad f \in B .$$

Proof. If u exists,

$$\Big\| \sup_j |T_j f_j| \Big\|_Y \le \Big\| \sup_j \Big| \frac{T_j f_j}{u} \Big| \Big\|_\infty \|u\|_Y \le C \sup_j \|f_j\|_B$$

and we get (2.13). Conversely if (2.13) holds, let

$$A = \{ \sup_{1 \le j \le n} |T_j f_j| : \sup_{1 \le j \le n} \|f_j\|_B \le 1 \} .$$

This is a directed subset of the closed ball of center 0 and radius C in Y. Then A has a least upper bound $G \in Y$ such that $\|G\|_Y \le C$ (see [24]). If we set $u = G/C$, we have

$$\sup_j |T_j f_j| \le C \sup_j \|f_j\|_B \cdot u$$

and, consequently,

$$\left\|\frac{Tf}{u}\right\|_\infty \le C\|f\|_B . \qquad \square$$

§ 3. Equivalence between weighted inequalities and vector-valued inequalities

Theorem 3.1. *Let* X *and* Y *be* p*-convex Banach lattices of functions on* $(\Sigma, d\sigma)$ *and* $(\Omega, d\omega)$ *respectively,* $1 \le p < \infty$ *. Let* $\mathcal{T}$ *be a family of sublinear operators* $T : X \to Y$ *. Then a sufficient condition for the vector-valued inequality*

$$(3.2)\qquad \left\|\left(\sum_j |T_j x_j|^p\right)^{1/p}\right\|_Y \le C\left\|\left(\sum_j |x_j|^p\right)^{1/p}\right\|_X ;\quad T_j \in \mathcal{T},\quad x_j \in X ,$$

to hold is that for every positive $u \in \tilde{Y}_p$ *there exists a positive* $v \in X_p$ *such that* $\|v\|_{\tilde{X}_p} \le \|u\|_{\tilde{Y}_p}$ *and*

$$\int_\Omega |Tx(\omega)|^p\, u(\omega)\, d\omega \le C^p \int_\Sigma |x(\sigma)|^p\, v(\sigma)\, d\sigma ,\quad T \in \mathcal{T},\quad x \in X .$$

If X *has absolutely continuous norm, the condition is also necessary.*

Proof. If the condition holds, let us prove (3.2)

$$\left\|\left(\sum_j |T_j x_j|^p\right)^{1/p}\right\|_Y^p = \left\|\sum_j |T_j x_j|^p\right\|_{Y^p} \le \int_\Omega \sum_j |T_j x_j|^p\, u\, d\omega$$

for some $u \in \tilde{Y}_p$, $u > 0$, and $\|u\|_{\tilde{Y}_p} \le 1$. We get the corresponding v and continue

$$\le C^p \int_\Sigma \sum_j |x_j|^p\, v\, d\sigma \le C^p \left\|\sum |x_j|^p\right\|_{X^p} \|v\|_{\tilde{X}_p} \le C^p \left\|\left(\sum_j |x_j|^p\right)^{1/p}\right\|_X .$$

Conversely, let us assume that (3.2) holds and that X has absolutely continuous norm. Let $u > 0$, $u \in \tilde{Y}_p$. We may assume $\|u\|_{\tilde{Y}_p} = 1$. We see that $Y \hookrightarrow L^p(u)$. Indeed, if $y \in Y$,

$$\int_\Omega |y(\omega)|^p\, u(\omega)\, d\omega \le \left\||y|^p\right\|_{Y^p} \|u\|_{\tilde{Y}_p} = \|y\|_Y^p .$$

Thus we have a family of $T : X \to L^p(u) \equiv B$ such that

$$\left(\sum_j \|T_j x_j\|_B^p\right)^{1/p} = \left\|\left(\sum_j |T_j x_j|^p\right)^{1/p}\right\|_B \le C\left\|\left(\sum_j |x_j|^p\right)^{1/p}\right\|_X .$$

Theorem 2.3 applies, and we get $v \in \tilde{X}_p$, $v > 0$ with $\|v\|_{\tilde{X}_p} \le 1$ such that

$$\int_\Omega |Tx(\omega)|^p\, u(\omega)\, d\omega = \|Tx\|_B^p \le C^p \int_\Sigma |x(\sigma)|^p\, v(\sigma)\, d\sigma\ , \quad T \in \mathcal{T}\ , \quad x \in X\ .$$

Thus, the condition is necessary. □

Note that in theorem 3.1, we actually get that T is bounded from $L^p(v)$ to $L^p(u)$ because X is dense in $L^p(v)$. Indeed, if $h \in (L^p(v))^* = L^{p'}(v^{-p'/p})$ is such that $\int_\Sigma x(\sigma)\, h(\sigma)\, d\sigma = 0$ for every $x \in X$, we have $h \in L^{p'}(v^{-p'/p}) \hookrightarrow X'$ and, consequently $h(\sigma) = 0$ a.e.

<u>Theorem 3.3.</u> *Let* X *and* Y *be p-concave Banach lattices of functions on* $(\Sigma, d\sigma)$ *and* $(\Omega, d\omega)$ *respectively,* $1 < p < \infty$. *Let* $\mathcal{T}$ *be a family of sublinear operators* $T : X \to Y$. *Then a sufficient condition for the vector-valued inequality* (3.2) *to hold is that for every positive* $v \in \hat{X}_p$ *there exists a positive* $u \in \hat{Y}_p$ *such that* $\|u\|_{\hat{Y}_p} \le \|v\|_{\hat{X}_p}$ *and*

$$\int_\Omega |Tx(\omega)|^p (u(\omega))^{-1}\, d\omega \le C^p \int_\Sigma |x(\sigma)|^p (v(\sigma))^{-1}\, d\sigma\ , \quad T \in \mathcal{T}\ .$$

If $(Y')^{p'}$ *is reflexive, the condition is also necessary.*

<u>Proof.</u> If the condition holds, let us prove (3.2). By (2.11) applied to the lattice X ,

$$\left\| \left(\sum_j |x_j|^p \right)^{1/p} \right\|_X^p = \int_\Sigma \sum_j |x_j(\sigma)|^p (v(\sigma))^{-1}\, d\sigma$$

for some $v \in \hat{X}_p$, $v > 0$ with $\|v\|_{\hat{X}_p} \le 1$. By considering the associated $u \in \hat{Y}_p$, we can continue writing:

$$\ge \frac{1}{C^p} \int_\Omega \sum_j |T_j x_j(\omega)|^p (u(\omega))^{-1}\, d\omega \ge \frac{1}{C^p} \left\| \left(\sum_j |T_j x_j|^p \right)^{1/p} \right\|$$

again by (2.11).

Conversely, suppose that (3.2) holds and $(Y')^p$ is reflexive. Given $v \in \hat{X}_p$, $v > 0$ with $\|v\|_{\hat{X}_p} = 1$ we have $B \equiv L^p(v^{-1}) \hookrightarrow X$ by (2.11). We may view $\mathcal{T}$ as a family of operators $T : B \to Y$ such that

$$\left\| \left(\sum_j |T_j f_j|^p \right)^{1/p} \right\|_Y \le C \left\| \left(\sum_j |f_j|^p \right)^{1/p} \right\|_B = \left(\sum_j \|f_j\|_B^p \right)^{1/p} .$$

We can apply theorem 2.9 to conclude that there exists $u \in \hat{Y}_p$, $u > 0$, with $\|u\|_{\hat{Y}_p} \le 1$ such that

$$\int_\Omega |Tx(\omega)|^p (u(\omega))^{-1}\, d\omega \le C^p \|x\|_B^p = C^p \int_\Sigma |x(\sigma)|^p (v(\sigma))^{-1}\, d\sigma\ .$$

□

The case $p = \infty$ is much simpler.

Theorem 3.4. *Let* $\mathcal{T}$ *be a family of sublinear operators* $T : X \to Y$, *where* X *and* Y *are Banach lattices of functions on* $(\Sigma,d\sigma)$ *and* $(\Omega,d\omega)$, *respectively. Then the inequality*

$$(3.5) \qquad \left\| \sup_j |T_j x_j| \right\|_Y \le C \left\| \sup_j |x_j| \right\|_X$$

holds if and only if for every $v \in X_+$, *there exists* $u \in Y_+$ *such that* $\|u\|_Y \le \|v\|_X$ *and the operators* $T \in \mathcal{T}$ *are uniformly bounded from* $v\cdot L^\infty$ *to* $u\cdot L^\infty$ *with* $\|T\| \le C$, *that is:*

$$\left\| \frac{Tx}{u} \right\|_\infty \le C \left\| \frac{x}{v} \right\|_\infty .$$

__Proof.__ To prove the sufficiency, let $v = \sup_j |x_j|$. Then we have the corresponding u , and since $\|x_j/v\|_\infty \le 1$, we get:

$$\left\| \sup_j |T_j x_j| \right\|_Y \le \left\| \sup_j \left| \frac{T_j x_j}{u} \right| \right\|_\infty \|u\|_Y \le C\|v\|_X$$

which is (3.5). Conversely, if (3.5) holds and we are given $v \in X_+$ with $\|v\|_X = 1$, we have $v\cdot L^\infty \hookrightarrow X$ since $\|x\|_X \le \|x/v\|_\infty \|v\|_X = \|x/v\|_\infty$. Let $B = v\cdot L^\infty$. Then (3.5) gives

$$\left\| \sup_j |T_j x_j| \right\|_Y \le C\cdot \left\| \sup_j \left| \frac{x_j}{v} \right| \right\|_\infty \le C \sup_j \|x_j\|_B .$$

Theorem 2.12 can be applied and we get $u \in Y_+$ with $\|u\|_Y \le \|v\|_X$ and

$$\left\| \frac{Tx}{u} \right\|_\infty \le C\|x\|_B = C\left\| \frac{x}{v} \right\|_\infty . \qquad \square$$

When we are dealing with a single Banach lattice X of functions on $(\Sigma,d\sigma)$, theorem 3.1 can be improved obtaining an inequality with the same weight in both sides. This unification of the weight is achieved by the Rubio de Francia algorithm ([19]) which we describe below.

__Theorem 3.6.__ *Let* X *be a* p-*convex Banach lattice of functions on* $(\Sigma,d\sigma)$, $1 \le p < \infty$, *and let* $\mathcal{T}$ *be a family of sublinear operators* $T : X \to X$. *Then a sufficient condition for the vector-valued inequality:*

$$(3.7) \qquad \left\| \left(\sum_j |T_j x_j|^p \right)^{1/p} \right\|_X \le C_1 \left\| \left(\sum_j |x_j|^p \right)^{1/p} \right\|_X ; \quad T_j \in \mathcal{T}, \quad x_j \in X,$$

to hold is that for every positive $u \in \tilde{X}_p$ *there exists a positive* $v \in X_p$ *such that* $u \le v$, $\|v\|_{\tilde{X}_p} \le 2\|u\|_{\tilde{X}_p}$ *and*

$$\int_\Sigma |Tx(\sigma)|^p \, v(\sigma)\, d\sigma \le C_2^p \int_\Sigma |x(\sigma)|^p \, v(\sigma)\, d\sigma \; ; \quad T \in \mathcal{T}, \quad x \in X .$$

If X *has absolutely continuous norm, the condition is also necessary. Moreover,*

$$2^{-1/p} \le C_1/C_2 \le 2^{1/p}.$$

Proof. The sufficiency is proved exactly as in theorem 3.1, yielding $C_1 \le 2^{1/p} C_2$.

Conversely, if X has absolutely continuous norm, let us prove the necessity. We assume that (3.7) holds, and that we have $u > 0$, $u \in \tilde{X}_p$. Theorem 3.1 gives us $U \in \tilde{X}_p$, $U > 0$ with $\|U\|_{\tilde{X}_p} \le \|u\|_{\tilde{X}_p}$ and such that

$$\int_\Sigma |Tx(\sigma)|^p\, u(\sigma)\, d\sigma \le C_1^p \int_\Sigma |x(\sigma)|^p\, U(\sigma)\, d\sigma\ , \quad T \in \mathcal{T}.$$

Let us call $u_0 = u$ and $u_1 = U$, and use theorem 3.1 again to obtain $U_1 \in \tilde{X}_p$, $U_1 > 0$ with

$$\|U_1\|_{\tilde{X}_p} \le \|u_1\|_{\tilde{X}_p} \le \|u\|_{\tilde{X}_p}$$

and such that

$$\int_\Sigma |Tx(\sigma)|^p\, u_1(\sigma)\, d\sigma \le C_1^p \int_\Sigma |x(\sigma)|^p\, U_1(\sigma)\, d\sigma\ , \quad T \in \mathcal{T}\ .$$

Now call $u_2 = U_1$ and continue. By induction we get $u_j \in \tilde{X}_p$, $u_j > 0$ with $\|u_j\|_{\tilde{X}_p} \le \|u\|_{\tilde{X}_p}$ and

$$\int_\Sigma |Tx(\sigma)|^p\, u_j(\sigma)\, d\sigma \le C_1^p \int_\Sigma |x(\sigma)|^p\, u_{j+1}(\sigma)\, d\sigma\ , \quad T \in \mathcal{T},\ \ j = 0,1,2,\dots\ .$$

Let $v = \sum_{j=0}^{\infty} 2^{-j} u_j$. Then $v \ge u_0 = u$, $\|v\|_{\tilde{X}_p} \le 2\|u\|_{\tilde{X}_p}$ and

$$\int_\Sigma |Tx(\sigma)|^p\, v(\sigma)\, d\sigma \le C_1^p \int_\Sigma |x(\sigma)|^p \sum_{j=0}^{\infty} 2^{-j}\, u_{j+1}(\sigma)\, d\sigma \le$$

$$\le 2C_1^p \int_\Sigma |x(\sigma)|^p\, v(\sigma)\, d\sigma$$

which is our condition with $C_2 \le 2^{1/p} C_1$. □

There does not seem to be a direct way to unify the weight in theorem 3.3 or 3.4 when $X = Y$. However, if we are dealing with linear operators, we can achieve the unification by using duality.

Theorem 3.8. *Let* $\mathcal{T}$ *be a family of linear operators* $T : X \to X$ *where* X *is a* p*-concave Banach lattice,* $1 < p < \infty$. *Then, a sufficient condition for* (3.7) *to hold is that for every positive* $u \in \hat{X}_p$ *there exists a positive* $v \in \hat{X}_p$ *such that* $u \le v$, $\|v\|_{\hat{X}_p} \le 2^{p/p'} \|u\|_{\hat{X}_p}$ *and*

$$\int_\Sigma |Tx(\sigma)|^p (v(\sigma))^{-1}\, d\sigma \le C_2^p \int_\Sigma |x(\sigma)|^p (v(\sigma))^{-1}\, d\sigma\ , \quad T \in \mathcal{T}\ .$$

If X *is reflexive, the condition is also necessary. Moreover,* $2^{-1/p'} \le C_1/C_2 \le 2^{1/p}$.

Proof. The sufficiency is proved as in theorem 3.3, giving $C_1 \le 2^{1/p} C_2$. Conversely, let us assume that X is reflexive and that (3.7) holds. The reflexivity implies (theorem 1.8), that $X' = X^*$. Thus, the family $\mathcal{T}^*$ consisting of the adjoint operators T^* of the operators $T \in \mathcal{T}$, is well defined as a class of operators on X' . Besides (3.7) implies

$$\Big\| \Big(\sum_j |T_j^* y_j|^{p'} \Big)^{1/p'} \Big\|_{X'} \le C_1 \Big\| \Big(\sum_j |y_j|^{p'} \Big)^{1/p'} \Big\|_{X'} \ ; \quad T_j^* \in \mathcal{T}.$$

Now X' is p'-convex and has absolutely continuous norm because X is reflexive. Thus, we can apply theorem 3.6. Observe that $(X')^{\sim}_{p'} = \big((X')^{p'}\big)' = \hat{X}_p^{p'/p}$. Now, given a positive $u \in \hat{X}_p$, let $U = u^{p'/p} \in (X')^{\sim}_{p'}$. We know that there is $V \in (X')^{\sim}_{p'}$ such that $U \le V$, $\|V\|_{(X')^{\sim}_{p'}} \le 2\|U\|_{(X')^{\sim}_{p'}}$ and

$$\int_\Sigma |T^*y(\sigma)|^{p'}\, V(\sigma)\, d\sigma \le 2C_1^{p'} \int_\Sigma |y(\sigma)|^{p'}\, V(\sigma)\, d\sigma\ , \quad T^* \in \mathcal{T}^*\ .$$

But this is equivalent to

$$\int_\Sigma |Tx(\sigma)|^p\, V(\sigma)^{-p/p'} d\sigma \le 2^{p/p'} C_1^p \int_\Sigma |x(\sigma)|^p\, V(\sigma)^{-p/p'} d\sigma\ .$$

If we write $V^{p/p'} = v \in \hat{X}_p$ we have:

$$u = U^{p/p'} \le V^{p/p'} = v\ ,$$

$$\|v\|_{\hat{X}_p} = \|V\|^{p/p'}_{(X')^{\sim}_{p'}} \le 2^{p/p'} \|U\|^{p/p'}_{(X')^{\sim}_{p'}} = 2^{p/p'} \|u\|_{\hat{X}_p}$$

and

$$\int_\Sigma |Tx(\sigma)|^p (v(\sigma))^{-1}\, d\sigma \le 2^{p/p'} C_1^p \int_\Sigma |x(\sigma)|^p (v(\sigma))^{-1}\, d\sigma$$

which is what we wanted with $C_2 \le 2^{1/p'} C_1$.

□

We can also use duality for the case $p = \infty$, improving theorem 3.4 when $X = Y$ and the operators are linear.

Theorem 3.9. *Let* $\mathcal{T}$ *be a family of linear operators* $T : X \to X$, *where* X *is a Banach lattice. Then a sufficient condition for the inequality*

(3.10) $$\Big\| \sup_j |T_j x_j| \Big\|_X \le C_1 \Big\| \sup_j |x_j| \Big\|_X\ , \quad T_j \in \mathcal{T},$$

to hold is that for every $u \in X_+$ *there exists* $v \in X_+$ *such that* $u \leq v$, $\|v\|_X \leq 2\|u\|_X$ *and*

$$\left\|\frac{Tx}{v}\right\|_\infty \leq C_2\left\|\frac{x}{v}\right\|_\infty \quad ; \quad T \in \mathcal{T} .$$

If X *is reflexive, the condition is also necessary. Moreover, we have:* $1/2 \leq C_1/C_2 \leq 2$.

Proof. The sufficiency is proved as in theorem 3.4 and gives $C_1 \leq 2C_2$. If X is reflexive, we have $X' = X^*$, and we may consider the class $\mathcal{T}^*$ formed by the adjoint operators $T^* : X' \to X'$. If (3.10) holds, we claim that

$$(3.11) \qquad \left\|\sum_j |T_j^* y_j|\right\|_{X'} \leq C_1\left\|\sum_j |y_j|\right\|_{X'} .$$

Indeed, the left hand side equals

$$\int_\Sigma \sum_j |T_j^* y_j|\ x = \int_\Sigma \sum_j (T_j^* y_j)\ a_j(\omega)\ x(\omega)\ d\omega$$

for some $x \in X$ with $\|x\| \leq 1$ and $\sup_j |a_j(\omega)| \leq 1$. But the last integral can be written as

$$\int_\Sigma \sum_j y_j T_j(a_j x) \leq \int_\Sigma \sum_j |y_j| \cdot \sup_j T_j(a_j x)\ d\sigma \leq$$

$$\leq \left\|\sum_j |y_j|\right\|_{X'} \left\|\sup_j T_j(a_j x)\right\|_X \leq C_1 \left\|\sup_j |a_j|\ |x|\right\|_X \left\|\sum_j |y_j|\right\|_{X'} \leq$$

$$\leq C_1 \left\|\sum_j |y_j|\right\|_{X'} .$$

Once we know (3.11), we can apply theorem 3.6. Note that X' has absolutely continuous norm because X is reflexive, and also $(X')_1^\sim = X'' = X$. Thus, given $u \in X_+$, we have $v \in X_+$ such that $u \leq v$, $\|v\|_X \leq 2\|u\|_X$ and

$$\int_\Sigma |T^* y(\sigma)|\ v(\sigma)\ d\sigma \leq 2C_1 \int_\Sigma |y(\sigma)|\ v(\sigma)\ d\sigma .$$

But, since $(L^1(v))^* = v \cdot L^\infty$ this is equivalent to

$$\left\|\frac{Tx}{v}\right\|_\infty \leq 2C_1\left\|\frac{x}{v}\right\|_\infty$$

which is what we wanted to prove with $C_2 \leq 2C_1$. □

Duality can also be used to obtain a variant of theorem 3.3 for a family of linear operators. For the necessity we require that X has absolutely continuous norm and also that Y is reflexive. These conditions are slightly different from those in theorem 3.3.

We shall end this section by writing versions of the previous theorems for Lebesgue spaces.

<u>Lemma 3.12.</u> a) *For* $q > p$, $(L^q)^{\sim}_p = L^\alpha$ *where* $\frac{1}{\alpha} = 1 - \frac{p}{q}$.

b) *For* $q < p$, $(L^q)^{\wedge}_p = L^\beta$ *where* $\frac{1}{\beta} = \frac{p}{q} - 1$.

<u>Proof.</u> a) L^q is q-convex and, consequently, p-convex. Now $(L^q)^{\sim}_p = (L^{q/p})' = L^{(q/p)'}$. But $\frac{1}{(q/p)'} = 1 - \frac{p}{q} = \frac{1}{\alpha}$.

b) L^q is q-concave and, consequently, p-concave $(L^q)^{\wedge}_p = L^{(q'/p')'(p'/p)}$.

But $\frac{1}{(q'/p')'(p'/p)} = \left(1 - \frac{p'}{q'}\right)\frac{p}{p'} = \left(\frac{1}{p'} - \frac{1}{q'}\right)p = \left(\frac{1}{q} - \frac{1}{p}\right)p = \frac{p}{q} - 1 = \frac{1}{\beta}$. □

It is interesting that in both cases the reciprocal of the exponent turns out to be $|1 - \frac{p}{q}|$. This allows us to combine in a single statement the versions of theorems 3.1 and 3.3 for Lebesgue spaces.

We have a family $\mathcal{T}$ of sublinear operators $T : L^q(\Sigma,d\sigma) \to L^r(\Omega,d\omega)$. We are interested in knowing when the following vector-valued inequality holds:

$$\left\|\left(\sum_j |T_jf_j|^p\right)^{1/p}\right\|_r \le C\left\|\left(\sum_j |f_j|^p\right)^{1/p}\right\|_q \; ; \quad T_j \in \mathcal{T}, \quad f_j \in L^q . \tag{3.13}$$

The answer is as follows:

<u>Theorem 3.14.</u> *Let* $1 \le p,q,r < \infty$, *and define* α *and* β *by* $\frac{1}{\alpha} = |1 - \frac{p}{r}|$, $\frac{1}{\beta} = |1 - \frac{p}{q}|$. *Then*

a) *If* $p < q,r$, (3.13) *holds if and only if for every* $u \in L^\alpha_+(d\omega)$, *there exists* $v \in L^\beta_+(d\sigma)$ *such that* $\|v\|_\beta \le \|u\|_\alpha$ *and*

$$\int_\Omega |Tf(\omega)|^p\, u(\omega)\, d\omega \le C^p \int_\Sigma |f(\sigma)|^p\, v(\sigma)\, d\sigma , \quad T \in \mathcal{T} .$$

b) *If* $p > q,r$, $r > 1$, (3.13) *holds if and only if for every* $v \in L^\beta_+(d\sigma)$, *there exists* $u \in L^\alpha_+(d\omega)$ *such that* $\|u\|_\alpha \le \|v\|_\beta$ *and*

$$\int_\Omega |Tf(\omega)|^p (u(\omega))^{-1}\, d\omega \le C^p \int_\Sigma |f(\sigma)|^p (v(\sigma))^{-1}\, d\sigma , \quad T \in \mathcal{T} .$$

Actually, theorem 3.14 is true even for $0 < p,q,r < \infty$. The same proofs work with some minor changes.

The version of theorem 3.6 for $X = L^q$ will be this:

<u>Theorem 3.15.</u> *Let* $1 \le p < q < \infty$ *and let* $\mathcal{T}$ *be a family of sublinear operators*

$T : L^q(\Sigma,d\sigma) \to L^q(\Sigma,d\sigma)$. *Define* α *by* $\frac{1}{\alpha} = 1 - \frac{p}{q}$. *Then the following conditions are equivalent:*

a) $$\Big\| \Big(\sum_j |T_j f_j|^p \Big)^{1/p} \Big\|_q \le C_1 \Big\| \Big(\sum_j |f_j|^p \Big)^{1/p} \Big\|_q \; ; \; T_j \in \mathcal{T}, \quad f_j \in L^q .$$

b) *For every* $u \in L^\alpha_+(d\sigma)$, *there exists* $v \in L^\alpha_+(d\sigma)$ *such that* $u \le v$, $\|v\|_\alpha \le 2\|u\|_\alpha$ *and*

$$\int_\Sigma |Tf(\sigma)|^p \, v(\sigma) \, d\sigma \le C_2^p \int_\Sigma |f(\sigma)|^p \, v(\sigma) \, d\sigma , \quad T \in \mathcal{T}.$$

Moreover, $2^{-1/p} \le C_1/C_2 \le 2^{1/p}$.

And here is the version of theorem 3.8:

<u>Theorem 3.16.</u> *Let* $1 < q < p < \infty$ *and let* $\mathcal{T}$ *be a family of linear operators* $T : L^q(\Sigma,d\sigma) \to L^q(\Sigma,d\sigma)$. *Define* α *by* $\frac{1}{\alpha} = \frac{p}{q} - 1$. *Then the following conditions are equivalent:*

a) $$\Big\| \Big(\sum_j |T_j f_j|^p \Big)^{1/p} \Big\|_q \le C_1 \Big\| \Big(\sum_j |f_j|^p \Big)^{1/p} \Big\|_q , \quad T_j \in \mathcal{T}, \quad f_j \in L^q .$$

b) *For every* $u \in L^\alpha_+(d\sigma)$, *there exists* $v \in L^\alpha_+(d\sigma)$ *such that* $u \le v$, $\|v\|_\alpha \le 2^{p/p'} \|u\|_\alpha$ *and*

$$\int_\Sigma |Tf(\sigma)|^p \big(v(\sigma)\big)^{-1} \, d\sigma \le C_2^p \int_\Sigma |f(\sigma)|^p \big(v(\sigma)\big)^{-1} \, d\sigma , \quad T \in \mathcal{T} .$$

Moreover, $2^{-1/p'} \le C_1/C_2 \le 2^{1/p}$.

<u>§ 4. Extrapolation theorems</u>

We are going to reformulate the theorems in section 3 for linear operators, replacing the conditions given there by seemingly weaker ones which do not assume any size relation between the weights. These conditions will suffice because we can apply the general principles of Linear Analysis.

We start by recalling the relation between the lattices and the weighted Lebesgue spaces.

<u>Lemma 4.1.</u> *Let* X *be a Banach lattice.*

a) *If* X *is* p-*convex, then* $X = \bigcap L^p(u)$, $u \in (\tilde{X}_p)_+$ *and*

$$\|x\|_X = \max \left\{ \left(\int_\Sigma |x(\sigma)|^p \, u(\sigma) \, d\sigma \right)^{1/p} , \quad \|u\|_{\tilde{X}_p} \le 1 , \quad u > 0 \right\} .$$

b) *If* X *is* p-*concave, then* $X = \bigcup L^p(v^{-1})$, $v \in (\hat{X}_p)_+$ *and*

$$\|x\|_X = \min \left\{ \left(\int_\Sigma |x(\sigma)|^p (v(\sigma))^{-1} \, d\sigma \right)^{1/p} , \; \|v\|_{\hat{X}_p} \le 1 , \; v > 0 \right\} .$$

Proof. a) $|x|^p u \in L^1 \quad \forall u \in (\tilde{X}_p)_+ \Leftrightarrow |x|^p \in (X^p)'' = X^p \Leftrightarrow x \in X$. The expression for the norm follows from $\|x\|_X = \| |x|^p \|_{X^p}^{1/p}$.

b) The identity for the norm has been proved already. It is a reformulation of (2.11). The expression of X as a union of L^p spaces is an immediate consequence. □

Here is the new formulation of theorem 3.1.

Theorem 4.2. *Let* $\mathcal{T}$ *be a family of linear operators* $T : X \to Y$, *where* X *and* Y *are* p-*convex Banach lattices. Then a sufficient condition for the inequality* (3.2) *to hold is that for every positive* $u \in \tilde{Y}_p$ *there exists a positive* $v \in \tilde{X}_p$ *such that all* $T \in \mathcal{T}$ *are uniformly bounded from* $L^p(v)$ *to* $L^p(u)$.

Proof. It follows from lemma 4.1 that $X(\ell^p) = \bigcap L^p(v)(\ell^p)$, $v \in (\tilde{X}_p)_+$ and $Y(\ell^p) = \bigcap L^p(u)(\ell^p)$, $u \in \tilde{Y}_{p+}$. What we have to prove is that all the operators $\tilde{T} : (x_j) \mapsto (T_j x_j)$ obtained by choosing $T_j \in \mathcal{T}$ are (uniformly) bounded from $X(\ell^p)$ to $Y(\ell^p)$. What we assume implies that for every $u \in (\tilde{Y}_p)_+$, there exists $v \in (\tilde{X}_p)_+$ such that $\tilde{T}$ maps $L^p(v)(\ell^p)$ to $L^p(u)(\ell^p)$. Thus, $X(\ell^p)$ is carried into all the $L^p(u)(\ell^p)$'s and, consequently, into $Y(\ell^p)$. Once we know that $\tilde{T}$ carries $X(\ell^p)$ to $Y(\ell^p)$, the fact that it is continuous follows from the closed graph theorem. Indeed, the graph of T is closed in $X(\ell^p) \times Y(\ell^p)$ because it is closed in $L^p(v)(\ell^p) \times L^p(u)(\ell^p)$ and $X(\ell^p) \times Y(\ell^p) \hookrightarrow L^p(v)(\ell^p) \times L^p(u)(\ell^p)$. □

Observation 4.3. Theorem 4.2 continues to hold for a family $\mathcal{T}$ of linearizable operators.

There are two ways to prove this:

1) If $T_j : X \to Y$ is given by

$$T_j x(\omega) = \|S_j x(\omega)\|_{B_j} \quad \text{with} \quad S_j : X \to Y(B_j)$$

linear, we consider the operator $\tilde{S} : (x_j) \to (S_j x_j)$ and we have to prove that $\tilde{S}$ is continuous from

$$X(\ell^p) \quad \text{to} \quad Y\big(\bigoplus_{\ell^p} B_j\big) .$$

What we are assuming implies that for every $u \in (\tilde{Y}_p)_+$, there exists $v \in (\tilde{X}_p)_+$ such that $\tilde{S}$ maps

$$L^p(v)(\ell^p) \quad \text{to} \quad L^p(u)\Big(\bigoplus_{\ell^p} B_j\Big) .$$

Now we just have to use the fact that

$$Y\Big(\bigoplus_{\ell^p} B_j\Big) = \bigcap L^p(u)\Big(\bigoplus_{\ell^p} B_j\Big) , \quad u \in (\tilde{Y}_p)_+$$

to conclude that $\tilde{S}$ takes

$$X(\ell^p) \quad \text{to} \quad Y\Big(\bigoplus_{\ell^p} B_j\Big) .$$

Since $\tilde{S}$ is linear, the theorem follows.

2) The other approach consists in associating to the operator T given by $Tx(\omega) = \|Sx(\omega)\|_B$, the family of linear operators $\{T_h\}$ where h ranges over all functions in $L^\infty(B^*)$ having $\|h\| \le 1$, and $T_h x(\omega) = \langle Sx(\omega), h(\omega)\rangle$. Now we consider the family $\mathcal{T}'$ which is the union of all the families $\{T_h\}$ when $T \in \mathcal{T}$. Since $|T_h x(\omega)| \le \|Sx(\omega)\|_B = |Tx(\omega)|$, the new family $\mathcal{T}'$, whose elements are linear operators, satisfies the same condition that $\mathcal{T}$ in terms of weights. Thus, the family $\mathcal{T}'$ satisfies (3.2) and, consequently, the family $\mathcal{T}$ also satisfies (3.2).

There is also a version for p-concave lattices:

<u>Theorem 4.4.</u> *Let $\mathcal{T}$ be a family of linear operators, uniformly bounded from* X *to* Y, *both* p-*concave lattices with absolutely continuous norm. Then, a sufficient condition for* (3.2) *to hold is that for every positive* $v \in \hat{X}_p$, *there exists a positive* $u \in \hat{Y}_p$ *such that all* T *are uniformly bounded from* $L^p(v^{-1})$ *to* $L^p(u^{-1})$.

<u>Proof.</u> It is clear that $\tilde{T}$ sends $X(\ell^p)$ to $Y(\ell^p)$. To prove that it is continuous, we can apply theorem 4.2 to the family of adjoint operators. □

<u>Observation 4.5.</u> Theorem 4.4 is also valid for a family $\mathcal{T}$ of linearizable operators. The second approach adopted in observation 4.3 works equally well in the p-concave case.

When we have one single lattice X, we get results with one single weight, which correspond to theorems 3.6 and 3.8. We write them together as follows:

<u>Theorem 4.6.</u> *Let* X *be a* p-*convex (resp.* p-*concave lattice),* $1 < p < \infty$, *with absolutely continuous norm (resp. reflexive) and let* $\mathcal{T}$ *be a family of linearizable operators. Then* (3.7) *holds if and only if for every* $u \in \tilde{X}_p$ *(resp.* $\hat{X}_p$ *),*

$u > 0$, *there exists* $v \in \tilde{X}_p$ *(resp.* $\hat{X}_p$ *), such that* $u \leq v$ *and all* $T \in \mathcal{T}$ *are uniformly bounded in* $L^p(v)$ *(resp.* $L^p(v^{-1})$ *).*

<u>Proof.</u> That (3.7) implies our condition follows from theorems 3.6 and 3.8. Note that even though theorem 3.8 is stated for linear operators, it can be applied to linearizable ones just as in observation 4.3 2). Conversely, let us see that our condition implies 3.7. For X p-convex, we just need to apply theorem 4.2, extended to linearizable operators. Indeed, $u \leq v$ implies $L^p(v) \hookrightarrow L^p(u)$ so that all $T \in \mathcal{T}$ are uniformly bounded from $L^p(v)$ to $L^p(u)$, and this is precisely the condition needed in theorem 4.2. For X p-concave, we apply theorem 4.4, extended to linearizable operators. This requires that for each $u \in (\hat{X}_p)_+$, we find $v \in (\hat{X}_p)_+$ such that all $T \in \mathcal{T}$ are uniformly bounded from $L^p(u^{-1})$ to $L^p(v^{-1})$. What we have now is $v \geq u$ and T uniformly bounded in $L^p(v^{-1})$. But $L^p(u^{-1}) \hookrightarrow L^p(v^{-1})$ and we actually have what we wanted. □

The case $p = \infty$ requires a special formulation, but it is proved with the same method:

<u>Theorem 4.7.</u> *Let* $\mathcal{T}$ *be a family of linearizable operators uniformly bounded in the reflexive Banach lattice* X . *Then* (3.10) *holds if and only if for every* $u \in X_+$ *there exists* $v \in X_+$ *such that* $u \leq v$ *and all* $T \in \mathcal{T}$ *are uniformly bounded in* $v \cdot L^\infty$.

The key to the extrapolation theorems is going to be a boundedness criterion for linear operators obtained from theorems 4.2, 4.4 and 4.6 by means of the following fundamental result due to Grothendieck and Krivine (see [13] and [14]).

<u>Theorem 4.8.</u> *Let* $T : X \rightarrow Y$ *be a linear operator, bounded from* X *to* Y , *both Banach lattices. Then*

$$\left\| \left(\sum_j |Tx_j|^2 \right)^{1/2} \right\|_Y \leq K_G \cdot \|T\| \cdot \left\| \left(\sum_j |x_j|^2 \right)^{1/2} \right\|_X , \quad x_j \in X ,$$

where K_G *is Grothendieck's universal constant whose value is still unknown although* $1 < K_G < 2$.

The corresponding theorem for Lebesgue spaces is due to Marcinkiewicz and Zygmund and it is much simpler (see [8], chapter V).

Here are the boundedness criteria we get, where for simplicity we write $\tilde{X}$ instead of $\tilde{X}_2$ and $\hat{X}$ instead of $\hat{X}_2$.

<u>Theorem 4.9.</u> *Let* X *and* Y *be reflexive Banach lattices of measurable functions*

on $(\Sigma,d\sigma)$ *and* $(\Omega,d\omega)$, *respectively and suppose we have a linear operator* T *sending measurable functions on* $(\Sigma,d\sigma)$ *to measurable functions on* $(\Omega,d\omega)$.

Then:

a) *If* X *and* Y *are 2-convex,* T *is bounded from* X *to* Y *if and only if for every* $u \in \tilde{Y}_+$ *there exists* $v \in \tilde{X}_+$ *such that* T *is bounded from* $L^2(v)$ *to* $L^2(u)$.

b) *If* X *and* Y *are 2-concave,* T *is bounded from* X *to* Y *if and only if for every* $v \in \hat{X}_+$ *, there exists* $u \in \hat{Y}_+$ *such that* T *is bounded from* $L^2(v^{-1})$ *to* $L^2(u^{-1})$.

Theorem 4.10. *Let* X *be a reflexive, 2-convex (resp. 2-concave) Banach lattice of measurable functions on* $(\Sigma,d\sigma)$ *and suppose* T *is a linear operator sending measurable functions on* $(\Sigma,d\sigma)$ *to measurable functions on* $(\Sigma,d\sigma)$ *. Then* T *is bounded in* X *if and only if for every* $u \in \tilde{X}_+$ *(resp.* $\hat{X}_+$ *) there exists* $v \in \tilde{X}_+$ *(resp.* $\hat{X}_+$ *) such that* $u \le v$ *and* T *is bounded in* $L^2(v)$ *(resp.* $L^2(v^{-1})$ *)*.

In theorems 4.9 and 4.10 the exponent 2 was crucial because of the Grothendieck-Krivine inequality which is false in general for $p \neq 2$. However, if T is linear and positive, the boundedness $T : X \to Y$ implies the boundedness $\tilde{T} : X(\ell^p) \to Y(\ell^p)$ for any $1 \le p \le \infty$, as we see next.

Proposition 4.11. *Let* $T : X \to Y$ *be bounded, linear and positive, where* X *and* Y *are Banach lattices. Then*

$$\left\| \left(\sum_j |Tx_j|^p \right)^{1/p} \right\|_Y \le \|T\| \left\| \left(\sum_j |x_j|^p \right)^{1/p} \right\|_X , \quad 1 \le p < \infty,$$

and

$$\left\| \sup_j |Tx_j| \right\|_Y \le \|T\| \left\| \sup_j |x_j| \right\|_X .$$

Proof. Assume $1 \le p < \infty$. The proof for $p = \infty$ is analogous, only the notation is different.

$$\left(\sum_{j=1}^n |x_j|^p \right)^{1/p} \ge \left| \sum_{j=1}^n a_j x_j \right| \quad \text{whenever} \quad \sum_{j=1}^n |a_j|^{p'} \le 1 .$$

Since T is positive,

$$T\left(\left(\sum_{j=1}^n |x_j|^p \right)^{1/p} \right) \ge \left| \sum_{j=1}^n a_j Tx_j \right| .$$

Hence

$$\Big(\sum_{j=1}^{n} |Tx_j|^p\Big)^{1/p} = \sup \Big\{ \Big|\sum_{j=1}^{n} a_j Tx_j\Big| : \sum_{j=1}^{n} |a_j|^{p'} \le 1 \Big\} \le T\Big(\Big(\sum_{j=1}^{n} |x_j|^p\Big)^{1/p}\Big)$$

and, consequently,

$$\Big\|\Big(\sum_{j=1}^{n} |Tx_j|^p\Big)^{1/p}\Big\|_Y \le \Big\|T\Big(\Big(\sum_{j=1}^{n} |x_j|^p\Big)^{1/p}\Big)\Big\|_Y \le \|T\| \, \Big\|\Big(\sum_{j=1}^{n} |x_j|^p\Big)^{1/p}\Big\|_X . \qquad \square$$

We can make the following

__Observation 4.12.__ If T is linear and positive, theorems 4.9 and 4.10 continue to hold for $p \neq 2$, $1 \le p < \infty$. Naturally we have to replace $\tilde{X}$ by $\tilde{X}_p$ and $\hat{X}$ by $\hat{X}_p$. Even $p = \infty$ is admissible in the concave case. We just have to use X instead of $\hat{X}_p$ and $v \cdot L^\infty$ instead of $L^p(v^{-1})$.

When the lattices are Lebesgue spaces, proposition 4.11 has the following version valid for linearizable operators.

__Proposition 4.13.__ *Let* $T : X \to Y$ *be bounded linearizable and positive where* $X = L^q(Ud\sigma)$ *and* $Y = L^q(Vd\omega)$, $U > 0$, $V > 0$. *Then, for every* $q \le p < \infty$, *we have* $\tilde{T}$ *bounded from* $X(\ell^p)$ *to* $Y(\ell^p)$ *with* $\|\tilde{T}\| = \|T\|$.

__Proof.__ When $p = q$ the result is obvious, so that we have $\tilde{T} : X(\ell^q) \to Y(\ell^q)$ bounded. By the positivity we also have $\tilde{T} : X(\ell^\infty) \to Y(\ell^\infty)$ bounded since $\sup |Tx_j| \le T(\sup |x_j|)$. Now the result follows by interpolation, which is perfectly legitimate since we are really interpolating a linear operator $\tilde{T}_0 : X(\ell^q) \to Y(\ell^q_B)$, $\tilde{T}_0 : X(\ell^\infty) \to Y(\ell^\infty_B)$ associated to T_0 such that $|Tx| = \|T_0 x\|_B$. $\square$

We are finally ready to obtain the extrapolation theorems. We shall use the following notations:
For an operator T sending measurable functions on $(\Sigma, d\sigma)$ to measurable functions on $(\Omega, d\omega)$ and $1 \le p < \infty$,

$$V_p(T) = \{(u,v) : u > 0 \text{ a.e. on } \Omega,\ v > 0 \text{ a.e. on } \Sigma \text{ and } T \text{ is bounded from } L^p(v) \text{ to } L^p(u)\}.$$

In particular for $p = 2$ we shall simply write $V(T)$ for $V_2(T)$. Also

$$V_\infty(T) = \{(u,v) : u > 0 \text{ a.e. on } \Omega,\ v > 0 \text{ a.e. on } \Sigma \text{ and } T \text{ is bounded from } v \cdot L^\infty \text{ to } u \cdot L^\infty\}.$$

__Observation 4.14.__ If T is sublinear and positive, $(u,v) \in V_\infty(T)$ if and only if $|Tv| \le Cu$ a.e. for some constant C.

Indeed, $T : v \cdot L^\infty \to u \cdot L^\infty$ bounded implies $|Tv/u| \le \|T\| \, \|v/v\|_\infty = \|T\|$. Conversely, if $|Tv| \le Cu$, we have

$$|f| = \left|\frac{f}{v}\cdot v\right| \le \left\|\frac{f}{v}\right\|_\infty v .$$

Since T is positive and sublinear,

$$|Tf| \le \left\|\frac{f}{v}\right\|_\infty |Tv| \le C\left\|\frac{f}{v}\right\|_\infty u .$$

Thus $T : v \cdot L^\infty \to u \cdot L^\infty$ is bounded with $\|T\| \le C$.

When $\Sigma = \Omega$, we shall consider

$$W_p(T) = \{w > 0 \text{ a.e. on } \Sigma = \Omega : T \text{ is bounded in } L^p(w)\},$$

$1 \le p < \infty$. Also in this case when $p = 2$ we shall write $W(T)$ for $W_2(T)$. Finally,

$$W_\infty(T) = \{w > 0 \text{ a.e. on } \Sigma = \Omega : T \text{ is bounded in } w \cdot L^\infty\}.$$

When T is positive sublinear we have, as before: $w \in W_\infty(T) \Leftrightarrow |Tw| \le Cw$ a.e.

Here is, first of all, the extrapolation theorem from L^2.

Theorem 4.15. *Let* S *and* T *be operators sending functions on* $(\Sigma, d\sigma)$ *to functions on* $(\Omega, d\omega)$, *such that:*

a) S *is linear*

b) T *is linearizable*

c) $V(S) \subset V(T)$.

If X *and* Y *are reflexive Banach lattices of functions on* $(\Sigma, d\sigma)$ *and* $(\Omega, d\omega)$, *respectively, both 2-convex or both 2-concave and if* S *is bounded from* X *to* Y, *then* T *is also bounded from* X *to* Y.

In case $X = Y$, c) *can be replaced by the weaker assumption*

c') $W(S) \subset W(T)$.

Proof. Suppose that X and Y are 2-convex. Let $u \in \tilde{Y}_+$. Since $S : X \to Y$ is bounded, theorem 4.9 implies that there exists $v \in \tilde{X}_+$ such that $(u,v) \in V(S) \subset V(T)$. Now we can proceed as in theorem 4.2 and observation 4.3, obtaining in particular that T is bounded from X to Y. The argument is similar in the case of 2-concave lattices or in the case $X = Y$. □

If S is linear and positive, extrapolation works from any $1 < p < \infty$.

Theorem 4.16. *Let* $1 < p < \infty$ *and suppose that* S *and* T *are operators sending functions on* $(\Sigma, d\sigma)$ *to functions on* $(\Omega, d\omega)$, *such that:*

a) S *is linear and positive*

b) T *is linearizable*

c) $V_p(S) \subset V_p(T)$.

If X *and* Y *are reflexive Banach lattices of functions on* $(\Sigma, d\sigma)$ *and* $(\Omega, d\omega)$, *respectively, both* p-*convex or both* p-*concave, and if* S *is bounded from* X *to* Y , *then* T *is also bounded from* X *to* Y .

In case X = Y *condition* c) *can be replaced by the weaker assumption*

c') $W_p(S) \subset W_p(T)$.

Proof. Suppose that X and Y are p-convex. Let $u \in (\tilde{Y}_p)_+$. Proposition 4.11 implies that S is bounded from $X(\ell^p)$ to $Y(\ell^p)$. Then theorem 3.1 implies that there exists $v \in (\tilde{X}_p)_+$ such that $(u,v) \in V_p(S) \subset V_p(T)$. Then observation 4.3 gives the boundedness of T .

Similar arguments work for p-concave lattices or for X = Y with c'). □

From theorem 4.7 we can obtain an extrapolation theorem from $p = \infty$. However, when $X \neq Y$ the result is trivial because $V_\infty(S) \subset V_\infty(T) \Rightarrow |Tf| \le CS(|f|)$. Indeed,

$$(Sv,v) \in V_\infty(S) \subset V_\infty(T) \Rightarrow T : v\cdot L^\infty \to (Sv)\cdot L^\infty \quad \text{i.e.} \quad \left|\frac{Tf}{Sv}\right| \le C \left|\frac{f}{v}\right| .$$

Putting $v = |f|$ we get $|Tf| \le CS(|f|)$. However for X = Y we get an interesting result.

Theorem 4.17. *Suppose that* X *is a reflexive Banach lattice and*

a) S *is linear and positive*

b) T *is linearizable*

c) $W_\infty(S) \subset W_\infty(T)$.

Then if S *is bounded in* X , T *is also bounded in* X .

For every fixed p_0 and p , and weights $u, v > 0$, the lattices $X = L^p(v)$ and $Y = L^p(u)$ are either p_0-convex if $p_0 \le p$ or p_0-concave if $p_0 \ge p$. Besides, they are reflexive provided $1 < p < \infty$. If we apply theorems 4.15, 4.16 and 4.17 to this case, we obtain

Theorem 4.18. *Let* S *be a linear operator and* T *a linearizable operator.*

a) *If* $V(S) \subset V(T)$, *then* $V_p(S) \subset V_p(T)$ *for every* $1 < p < \infty$.

b) *If we have a single measure space and* $W(S) \subset W(T)$, *then* $W_p(S) \subset W_p(T)$ *for every* $1 < p < \infty$.

c) *If* S *is positive and* $V_{p_0}(S) \subset V_{p_0}(T)$ *for some* $1 < p_0 \le \infty$, *then* $V_p(S) \subset V_p(T)$ *for every* $1 < p < \infty$.

d) *If we have a single measure space,* S *is positive and* $W_{p_0}(S) \subset W_{p_0}(T)$ *for some* $1 < p_0 \leq \infty$ *, then* $W_p(S) \subset W_p(T)$ *for every* $1 < p < \infty$.

If the operator S is not linear, but still linearizable and positive, we can get an extrapolation theorem from above, by making use of proposition 4.13.

Theorem 4.19. *Let* S *and* T *be linearizable operators. Assume also that* S *is positive.*

a) *If* $V_{p_0}(S) \subset V_{p_0}(T)$ *for some* $1 < p_0 \leq \infty$ *, then* $V_p(S) \subset V_0(T)$ *for every* $1 < p \leq p_0$.

b) *If we have a single measure space and* $W_{p_0}(S) \subset W_{p_0}(T)$ *for some* $1 < p_0 \leq \infty$*, then* $W_p(S) \subset W_p(T)$ *for every* $1 < p \leq p_0$.

Proof. Let us see, for example, how to prove b). $w \in W_p(S)$ implies that S is bounded in $X = L^p(w)$. Since $p < p_0$, proposition 4.13 applies to S , which is linearizable and positive. Thus, $\tilde{S}$ is bounded in $X(\ell^{p_0})$. Now theorems 4.6 or 4.7 can be used. Note that X is p_0-concave and reflexive. From the boundedness of $\tilde{S}$, we get that for every $U \in \hat{X}_{p_0}$ if $p_0 < \infty$ (or $U \in X$ if $p_0 = \infty$) with $U > 0$, there exists $V \in \hat{X}_{p_0}$ if $p_0 < \infty$ (or $V \in X$ if $p_0 = \infty$) such that $U \leq V$ a.e. and $V^{-1} \in W_{p_0}(S)$ if $p_0 < \infty$ (or $V \in W_\infty(S)$ if $p_0 = \infty$). But $W_{p_0}(S) \subset W_{p_0}(T)$ and we can apply again theorems 4.6 or 4.7, this time to T and in the opposite direction. We conclude that, in particular, T is bounded in X , that is: $w \in W_p(T)$. □

Sometimes the classes $V_p(S)$ or $W_p(S)$ behave well under duality, and this can be used to extrapolate from a given p_0 to any other p . Here is a result in this direction:

Theorem 4.20. *Let* S *be an operator linearizable and positive such that for every* $1 < p < \infty$,

(4.21) $$w \in W_p(S) \leftrightarrow w^{-p'/p} \in W_{p'}(S) .$$

Let T *be a linearizable operator such that* $W_{p_0}(S) \subset W_{p_0}(T)$ *for some* $1 < p_0 \leq \infty$ *. Then* $W_p(S) \subset W_p(T)$ *for every* $1 < p < \infty$.

Proof. We just need to consider $p_0 < p < \infty$, since the remaining cases are covered by the previous theorem. Assume T is linear, $w \in W_p(S) \rightarrow w^{-p'/p} \in W_{p'}(S)$. But $p' < p_0'$ and

$$W_{p_0'}(S) = \left(W_{p_0}(S)\right)^{-p_0'/p_0} \subset \left(W_{p_0}(T)\right)^{-p_0'/p_0} = W_{p_0'}(T^*) .$$

Theorem 4.19 yields $W_{p'}(S) \subset W_{p'}(T^*)$. Thus $w^{-p'/p} \in W_{p'}(T^*)$ and this is equivalent to $w \in W_p(T)$.

For T non-linear, we just need to consider the linearizations $\{T_h\}$ as in observation 4.3 2) and apply theorem 4.19 to the collection of the adjoints $\{T_h^*\}$. □

The last six theorems are variants of an abstract extrapolation theorem. By making specific choices of the operator S , we get concrete extrapolation theorems. Here are some candidates for S .

1) Let M be the Hardy-Littlewood maximal operator, sending a function $f \in L^1_{loc}(\mathbb{R}^n)$ to

$$Mf(x) = \sup_{Q \ni x} \frac{1}{|Q|} \int_Q |f(y)| \, dy ,$$

where the supremum is taken over all the cubes Q with sides parallel to the coordinate axes and containing the point x . It was proved by Muckenhoupt [16] (see also [4]) that for $1 < p < \infty$ $W_p(M) = A_p$, the class of weights defined by

$$(4.22) \qquad w \in A_p \Leftrightarrow \sup_Q \left(\frac{1}{|Q|} \int_Q w\right)^{1/p} \left(\frac{1}{|Q|} \int_Q w^{-p'/p}\right)^{1/p'} < \infty ,$$

where the sup is taken over all the cubes with sides parallel to the coordinate axes. Also

$$W_\infty(M) = A_1 = \{w \ge 0 : Mw \le Cw \quad \text{a.e.}\} .$$

2) Let M^* be the strong maximal operator, sending a function $f \in L^1_{loc}(\mathbb{R}^n)$ to

$$M^*f(x) = \sup_{R \ni x} \frac{1}{|R|} \int_R |f(y)| \, dy ,$$

where the supremum is now taken over all the intervals (i.e.: Cartesian products of intervals) containing x .
It can be seen (for example in [8] IV.6) that for $1 < p < \infty$ $W_p(M^*) = A_p^*$, the class of weights defined by

$$(4.23) \qquad w \in A_p^* \Leftrightarrow \sup_R \left(\frac{1}{|R|} \int_R w\right)^{1/p} \left(\frac{1}{|R|} \int_R w^{-p'/p}\right)^{1/p'} < \infty ,$$

where the sup is taken over all intervals. Also

$$W_\infty(M^*) = A_1^* = \{w \ge 0 : M^*w \le Cw \quad \text{a.e.}\} .$$

Alternatively, for $1 \le p < \infty$, $w \in A_p^*$ if and only if it is A_p in each variable, with uniform constant, for a.e. determination of the other variables.

3) The operator M is linearizable and positive but it is very interesting to know that there is a linear operator giving rise to the same classes of weights.

If $n = 1$, we can take the Hilbert transform H given by:

$$Hf(x) = \text{p.v.}\ \frac{1}{\pi} \int_{-\infty}^{\infty} \frac{f(y)}{x-y}\, dy$$

and we have ([8]): $W_p(H) = A_p$, $1 < p < \infty$.

If $n > 1$, we can take the Riesz transforms $R_1, R_2, \ldots, R_n$ given by:

$$R_j f(x) = \text{p.v.}\ c_n \int_{\mathbb{R}^n} \frac{x_j - y_j}{|x-y|^{n+1}} f(y)\, dy \ ; \quad c_n = \frac{\Gamma((n+1)/2)}{\pi^{(n+1)/2}}$$

If we consider $R = R_1 + R_2 + \ldots + R_n$, we have (see [4] and [8])

$$W_p(R) = A_p, \quad 1 < p < \infty .$$

4) Also for M^* we can find a linear substitute which turns out to be the multiple Hilbert transform H^* given by:

$$H^* f(x) = \lim_{\varepsilon_1, \varepsilon_2, \ldots, \varepsilon_n \to 0} \int_{|x_j - y_j| > \varepsilon_j} \frac{f(y)\, dy}{(x_1 - y_1)(x_2 - y_2)\ldots(x_n - y_n)} .$$

We have (see [8] IV.6.) $W_p(H^*) = A_p^*$, $1 < p < \infty$

For these particular examples we get the following extrapolation theorem:

<u>Theorem 4.24.</u> *Let* T *be a linearizable operator whose domain and range consist of measurable functions on* $\mathbb{R}^n$. *Suppose that either*

(i) *for some* $1 < p_0 < \infty$, T *is bounded in* $L^{p_0}(w)$ *for every* $w \in A_{p_0}$ *(resp.* $A_{p_0}^*$ *) or*

(ii) T *is bounded in* $w \cdot L^\infty$ *for every* $w \in A_1$ *(resp.* A_1^* *).*

Then

a) *for every* $1 < p < \infty$ *and every* $w \in A_p$ *(resp.* A_p^* *),* T *is bounded in* $L^p(w)$.

Not only that, but in general

b) *f* X *is a reflexive Banach lattice 2-convex or 2-concave and* R *(resp.* H^* *) is bounded in* X, *then* T *also is bounded in* X.

<u>Proof.</u> Note that the classes A_p and A_p^* satisfy condition (4.21), so that theorem 4.20 can be applied with $S = M$ or M^*, obtaining a). In particular, a)

holds with $p = 2$ and we can apply theorem 4.15 with $S = R$ or H^* to obtain b) □

Observation 4.25. Condition (ii) can be replaced by this weaker one:

T maps $w \cdot L^\infty$ into $B.M.O.(w)$ for every $w \in A_1$, where

$$g \in B.M.O.(w) \Leftrightarrow \sup_Q \frac{1}{w(Q)} \int_Q |g(x) - g_Q| \, dx < \infty$$

with $g_Q = \frac{1}{|Q|} \int_Q g$.

If we define $g^\#(x) = \sup_{Q \ni x} \frac{1}{|Q|} \int_Q |g(y) - g_Q| \, dy$,

we observe that for $w \in A_1$,

$$g \in B.M.O.(w) \rightarrow g^\#(x) = \sup_{Q \ni x} \frac{w(Q)}{|Q|} \frac{1}{w(Q)} \int_Q |g(y) - g_Q| \, dy \le$$

$$\le C.Mw(x) \le Cw(x) \rightarrow g^\# \in w \cdot L^\infty$$

If we define $T^\# f = (Tf)^\#$ we see that $T^\#$ satisfies (ii). Then if $w \in A_p$, $1 < p < \infty$, we have

$$\int_{\mathbf{R}^n} |Tf|^p w \le C \int_{\mathbf{R}^n} |(Tf)^\#|^p w \le C \int_{\mathbf{R}^n} |f|^p w .$$

For the first inequality see [8], chapter IV, theorem 2.20. The weaker condition is sometimes easy to check whereas (ii) fails for many natural operators satisfying a).

We shall finish this section by recalling briefly the history of the extrapolation theorem. It was first discovered by Rubio de Francia [19] with a non-constructive proof of the type we have given, but only for the A_p classes. Then García-Cuerva [6] gave a constructive proof that used the particular definition of the A_p classes plus the Rubio de Francia algorithm. Jawerth [12] proved a general theorem for L^p spaces by using the Rubio de Francia algorithm plus interpolation. Then it came the general formulation of Rubio de Francia [20], [21] in lattices, which is the one we have presented. The case $p_0 = \infty$ was treated in [10] and [7].

§ 5. Applications

We shall give a characterization of U.M.D. Banach lattices in terms of A_p weights due to José Luis Rubio de Francia [22]. U.M.D. stands for unconditional martingale differences. Here is the original definition of this condition:

Definition 5.1. We shall say that a Banach space B is U.M.D. if and only if the following inequality holds:

$$\left\|\sum_{k=1}^{n} \varepsilon_k d_k\right\|_{L^p(B)} \leq C_{p,B}\left\|\sum_{k=1}^{n} d_k\right\|_{L^p(B)}$$

for all n, all $\varepsilon_k = \pm 1$, and all B-valued martingale differences (that is: $d_k = f_k - f_{k-1}$ for a B-valued martingale $\{f_k\}$), p is some fixed exponent with $1 < p < \infty$.

Even though this definition seems to depend on p, it does not. Actually Burkholder [3] gave a geometrical characterization called ρ-convexity, which is obviously independent of p.

We shall not use any of these characterizations. Instead, we shall base our discussion upon the following

Theorem 5.2. *The Banach space* B *is U.M.D. if and only if the* B*-valued extension* H_B *of the Hilbert transform* H *(say, on the torus* $\mathbb{T}$ *) is bounded in* $L^p(B)$ *for some* $1 < p < \infty$. *This means, that the operator* H_B *defined on*

$$L^p \otimes B = \{\sum_j \phi_j(t)b_j : \phi_j \in L^p,\ b_j \in B\} \text{ by } H_B\left(\sum_j \phi_j b_j\right) = \sum_j (H\phi_j)b_j$$

satisfies: $\|H_B(f)\|_{L^p(B)} \leq C\|f\|_{L^p(B)}$

so that it extends continuously to $L^p(B)$.

That the boundedness of H_B is necessary was proved by Burkholder [3] and that it is sufficient by Bourgain [2]. When X is a Banach lattice of functions on $(\Sigma, d\sigma)$ we can view $L^p(X) = L^p(\mathbb{T})(X)$ as a lattice of functions of two variables $f(t,\sigma)$, $t \in \mathbb{T}$, $\sigma \in \Sigma$. If we have an operator A bounded in L^p, we can define $\tilde{A}$ at least in $L^p \otimes X$ by

$$\tilde{A}f(t,\sigma) = A\big(f(\cdot,\sigma)\big)(t) .$$

Note that $\tilde{H} = H_X$ on $L^p \otimes X$.

Here is the characterization of U.M.D. Banach lattices in terms of A_p weights:

Theorem 5.3. *Let* X *be a Banach lattice, reflexive and* p_0*-convex for some* $p_0 > 1$. *Fix* p *such that* $1 < p < p_0$. *Then* X *is U.M.D. if and only if for every* $u \in L^2(\tilde{X}_p)$, $u \geq 0$, *there exists* $w \in L^2(\tilde{X}_p)$ *such that* $u \leq w$, $\|w\| \leq 2\|u\|$ *and* $w(\cdot,\sigma) \in A_p$ *uniformly in* σ *for a.e.* $\sigma \in \Sigma$.

Proof. Let X be U.M.D. and take $q = 2p$. Then H_X is bounded in $L^q(X)$.

Not only that. It is easy to see that $X(\ell^p)$ is also U.M.D., so that, H_X has an ℓ^p extension. We can apply theorem 3.6 with $Y = L^q(X)$ in place of the lattice X appearing there. Observe that

$$\tilde{Y}_p = (L^{q/p}(X^p))'' = L^2(\tilde{X}_p) .$$

Thus, given $u \in L^2(\tilde{X}_p)$, $u \geq 0$, we have $w \in L^2(\tilde{X}^p)$ such that $u \leq w$, $\|w\| \leq 2\|u\|$ and

$$(5.4) \qquad \int_\Sigma \int_{\mathbb{T}} |H_X f(t,\sigma)|^p \, w(t,\sigma) \, dt \, d\sigma \leq C \int_\Sigma \int_{\mathbb{T}} |f(t,\sigma)|^p \, w(t,\sigma) \, dt \, d\sigma .$$

Let us apply this inequality to $f(t,\sigma) = \phi(t) \, \chi_E(\sigma)$ where ϕ is a trigonometric polynomial with rational coefficients and E is a subset of Σ with $|E| < \infty$. Since $H_X f(t,\sigma) = H\phi(t) \cdot \chi_E(\sigma)$, we have

$$\int_E \int_{\mathbb{T}} |H\phi(t)|^p \, w(t,\sigma) \, dt \, d\sigma \leq C \int_E \int_{\mathbb{T}} |\phi(t)|^p \, w(t,\sigma) \, dt \, d\sigma$$

and, consequently,

$$(5.5) \qquad \int_{\mathbb{T}} |H\phi(t)|^p \, w(t,\sigma) \, dt \leq C \int_{\mathbb{T}} |\phi(t)|^p \, w(t,\sigma) \, dt$$

for every $\sigma \notin E_0$, a set independent of ϕ and having $|E_0| = 0$.

This implies that $w(\cdot,\sigma)$ is an A_p-weight with uniform constant for every $\sigma \notin E_0$. The converse is even easier. If given $u \in L^2(\tilde{X}_p)$, $u \geq 0$, we have $w \in L^2(\tilde{X}_p)$ such that $u \leq w$, $\|w\| \leq 2\|u\|$ and (5.5) holds, we can obtain (5.4) by a limiting process. Then, by the easy part of theorem 3.6, H_X is bounded in $L^q(X)$ so that X is U.M.D. □

The condition in theorem 5.3 will be abreviated by saying that " $L^2(\tilde{X}_p)$ has enough A_p weights ".

Incidentally, note that in theorem 5.3 the exponent 2 does not play any role and we can use $L^s(\tilde{X}_p)$ for any $1 < s < \infty$. Theorem 5.3 can be used to prove the following

<u>Theorem 5.6.</u> *Let* X *be a U.M.D. lattice. Then there exists* $\varepsilon > 0$ *such that* X^a *is U.M.D. for every* $0 < a < 1 + \varepsilon$.

<u>Proof.</u> There are two different parts in this result. The deepest one is for $1 < a < 1 + \varepsilon$. It is proved in the following way. If X is U.M.D., it is known that X is superreflexive, and consequently p_0-convex for some $p_0 > 1$. Fix $1 < p < p_0$. By theorem 5.3, $L^2(\tilde{X}_p)$ has enough A_p weights. It is a well known fact in A_p theory that A_p weights satisfy a so called reverse Hölder's inequality (see [8]) and, consequently, every A_p weight is an $A_{p-\varepsilon}$ weight for

some $\varepsilon > 0$ depending only on the A_p constant. Thus $L^2(\tilde{X}_p)$ has enough $A_{p-\varepsilon}$ weights for some ε depending only on p and the U.M.D. constant of X. Let $Y = X^{p/(p-\varepsilon)}$, so that $Y^{p-\varepsilon} = X^p$. Then Y satisfies the condition in theorem 5.3 with $p - \varepsilon$ instead of p. It follows that $Y = X^a$ for some $a > 1$ is a U.M.D. lattice.

The second part of the result, the one for $0 < a < 1$, is based upon the so called "magical identity". This identity works for the conjugate function operator which sends the trigonometric polynomial $\sum a_n e^{int}$ to the trigonometric polynomial $\sum -i(\operatorname{sgn} n)a_n e^{int}$. The difference between this operator and the Hilbert transform H is dominated by the Hardy-Littlewood maximal operator so that its associated weights are again the A_p weights, and it is also good to characterize U.M.D. This justifies that we call H to the conjugate function also. With this notation we have the magic formula

$$(Hf)^2 - f^2 = 2H(fHf)$$

simply because $(f + iHf)^2$ is analytic. Now the magic formula works equally well for $\tilde{H}$, so that we have

$$\tilde{H}f(t,\sigma)^2 = f(t,\sigma)^2 + 2\tilde{H}(f\tilde{H}f)(t,\sigma)$$

and this implies:

$$\begin{aligned}
\|\tilde{H}f\|_{L^4(X^{1/2})} &= \int \|\tilde{H}f\|^4_{X^{1/2}} = \int \| |\tilde{H}f|^2 \|^2_X \le \\
&\le \int \left(\|f(t,\sigma)^2\|_X + 2\|\tilde{H}(f\tilde{H}f)\|_X \right)^2 \le \\
&\le 2 \int \|f(t,\sigma)^2\|^2_X + 8 \int \|\tilde{H}(f\tilde{H}f)\|^2_X \le \\
&\le 2\|f\|^4_{L^4(X^{1/2})} + 8C \int \|f\tilde{H}f\|^2_X \le \\
&\le 2\|f\|^4_{L^4(X^{1/2})} + C \int \|f\|^2_{X^{1/2}} \|\tilde{H}f\|^2_{X^{1/2}} \le \\
&\le 2\|f\|^4_{L^4(X^{1/2})} + C\|f\|^2_{L^4(X^{1/2})} \|\tilde{H}f\|^2_{L^4(X^{1/2})} \le \\
&\le C_\varepsilon \|f\|^4_{L^4(X^{1/2})} + \varepsilon\|\tilde{H}f\|^4_{L^4(X^{1/2})}
\end{aligned}$$

This implies that $X^{1/2}$ is U.M.D. By iteration we get that X^a is U.M.D. if $a = 2^{-k}$, $k = 1,2,\ldots$. We obtain the result for every $0 < a < 1$ by interpolation. □

In order to present the next result, we shall use the notation $[\ ,\ ]_\Theta$, $0 < \Theta < 1$, for Calderon's complex interpolation method, and we shall also recall Calderon's identity for Banach lattices. This holds provided at least one of the lattices X_0, X_1 is reflexive:

$$[X_0,X_1]_\theta = X_0^{1-\theta}X_1^\theta = \{x : |x| = x_0^{1-\theta}x_1^\theta ,\ x_0 \in X_0 ,\ x_1 \in X_1\}$$

Corollary 5.7. *Let* X *be a U.M.D. lattice in* $(\Sigma,d\sigma)$. *Then there exist* θ *with* $0 < \theta < 1$ *and another U.M.D. lattice* X_0 *such that* $X = [L^2(d\sigma),X_0]_\theta$.

Proof. Take $a > 1$ such that X^a and $(X^*)^a$ are U.M.D. Call $Y = ((X^*)^a)^*$, also U.M.D. The key observation is that $X = [L^1,Y]_{1/a}$. In order to verify this, we just need to check that the duals coincide. But

$$([L^1,Y]_{1/a})^* = [L^\infty,(X^*)^a]_{1/a} = (L^\infty)^{1/a'}((X^*)^a)^{1/a} = X^*$$

Then

$$X = X^{1/2}X^{1/2} = X^{1/2}([L^1,Y]_{1/a})^{1/2} = X^{1/2}((L^1)^{1/a'}Y^{1/a})^{1/2} =$$

$$= X^{1/2}(L^1)^{1/2a'}Y^{1/2a} = (L^2)^{1/a'}((X^a)^{1/2}Y^{1/2})^{1/a} =$$

$$= [L^2,[X^a,Y]_{1/2}]_{1/a} = [L^2,X_0]_{1/a}$$

which is what we wanted because $X_0 = [X^a,Y]_{1/2}$ is U.M.D. □

Corollary 5.7 extends a previous result of Pisier [17] in which X_0 was just a Banach lattice, not necessarily U.M.D.

The next theorem will allow us to obtain the boundedness of the vector-valued extension to a U.M.D. Banach lattice of a huge class of operators.

Theorem 5.8. *Let* X *be a U.M.D. Banach lattice and let* T *be a linearizable operator which is bounded in* $L^p(w) = L^p(\mathbb{T},w(t)dt)$ *for every* $w \in A_p$ *and every* $1 < p < \infty$. *Then* $\tilde{T}f(t,\sigma) = T(f(\cdot,\sigma))(t)$ *is bounded in* $L^p(X) = L^p(\mathbb{T})(X)$ *for every* $1 < p < \infty$.

Proof. If $L^p(X)$ is 2-convex or 2-concave, we just need to apply theorem 4.15 to the lattice $L^p(X)$, the linear operator $\tilde{H}$ and the linearizable operator $\tilde{T}$. We need to see that $W(\tilde{H}) \subset W(\tilde{T})$. In order to do that we proceed as in the proof of theorem 5.3. We realize that if a weight $w(t,\sigma)$ belongs to $W(\tilde{H})$, then $w(\cdot,\sigma)$ is an A_2 weight with uniform constant for a.e. $\sigma \in \Sigma$. Now by the hypothesis we have assumed about T, we get an inequality like (5.5) with T in place of H. Integrating on Σ and completing gives $w \in W(\tilde{T})$. In the general case, since X is U.M.D., we know that $L^p(X)$ is p_0-convex for some $1 < p_0 < \infty$. We use the fact that $\tilde{M}$ is bounded in $L^p(X)$ (see [22]) and apply theorem 4.16 exactly as we did before. □

Corollary 5.9. *Let us denote by*

$$S_n f(t) = \sum_{-n}^{n} \hat{f}(k) e^{ikt} ,$$

the n-*th partial sum of the Fourier series of* f . *Then if* X *is a U.M.D. Banach lattice, we have the following convergence a.e. of the* X-*valued Fourier series*

$$\|\tilde{S}_n f(t) - f(t)\|_X \rightarrow 0 \quad a.e., \quad f \in L^p(X) , \quad 1 < p < \infty .$$

Proof. Let $Tf(t) = \sup_n |S_n f(t)|$ be Carleson's maximal partial sum operator. Theorem 5.8 can be applied to it (see [11]), and concludes that $\tilde{T}$ is bounded in $L^p(X)$, $1 < p < \infty$.

But then

$$\int_{\mathbb{T}} \sup_n \|S_n f(t)\|_X^p \, dt \le \int_{\mathbb{T}} \|\sup_n |S_n f(t,\cdot)|\|_X^p \, dt =$$

$$= \int_{\mathbb{T}} \|\tilde{T}f(t,\cdot)\|_X^p \, dt = \|\tilde{T}f\|_{L^p(X)}^p \le C\|f\|_{L^p(X)}^p$$

which is enough for our purposes. □

References

[1] BENNET, C. and SHARPLEY, R. : *Interpolation of Operators*. Academic Press 1988.

[2] BOURGAIN, J.: *Some remarks on Banach spaces in which martingale differences are unconditional*. Arkiv Mat. 21 (1983), 163 - 168.

[3] BURKHOLDER, D. L.: *A geometric condition that implies the existence of certain singular integrals of Banach-space valued functions*. Conf. Harmonic Analysis in honor of A. Zygmund. Wadsworth Inc (1981), 270 - 286.

[4] COIFMAN, R. and FEFFERMAN, C.: *Weighted norm inequalities for maximal functions and singular integrals*. Studia Math. 51 (1974), 241 - 250.

[5] DUOANDIKOETXEA, J. and RUBIO DE FRANCIA, J. L.: *Maximal and singular integral operators via Fourier transform estimates*. Invent. Math. 84 (1986), 541 - 561.

[6] GARCÍA-CUERVA, J.: *An extrapolation theorem in the theory of* A_p *weights*. Proc. Amer. Math. Soc. 83 (1983), 422 - 426.

[7] GARCÍA-CUERVA, J.: *General endpoint results in extrapolation*. In Analysis and P.D.E. ed. Cora Sadosky Lect. Notes in Pure and Applied Math. Marcel Dekker, vol. 122 (1990), 161 - 169.

[8] GARCÍA-CUERVA, J. and RUBIO DE FRANCIA, J. L.: *Weighted norm inequalities and related topics*. Nort-Holland Math. Studies 116 (1985).

[9] GILBERT, J. E.: *Nikishin-Stein theory and factorization with applications.* Proc. Symp. Pure Math. XXXV (2) (1979), 233 - 267.

[10] HARBOURE, E., MACIAS, R.A. and SEGOVIA, C.: *Extrapolation results for classes of weights.* Amer. Jour. of Math. 110 (1988), 383 - 397.

[11] HUNT, R. A. and YOUNG, W. S.: *A weighted norm inequality for Fourier series.* Bull. Amer. Math. Soc. 80 (1974), 274 - 277.

[12] JAWERTH, B.: *Weighted inequalities for maximal operators: linearization, localization and factorization.* Amer. Jour. of Math. 108 (1986), 361 - 414.

[13] KRIVINE, J. L.: *Theorems de factorisation dans les espaces reticules.* Sem. Maurey-Schwartz 1973/74. Palaiseau (France) exp XXII - XXIII.

[14] LINDENSTRAUSS, J. and TZAFRIRI, L.: *Classical Banach Spaces II.* Springer-Verlag (1979).

[15] MAUREY, B.: *Théorèmes de Factorisation pour les Operateurs Lineaires a Valeurs dans les Espaces* L^p. Asterisque No.11, Soc. Math. France (1974).

[16] MUCKENHOUPT, B.: *Weighted norm inequalities for the Hardy maximal function.* Trans. Amer. Math. Soc. 165 (1972), 207 - 226.

[17] PISIER, G.: *Some applications of the complex interpolation method to Banach lattices.* Journal d'Anal. Math. 35 (1979), 264 - 281.

[18] RUBIO DE FRANCIA, J. L.: *Weighted norm inequalities and vector-valued inequalities.* Harmonic Analysis (Ricci, F. and Weiss, G. editors) Springer-Verlag (1982), 86 - 101.

[19] RUBIO DE FRANCIA, J. L.: *Factorization theory and* A_p *weights.* Amer. Jour. Math. 106 (1984), 533 - 547.

[20] RUBIO DE FRANCIA, J. L.: *Acotación de operadores en retículos de Banach y desigualdades con peso.* Memorias de la Real Academia de Ciencias de Madrid. Serie de Ciencias Exactas Tomo XVIII (1985).

[21] RUBIO DE FRANCIA, J. L.: *Linear Operators in Banach lattices and Weighted* L^2 *inequalities.* Math. Nachr. 133 (1987), 197 - 209.

[22] RUBIO DE FRANCIA, J. L.: *Martingale and integral transforms of Banach space valued functions.* Probability and Banach Spaces. (J. Bastero and M. San Miguel editors). Lecture Notes in Mathematics 1221. Springer-Verlag (1986), 195 - 222.

[23] RUDIN, W.: *Functional Analysis.* McGraw-Hill New York etc. (1973).

[24] ZAANEN, A. C.: *Integration.* North-Holland Amsterdam (1967).

Weighted Inequalities in Fourier Analysis

H. P. Heinig

Hamilton, Ontario

<u>Introduction.</u> Inequalities involving the Fourier transform and its variants have been an essential part of Fourier analysis from its early beginnings. This is not surprising because the size and estimate of the Fourier transform in various function spaces is significant in the development of harmonic analysis and is underscored by the numerous applications they yield in science and engineering. The first L^p-estimate of the Fourier transform involving power weights are the results of Paley–Titchmarsh (1934) with extensions by Pitt (1937), Stein (1956) and Rooney (1966). Somewhat more general weights were considered by Hirschman (1957) and Flett (1973). Only in 1978, Muckenhoupt ([22]) formally posed the problem of characterizing for given indices p and q those non-negative weight functions u and v for which the inequality

$$\|\hat{f}\|_{q,u} \le C\|f\|_{p,v}$$

holds, for all $f \in L^1$. This problem has been studied by a number of workers, including Muckenhoupt, and, although the complete solution is still illusive, much progress has been made.

This research was supported by the Natural Sciences and Engineering Research Council of Canada grant A-4837

Many results regarding the weighted Fourier transform inequality follow from interpolation arguments and those in turn depend on results about the Hardy operator, its dual as well as variants and generalizations. It is therefore also natural to study in this connection weighted estimates for the Hardy operator. In 1986 a survey article on these subjects appeared in the Proceedings of the Spring School held in Litomyšl, Czechoslovakia (Nonlinear Analysis, Function Spaces and Applications v.3; M. Krbec, A. Kufner, J. Rákosnik, ed., Teubner Texte zur Math. Bd. 43, Leipzig 1986). The article which follows, is an attempt to describe the progress which has been made in this area since the Litomyšl paper was written.

This work is divided into three parts. The first describes recent generalizations and extensions of Hardy's inequality. We describe how weighted gradient inequalities follow in a natural way from Hardy's inequality. In a limiting case of the weighted L^p inequality for the Hardy averaging operator, we obtain weighted characterizations of exponential inequalities of this operator. These results have higher dimensional analogues which together with their discrete versions seem to be new. Then we describe results on weighted Hardy inequalities for decreasing functions. These characterizations lead to weight characterizations of more general operators, such as the Laplace transform and the Riemann–Liouville operator, on classical Lorentz spaces. Section 2 contains various weighted and measure weighted Fourier inequalities. In particular, we point out that certain weighted Fourier inequalities are equivalent to weighted Hardy inequalities. The last section describes some application of these results to Fourier restriction and extensions theorems as well as uncertainty inequalities.

It is a pleasure to thank the organizers, Professors M. Krbec, A. Kufner and J. Rákosnik for the invitation to participate in this conference, for their fine organization and generous hospitality which made this visit to Czechoslovakia especially enjoyable.

Before we begin our discussion, we collect some notations and definitions used in the sequel.

As usual we write for $(-\infty, \infty) = \mathbb{R}$, $\mathbb{R}^+ = (0, \infty)$, $\mathbb{R}^n = \{x = (x_1,...,x_n),\ x_i \in \mathbb{R},\ i = 1,2,...,n\}$ and $dx = dx_1 dx_2 ... dx_n$. The Fourier transform of f on $\mathbb{R}^n$ is defined by

$$\hat{f}(x) = \int_{\mathbb{R}^n} e^{-ix\cdot y} f(y)dy, \quad x \in \mathbb{R}^n,$$

whenever the integral converges. L^p denote the usual Lebesgue spaces of functions with

$$\|f\|_p^p \equiv \int_{\mathbb{R}^n} |f(x)|^p dx < \infty, \quad 0 < p < \infty.$$

If u is a non-negative weight function, then we write $f \in L^p_u$, if $\|f\|_{p,u} = \|fu^{1/p}\|_p < \infty$. Similarly, if μ is a measure, then $f \in L^p_\mu$, if $\|f\|_{p,\mu} < \infty$. The conjugate index p' of p, $0 < p < \infty$, is defined by $\frac{1}{p} + \frac{1}{p'} = 1$ with $p' = \infty$ if $p = 1$.

$C_c(\mathbb{R})$ denotes the space of continuous functions of compact support on $\mathbb{R}$, while $C_0(\mathbb{R})$ is the space of continuous functions $f(x)$ vanishing as $|x| \to \infty$. A measure μ on $\mathbb{R}$ is a linear functional on $C_c(\mathbb{R})$. μ is positive if $<\mu, f> \geq 0$ for all $f \in C_c(\mathbb{R})$, $\mathscr{S}(\mathbb{R})$, respectively, $\mathscr{S}(\mathbb{R}^n)$ are the spaces of Schwartz functions on $\mathbb{R}$, respectively $\mathbb{R}^n$.

We write $f\downarrow$ if f is decreasing on $\mathbb{R}^+$ and decreasing means non-increasing.

$B_r(x_0) \subset \mathbb{R}^n$ is the n-ball centered at x_0 with radius $r > 0$. $\Sigma_{n-1}(\rho) = \{x \in \mathbb{R}^n\colon |x| = \rho\}$ is the n-sphere of radius ρ, while $\Sigma = \Sigma_{n-1}(1)$. $d\sigma$ or $d\tau$ are the surface measures on these spheres, and χ_E is the characteristic function of the set E. We shall write $G \approx H$, if there are positive constants C_0, C_1 such that $C_0 \leq G/H \leq C_1$. Constants will be denoted by C,B and A (at times with subscripts), which may be different from place to place. Finally, inequalities $\|Tf\| \leq C\|f\|$ are interpreted to mean that if the right side is finite, so is the left, and the inequality holds.

1. <u>Recent generalisations and variants of Hardy's inequality.</u> The classical Hardy inequality states that if P is the averaging operator

$$(Pf)(x) = \frac{1}{x}\int_0^x f(t)dt, \quad x > 0, \tag{1.1}$$

then P is bounded on $L^p(\mathbb{R}^+)$, $p > 1$. This results and its generalizations and variants find numerous applications in many branches of mathematics. For example in the theory of differential equations, approximation theory, interpolation theory, the study of function spaces, and Fourier analysis to mention a few. During the last twenty years, Hardy's inequality was extended to general measures and in particular to measures generated by weight functions. The general result in terms of weights may be summarized in Theorem 1.1, which will be the starting point of our discussion.

<u>Theorem 1.1.</u> Suppose $0 < p, q \leq \infty$, $p \geq 1$. If f is non-negative and u,v are weights on $(0, \infty)$, then the inequality

$$\left[\int_0^\infty u(x)\left(\int_0^x f(t)dt\right)^q dx\right]^{1/q} \leq C\left[\int_0^\infty v(x)\, f(x)^p dx\right]^{1/p}$$

holds, if and only if

(i) for $1 \leq p \leq q \leq \infty$, $\sup_{r>0} \left(\int_r^\infty u(x)dx\right)^{1/q} \left(\int_0^r v(x)^{1-p'}dx\right)^{1/p'} < \infty$;

(ii) for $0 < q < p < \infty$, $p \geq 1$, $\left\{\int_0^\infty \left[\left(\int_x^\infty u(t)dt\right)^{1/q}\left(\int_0^x v(t)^{1-p'}dt\right)^{1/q'}\right]^r v(x)^{1-p'}dx\right\}^{1/r} < \infty$,

where $\frac{1}{r} = \frac{1}{q} - \frac{1}{p}$.

There is of course also an obvious dual result which is obtained easily via changes of variables. For a proof of this result together with an excellent historical survey and many applications we refer to the forthcoming book by Kufner and Opic [18].

It may be instructive to illustrate how Theorem 1.1 can be applied to a gradient type differential inequality in $\mathbb{R}^n$. Note that the averaging operator P of (1.1) and its dual Q can be written in the form

$$(Pf)(x) = \int_0^1 f(xt)dt \qquad (Qf)(x) = \int_1^\infty f(xt) \frac{dt}{t}$$

But in this form the operator is also well defined for $x \in \mathbb{R}^n$. Now if Q^* is defined by $(Q^*f)(x) = -x \cdot (\nabla f)(x)$, $x \in \mathbb{R}^n$ then for any $f \in C_0^1(\mathbb{R}^n)$ we get

$$Q(Q^*f)(x) = -\int_1^\infty (tx)\cdot(\nabla f)(tx) \frac{dt}{t} = -\int_1^\infty \frac{d}{dt} f(tx)dt = f(x).$$

Now using polar coordinates it follows from this, that

$$\int_{\mathbb{R}^n} |f(x)|^q u(x)dx = \int_{\mathbb{R}^n} u(x)|\int_1^\infty (Q^*f)(xt) \frac{dt}{t}|^q dx = \int_\Sigma \int_0^\infty u(s\sigma)s^{n-1}|\int_1^\infty (Q^*f)(\sigma st) \frac{dt}{t}|^q ds d\sigma$$

$$= \int_\Sigma \int_0^\infty u(s\sigma)s^{n-1}|\int_s^\infty (Q^*f)(\sigma y) \frac{dy}{y}|^q ds d\sigma.$$

Since the inner integral corresponds to the one dimensional dual operator one obtains on applying the dual of Theorem 1.1 (i) the following gradient inequality of Sinnamon [28]:

<u>Theorem 1.2.</u> Suppose $1 < p < \infty$, $0 < q < \infty$, then for all $f \in C_0^\infty(\mathbb{R}^n)$

$$\left(\int_{\mathbb{R}^n} |f(x)|^q v(x)dx\right)^{1/q} \le C\left(\int_{\mathbb{R}^n} |x\cdot(\nabla f)(x)|^p u(x)dx\right)^{1/p}$$

holds, if and only if

(i) for $p = q$, $\sup_{x \in \mathbb{R}^n} \left(\int_0^1 v(tx)t^{n-1}dt\right)^{1/p} \left(\int_1^\infty [u(tx)t^n]^{1-p'} \frac{dt}{t}\right)^{1/p'} < \infty$

(ii) for $0 < q < p$, $p > 1$

$$\left\{\int_{\mathbb{R}^n} \left[\left(\int_0^1 v(tx)t^{n-1}dt\right)^{1/q} \left(\int_1^\infty [u(tx)t^n]^{1-p'} \frac{dt}{t}\right)^{1/q'}\right]^r u(x)^{1-p'} dx\right\}^{1/r} < \infty$$

where $\frac{1}{r} = \frac{1}{q} - \frac{1}{p}$

It should be remarked here that in the case $1 < p < q < \infty$, not only does the proof of this result fail but also the inequality (c.f. [28, Theorem 3.4]).

If the indices $p, q \in (0, 1)$ then either the domain spaces of the Hardy operator must be replaced by the boundary values of functions in weighted Hardy spaces in order to get weighted inequalities ([12]) or the inequalities are in the opposite direction ([3]). For example in case $q \leq p < 0$ the following holds:

Theorem 1.3. ([3, Theorem 1]). Let $q \leq p < 0$, u,v and f positive a.e. and suppose

$$K(r) \equiv \left(\int_0^r u(x)dx\right)^{1/q} \left(\int_0^r v(x)^{1-p'} dx\right)^{1/p'}$$

Then

$$\left(\int_0^\infty v(x)f(x)^p dx\right)^{1/p} \leq C \left(\int_0^\infty u(x)\left(\int_0^\infty f(t)dt\right)^q dx\right)^{1/q} \tag{1.2}$$

provided $K(r)$ is non-decreasing and $\inf_{r>0} K(r) \equiv B > 0$. Conversely, (1.2) implies $\inf K(r) \equiv B > 0$.

1.1. Inequalities related to those of Hardy's. We now discuss weighted inequalities which may be considered limiting cases of Hardy's inequality. Indeed we characterize the weights for which certain exponential inequalities hold. Unlike Hardy's inequality the higher dimensional generalizations of these results follow in a straight–forward manner. We give in addition a discrete form which yields a weighted characterization of Carleman's inequality. The results of this section have been obtained jointly with R. Kerman and M. Krbec.

Theorem 1.1 with $p = q$ shows (with explicit constant) that

$$\int_0^\infty u(x) \left(x^{-1} \int_0^x f(t)^{1/p}\, dt\right)^p dx \le B_p^p\, p(p')^{p-1} \int_0^\infty v(x)f(x)dx \tag{1.3}$$

holds, if and only if

$$B_p^p \equiv \sup_{r>0} \left(\int_r^\infty x^{-p}u(x)dx\right) \left(\int_0^r v(x)^{1-p'}dx\right)^{1/p'} < \infty.$$

Now if $p \to \infty$, then an application of Fatou's lemma shows that (1.3) implies

$$\int_0^\infty u(x)\exp\left(x^{-1} \int_0^x \ln f(t)dt\right)dx \le C \int_0^\infty v(x)f(x)dx, \tag{1.4}$$

provided $\lim B_p^p p(p')^{p-1} = C$ as $p \to \infty$. In this sense (1.4) might be considered the limiting case of Hardy's inequality. Of course there is no reason to assume that $B_p^p p(p')^{p-1} \to C$ as $p \to \infty$.

Besides the operator P of (1.1) we define Q_α, $\alpha > 0$, as follows:

$$(Q_\alpha f)(x) = \alpha x^\alpha \int_x^\infty t^{-\alpha-1} f(t)dt.$$

Theorem 1.4. Let u, v and f be positive a.e. and

$$w(x) = u(x)\exp[(P \ln \tfrac{1}{v})(x)], \quad x > 0.$$

Then, the following are equivalent:

(i) $\int_0^\infty u(x)\exp(P\,\ln f)(x)dx \le C_1 \int_0^\infty v(x)f(x)dx$

(ii) $(Pw)(x) + \alpha^{-1}(Q_\alpha w)(x) \le C_2, \quad \alpha > 0$

(iii) $(Pw) \in L^\infty$

(iv) $\int_0^\infty w(x)(Pf^{1/p})(x)dx \le C_3 \int_0^\infty f(x)dx, \quad (p > 2)$

where C_3 is independent of p.

Proof. (i) ⇒ (ii). Write $f = \frac{g}{v}$, where $g(x) = t^{-1}\chi_{(0,t)}(x) + x^{-\alpha-1}e^{-1-\alpha}t^{\alpha}\chi_{(t,\infty)}(x)$, $t > 0$, $\alpha > 0$. Substituting into (i) the right side becomes

$$C_1\left[t^{-1}\int_0^t dx + e^{-1-\alpha}t^{\alpha}\int_t^\infty x^{-\alpha-1}dx\right] = C_1[1 + e^{-1-\alpha}/\alpha],$$

while the left side takes the form

$$\int_0^\infty u(x)\exp(P\,\ln \tfrac{g}{v})(x)dx = \int_0^\infty u(x)\exp(P\,\ln \tfrac{1}{v})(x)\exp(P\,\ln g)dx$$

$$= \int_0^\infty w(x)\exp\left(x^{-1}\int_0^x \ln g(y)dy\right)dx = \left(\int_0^t + \int_t^\infty\right) \equiv I_1 + I_2$$

respectively. With the above defined g

$$I_1 = \int_0^t w(x)\exp\left[\frac{1}{x}\int_0^x \ln\left(\frac{1}{t}\right)dy\right]dx = \frac{1}{t}\int_0^t w(x)dx = (Pw)(t)$$

and

$$I_2 = \int_t^\infty w(x)\exp\left(x^{-1}\int_0^t \ln\left(\frac{1}{t}\right)dy\right)\exp\left(x^{-1}\int_t^x \ln\left(e^{-1-\alpha}t^\alpha y^{-\alpha-1}\right)dy\right)dx$$

$$= \int_t^\infty w(x)\exp\left[x^{-1}t\ \ln\left(\frac{1}{t}\right) + \left(1-\frac{t}{x}\right)\ln\left(e^{-1-\alpha}t^\alpha\right) - (\alpha+1)x^{-1}\int_t^x \ln y\,dy\right]dx$$

$$= \int_t^\infty w(x)\exp\left[x^{-1}t\ \ln\left(\frac{1}{t}\right) + \left(1-\frac{t}{x}\right)\left(-1-\alpha\right) + \alpha\left(1-\frac{t}{x}\right)\ln t - x^{-1}(\alpha+1)(x\ln x - x + t - t\ln t)\right]dx$$

$$= \int_t^\infty w(x)\exp\left[x^{-1}t\ \ln\left(\frac{1}{t}\right) + \left(1-\frac{t}{x}\right)\ln t - \left(\alpha+1\right)\ln x + x^{-1}\,t(\alpha+1)\ln t\right]dx$$

$$= \int_t^\infty w(x)\exp\left[\alpha \ln t - (\alpha+1)\ln x\right]dx = t^\alpha\int_t^x x^{-\alpha-1}\,w(x)dx$$

$$= \alpha^{-1}(Q_\alpha w)(t).$$

Therefore, $(Pw)(t) + \alpha^{-1}(Q_\alpha w)(t) \le C_1[1 + e^{-1-\alpha}/\alpha] \equiv C_2$ and (ii) holds.

(ii) ⇒ (iii) is obvious. To prove (iii) ⇒ (iv) note that by Hölder's inequality with $q > 1$

$$(Ph)(x) \le [(Ph^q)(x)]^{1/q} \le \|h\|_q/x^{1/q},$$

so that

$$\int_{\{x:(Ph)(x)>\lambda\}} w(x)dx \le \int_{\{x:(\|h\|_q/\lambda)^q>x\}} w(x)dx = \int_0^{(\|h\|_q/\lambda)^q} w(x)dx \le C(\|h\|_q/\lambda)^q .$$

Here the last inequality is implied by (iii) with C independent of q. Hence the operator P is bounded from L^q to weighted "weak" L^q_w, $q > 1$ with norm independent of q. Now let $p > 2$, then we apply the Marcinkiewicz interpolation theorem with $q_0 = p/2$ and $q_1 = 3p/2$, $(\theta = 3/4)$ to obtain

$$\int_0^\infty w(x)\,(Ph)(x)^p dx \le C_3 \int_0^\infty h(x)^p dx$$

where C_3 is independent of p. With $h^p = f$ this implies (iv).

Finally (iv) $\Rightarrow$ (i) follows from Fatou's lemma and the fact that

$$\lim_{p\to\infty} (Pf^{1/p})^p(x) = \exp(P \ln f)(x) \qquad ([33;\ p.\ 344,\ 5c]).$$

Remark 1.5. If (iii) holds, then $(Pw)(x) \le C$, so that $(Q_\alpha(Pw))(x) \le (Q_\alpha C)(x) = C$. Thus

$$C \ge (Q_\alpha(Pw))(x) = \alpha x^\alpha \int_x^\infty y^{-\alpha-2}\left(\int_0^y w(t)dt\right)dy$$

$$= \alpha x^\alpha \int_x^\infty y^{-\alpha-2}\left[\int_0^x w(t)dt + \int_x^y w(t)dt\right]dy = \frac{\alpha}{\alpha+1}(Pw)(x) + \alpha x^\alpha \int_x^\infty w(t) \int_t^\infty y^{-\alpha-2}dydt$$

$$= \frac{\alpha}{\alpha+1}(Pw)(x) + \frac{1}{\alpha+1}(Q_\alpha w)(x).$$

Therefore, $Pw \in L^\infty$, if and only if $Q_\alpha w \in L^\infty$ and we obtain

Corollary 1.6. If u, v and f are positive and

$$w(x) = u(x)\exp(P \ln 1/v)(x), \quad x > 0,$$

then

$$\int_0^\infty u(x)\exp\left(x^{-1}\int_0^x f(y)dy\right)dx \le C\int_0^\infty v(x)\exp(f(x))dx,$$

if and only if for any $\alpha > 0$, $\sup_{x>0} \alpha^{-1}(Q_\alpha w)(x) < \infty$.

A simple calculation shows that $u(x) = v(x) = x^\beta$, β real, satisfy the weight condition and so do the functions

$$u(x) = e^{\beta x/2},\ v(x) = e^{\beta x},\ \beta \text{ real}.$$

The characterization of the weights for the exponential inequality for the averaging operator in higher dimensions carries over in a straight forward way unlike the corresponding characterizations in L^p–spaces, where the n–dimensional result, $n > 2$, is still open. Here we simply state the two dimensional case of the exponential inequality.

Let $x = (x_1, x_2)$, $y = (y_1, y_2)$, and P^2, Q_α^2 be defined by

$$(P^2f)(x) = x_1^{-1}x_2^{-1}\int_0^{x_1}\int_0^{x_2} f(y)dy,\ (Q_\alpha^2 f)(x) = \alpha_1\alpha_2 x_1^{\alpha_1}x_2^{\alpha_2}\int_{x_1}^\infty\int_{x_2}^\infty y_1^{-\alpha_1}y_2^{-\alpha_2-1}f(y)dy$$

$\alpha_1 > 0$, $\alpha_2 > 0$, then the following holds:

Theorem 1.7. Let u, v and f be positive a.e. on $\mathbb{R}^2_+$, then

$$\int_{\mathbb{R}^2_+} u(x) \exp(P^2 \ln f)(x)dx \le C \int_{\mathbb{R}^2_+} v(x) f(x)\, dx$$

if and only if $\sup_{x_1>0, x_2>0} (Q^2 w)(x) < \infty$, where $w(x) = u(x) \exp(P^2 \ln 1/v)(x)$.

There are further extensions of these results to more general operators with positive kernels satisfying certain homogeneity conditions. The Laplace transform is a special case of these. We shall not discuss these generalizations here, instead we consider the discrete case and give a variant of the arithmetic–geometric mean inequality.

Recall that if $a_1, a_2, \dots$; are positive real numbers, then Carleman's inequality asserts that

$$\sum_{n=1}^{\infty} [a_1 a_2 \dots a_n]^{1/n} \le e \sum_{n=1}^{\infty} a_n, \tag{1.5}$$

where the constant e is sharp. There are a number of weighted generalizations of this inequality, some with sharp constants (c.f. [10], [13], [19]). The next result characterizes weights for which a weighted inequality corresponding to (1.5) holds.

Theorem 1.8. Let $\{u_n\}$, $\{v_n\}$, $\{a_n\}$ be sequences of positive numbers and suppose

$$w_n = u_n \exp\left[\frac{1}{n} \sum_{k=1}^{n} \ln(1/v_k)\right] \quad n = 1,2,\dots ;$$

Then

$$\sum_{n=1}^{\infty} u_n [a_1 a_2 \dots a_n]^{1/n} \le C \sum_{n=1}^{\infty} v_n a_n \tag{1.6}$$

if and only if

$$\sup_{k\geq 1} k^{\alpha} \sum_{n=k}^{\infty} n^{-\alpha-1} w_n \equiv A < \infty, \alpha > 0. \tag{1.7}$$

Moreover

$$\alpha A/(\alpha + e^{-1-\alpha}) \leq C \leq e^{\alpha} A.$$

Proof. Let $b_n = a_n v_n$, $n = 1,2,\ldots$; then (1.6) is equivalent to

$$\sum_{n=1}^{\infty} w_n \exp\left[\frac{1}{n} \sum_{k=1}^{n} \ln b_k\right] \leq C \sum_{n=1}^{\infty} b_n. \tag{1.8}$$

Let $f(t) = b_k$, if $k - 1 < t \leq k$, $k = 1,2,\ldots$; and zero otherwise. Since

$$\int_0^1 \ln(y^{\alpha})dy = -\alpha$$

the left side of (1.8) becomes

$$\sum_{n=1}^{\infty} w_n \exp\left[\frac{1}{n} \sum_{n=1}^{n} \int_{k-1}^{k} \ln f(t)dt\right] = \sum_{n=1}^{\infty} w_n \exp\left[\frac{1}{n} \int_0^n \ln f(t)dt\right]$$

$$= \sum_{n=1}^{\infty} w_n \exp\left[\int_0^1 \ln f(ny)dy\right] = e^{\alpha} \sum_{n=1}^{\infty} w_n \exp\left[\int_0^n \ln(y^{\alpha} f(ny))dy\right]$$

$$\leq e^{\alpha} \sum_{n=1}^{\infty} w_n \int_0^1 y^{\alpha} f(ny)dy = e^{\alpha} \sum_{n=1}^{\infty} w_n n^{-1-\alpha} \int_0^n x^{\alpha} f(x)dx$$

$$= e^{\alpha} \sum_{n=1}^{\infty} w_n n^{-1-\alpha} \sum_{k=1}^{n} \int_{k-1}^{k} x^{\alpha} f(x)dx = e^{\alpha} \sum_{n=1}^{\infty} w_n n^{-1-\alpha} \sum_{k=1}^{n} k^{\alpha} b_k$$

$$= e^{\alpha} \sum_{k=1}^{\infty} b_k \Big[k^{\alpha} \sum_{n=k}^{\infty} n^{-1-\alpha} w_n \Big] \le A e^{\alpha} \sum_{k=1}^{\infty} b_k .$$

Here the first inequality follows from Jensen's inequality. Hence (1.8) follows.

To show that (1.8) implies (1.7) let

$$b_k = \begin{cases} 1/k, & \text{if } n \le k \\ e^{-1-\alpha} k^{\alpha} n^{-1-\alpha}, & \text{if } n > k ; \end{cases}$$

$n = 1,2,\dots$; k fixed. Then

$$\sum_{n=1}^{\infty} b_n = \frac{1}{k} \sum_{n=1}^{k} 1 + e^{-1-\alpha} k^{\alpha} \sum_{n=k+1}^{\infty} n^{-\alpha-1} \le 1 + e^{-1-\alpha}/\alpha$$

and by (1.8)

$$C(1 + e^{-1-\alpha}/\alpha) \ge \sum_{n=1}^{k} w_n \exp\Big[\frac{1}{n} \sum_{j=1}^{n} \ln b_j\Big] + \sum_{n=k+1}^{\infty} w_n \exp\Big[\frac{1}{n} \sum_{j=1}^{n} \ln b_j\Big] \equiv I_1 + I_2,$$

respectively. But $b_j = 1/k$ if $j \le k$, so $I_1 = \frac{1}{k} \sum_{n=1}^{k} w_n$ and

$$I_2 = \sum_{n=k+1}^{\infty} w_n \exp\Big[\frac{1}{n} \Big(\sum_{j=1}^{k} \ln 1/k + \sum_{j=k+1}^{n} \big(e^{-1-\alpha} k^{\alpha}/j^{\alpha+1}\big)\Big)\Big]$$

$$= \sum_{n=k+1}^{\infty} w_n \exp(n^{-1} k \ln 1/k) \exp[n^{-1}(n-k-1) \ln(e^{-1-\alpha} k^{\alpha})] \cdot \exp\Big[-n(\alpha + 1) \sum_{j=k+1}^{n} \log j\Big].$$

But

$$\sum_{j=k+1}^{n} \log j \leq -n + k - 1 + (n+1)\ln n - (k+1)\ln k$$

so that

$$I_2 \geq \sum_{n=k+1}^{\infty} w_n \exp[\ln(k^{\alpha}) + \ln(n^{-\alpha-1}) + n^{-1} \ln(k/n^{\alpha+1})]$$

$$\geq k^{\alpha} e^{-1-\alpha} \sum_{n=k+1}^{\infty} w_n\, n^{-\alpha-1}$$

since $n^{-1} \ln(k/n^{\alpha+1}) \geq -n^{-1}(\alpha+1) \ln n \geq -\alpha-1$. It follows then that

$$C[1 + e^{-1-\alpha}/\alpha] \geq k^{-1} \sum_{n=1}^{k} w_n + e^{-\alpha-1} k^{\alpha} \sum_{n=k}^{\infty} w_n\, n^{-\alpha-1} - e^{-\alpha-1} w_k k^{-1}$$

$$\geq e^{-\alpha-1} k^{\alpha} \sum_{n=k}^{\infty} w_n\, n^{-\alpha-1}.$$

Hence $C \geq \frac{\alpha}{\alpha + e^{-1-\alpha}} A$, which proves the theorem.

Extensions of this results to higher dimensions follow as in the continuous case. In addition characterizations of weights in the directions of the results in [10], [13] and [19] are possible.

We now consider briefly more general operators than the averaging operator and their boundedness in weighted L^p spaces. Specifically, we consider the Riemann–Liouville fractional integral operator defined by

$$(P_{\alpha}f)(x) = \frac{1}{\Gamma(\alpha)} \int_0^x (x-t)^{\alpha-1} f(t)\,dt, \qquad 0 < \alpha < \infty. \tag{1.9}$$

Recently Martin–Reyes and Sawyer [20] and independently Stepanov [31] proved that $P_\alpha: L^p_v \longrightarrow L^q_u$, $1 \le p \le q < \infty$, $\alpha \ge 1$ is bounded, if and only if

$$\sup_{r>0} \left(\int_r^\infty u(t)(t/r)^{q(\alpha-1)}dt\right)^{1/q} \left(\int_0^r (r-t)^{p'(\alpha-1)}\, v(t)^{1-p'}dt\right)^{1/p'} \equiv A < \infty$$

and

$$\sup_{r>0} \left(\int_r^\infty u(t)(t-r)^{q(\alpha-1)}dt\right)^{1/q} \left(\int_0^r v(t)^{1-p'}dt\right)^{1/p'} \equiv B < \infty.$$

It should be noted that Stepanov's (equivalent) conditions are somewhat different, namely

$$\max_{\gamma=0,1} \sup_{r>0} \left(\int_r^\infty (t-x)^{(\alpha-1)(1-\gamma)}\, u(t)dt\right)^{1/q} \left(\int_0^r (r-t)^{(\alpha-1)\gamma p'} v(t)dt\right)^{1/p'} < \infty.$$

Stepanov further characterized the weights u, v for which $P_\alpha: L^p_v \longrightarrow L^q_u$ is bounded in the index range $0 < q < p < \infty$, $p > 1$.

The corresponding characterizations of the weights for P_α when $0 < \alpha < 1$ has not yet been solved, however, sufficient conditions, similar to those given above are known in this case ([1], [14]). We shall return to this question shortly.

For far more general integral operators with positive kernels, weight characterization of a different kind were given in [16] (see also the literature cited there).

1.2. Weighted inequalities for operators of monotone functions.

If $1 < p = q < \infty$ and $u(t) = v(t)/t^p$, where $v(t)$ is defined by

$$v(t) = \begin{cases} 0 & \text{if } 1 < t < 2 \\ t^{-1/2} & \text{if } 0 < t \le 1 \text{ or } t \ge 2 \end{cases},$$

then clearly (i) of Theorem 1.1. fails. However, if f is a decreasing function, then as shown in [2]

$$\int_0^\infty v(x)\,(Pf)(x)^p dx \le C \int_0^\infty v(x)\, f(x)^p dx, \quad f \ge 0.$$

It is therefore natural to consider the analogue of Theorem 1.1 for decreasing functions. The first characterization of weights in this direction is due to Ariño and Muckenhoupt. Their result is

Theorem 1.9. ([2; Theorem 1.7]). If $1 \le p < \infty$ and $u(x) \ge 0$ then $\|Pf\|_{p,u} \le C\|f\|_{p,u}$ holds for all non-negative decreasing f, if and only if for all $r > 0$

$$\int_r^\infty x^{-p} u(x)dx \le C\, r^{-p} \int_0^r u(x)dx. \tag{1.10}$$

Let $\Lambda_p(u)$ denote the classical Lorentz spaces, that is, the set of measurable functions g on $\mathbb{R}^n$, such that

$$\|g\|_{\Lambda_p(u)} \equiv \left\{\int_0^\infty g^*(x)^p u(x)dx\right\}^{1/p} < \infty,\ 1 \le p < \infty,$$

where g^* is the rearrangement of $|g|$, namely

$$g^*(x) = \inf\{s > 0;\ \mu(\{t\colon |g(t)| > s\}) \le x\},$$

here μ denotes Lebesgue measure. It follows then from Theorem 1.9 that the Hardy–Littlewood maximal operator is bounded on $\Lambda_p(u)$ if and only if u satisfies (1.10).

There are several generalizations of Theorem 1.9 due to Braverman ([8]), Neugebauer ([23]), Sawyer ([27]) and Stepanov ([32]). Braverman defined the operator T_φ by

$$(T_\varphi f)(x) = x^{-1} \int_0^x f(t)\, \varphi(t/x)dt \quad x > 0$$

where $\varphi\downarrow$, $\varphi:(0, 1) \to \mathbb{R}^+$ and satisfies $\varphi(xy) \le C\, \varphi(x)\, \varphi(y)$. His result may then be formulated as follows:

Theorem 1.10. If $1 \le p < \infty$, then $\|T_\varphi f\|_{\Lambda_p(u)} \le C\|f\|_{\Lambda_p(u)}$, if and only if for each $r > 0$

$$\int_r^\infty u(x) \left(\int_0^{r/x} \varphi(t)dt\right)^p dx \le A \int_0^r u(x)dx.$$

While this result reduces to Theorem 1.9 with $\varphi \equiv 1$, it also shows with $\varphi(x) = (1 - x)^{\alpha-1}$, $0 < \alpha \le 1$, a one weight characterization, of the boundedness of the Riemann–Liouville fractional integral operator on $\Lambda_p(u)$ (c.f. [20], [31] and the previous remarks).

The result of Stepanov yields a two weight characterization for a wide range of indices for the Hardy operator:

Theorem 1.11. ([32]). Necessary and sufficient conditions for

$$\left(\int_0^\infty (Pf)(x)^q\, u(x)dx\right)^{1/q} \le C\left(\int_0^\infty v(x)f(x)^p dx\right)^{1/p},$$

$0 < p, q < \infty$ to be satisfied for all non-negative decreasing f is that

(a) for $1 < p \leq q < \infty$

$$A_0 \equiv \sup_{r>0} \left(\int_0^r u(t)dt\right)^{1/q} \left(\int_0^r v(x)dx\right)^{-1/p} < \infty$$

and

$$A_1 \equiv \sup_{r>0} \left(\int_r^\infty t^{-q} u(t)dt\right)^{1/q} \left(\int_0^r t^{p'} V(t)^{-p'} v(t)dt\right)^{1/p'} < \infty,$$

where $V(t) = \int_0^t v$, holds.

(b) For $0 < q < p < \infty$, $1/r = 1/q - 1/p$, $p > 1$,

$$B_0 \equiv \left\{\int_0^\infty \left[\left(\int_0^t u(x)dx\right)^{1/p} \left(\int_0^t v(x)dx\right)^{-1/p}\right]^r u(t)dt\right\}^{1/r} < \infty$$

and

$$B_1 \equiv \left\{\int_0^\infty \left[\left(\int_t^\infty x^{-q} u(x)dx\right)^{1/q} \left(\int_0^t x^{p'} V(x)^{-p'} v(x)dx\right)^{-1/q'}\right]^r t^{p'} V(t)^{-p'} v(t)dt\right\}^{1/r} < \infty$$

holds.

(c) For $0 < p \leq q < \infty$, $0 < p < 1$, that $A_0 < \infty$ and

$$A_1^{\#} \equiv \sup_{r>0} r\left(\int_r^\infty x^{-q} u(x)dx\right)^{1/q} \left(\int_0^r v(x)dx\right)^{-1/p} < \infty .$$

It should be noted here that the index range in (c) shows that for decreasing f the mapping properties of the Hardy operator are fundamentally different than those for arbitrary f. In the index range $1 < p, q < \infty$, these results can also be obtained from Sawyer's reverse Hölder inequality ([17]):

$$\sup_{g\geq 0,\, g\downarrow} \frac{\int_0^\infty gf}{\left(\int_0^\infty g^p v\right)^{1/p}} \approx \left\{\int_0^\infty \left(\int_0^x f\right)^{p'} V(x)^{-p'} v(x)dx\right\}^{1/p'} + \frac{\int_0^\infty f}{\left(\int_0^\infty v\right)^{1/p}}$$

where as before $V(x) = \int_0^x v$. This results is very useful in establishing mapping properties of more general integral operators. Thus, if

$$(Tf)(x) = \int_0^\infty k(x, y)f(y)dy, \quad k(x, y) \geq 0,$$

then (with $\int_0^\infty v = \infty$) the above estimates imply that

$$\left(\int_0^\infty (Tg)(x)^q \, u(x)dx\right)^{1/q} \leq C\left(\int_0^\infty g(x)^p \, v(x)dx\right)^{1/p}$$

holds for all $g\downarrow$ and non-negative, if and only if T^*, the dual of T satisfies

$$\left[\int_0^\infty \left(\int_0^x T^*f\right)^{p'} V(x)^{-p'} v(x)dx\right]^{1/p'} \le C\left(\int_0^\infty f(x)^{q'} u(x)^{1-q'} dx\right)^{1/q'} \qquad (1.11)$$

for all $f \ge 0$.

Now if $T = P$, the Hardy averaging operator, then

$$\int_0^x T^*f(t)dt = \int_0^x \left(\int_t^\infty \frac{f(y)}{y} dy\right)dt = \int_0^x \int_t^x \frac{f(y)}{y} dydt + \int_0^x \left(\int_x^\infty \frac{f(y)}{y} dy\right)dt$$

$$= \int_0^x f(y)dy + x \int_x^\infty \frac{f(y)}{y} dy$$

so that for $1 < p, q < \infty$, Theorem 1.11 follows from (1.11) and Theorem 1.1.

It should be noted that Sawyer's result also permits one to characterize weights for which the Hilbert transform and Riesz potential is bounded from $\Lambda_p(v)$ to $\Lambda_q(u)$, $1 \le p, q \le \infty$ (See e.g. [17]).

Let L denote the Laplace transform

$$(Lf)(x) = \int_0^\infty e^{-xt} f(t)dt \qquad x > 0,$$

then there is a simple characterization of weights for which this operator defined on decreasing functions is bounded from L_v^p to L_u^q. Since this result does not seem to follow from Sawyer's work we give the simple argument next.

Theorem 1.12. Let f be a non–negative decreasing function and $1 < p \le q < \infty$, then

$$\left(\int_0^\infty u(x)|x^{-1}(Lf)(x^{-1})|^q dx\right)^{1/q} \le C\left(\int_0^\infty v(x)f(x)^p dx\right)^{1/p} \qquad (1.12)$$

and if and only if A_0 and A_1 of Theorem 1.11 are finite.

Proof. By the second mean value theorem and the fact that f is decreasing

$$(Lf)(x) = \int_0^{1/x} e^{-xy} f(y)dy + \int_{1/x}^\infty e^{-xy} f(y)dy \le \int_0^{1/x} f(y)dy + e^{-1}f(1/x)/x$$

$$\le (1+e^{-1}) \int_0^{1/x} f(y)dy .$$

But then by Theorem 1.11 the sufficiency part follows.

For the converse, assume first that $V(\infty) = \infty$ (recall $V(x) = \int_0^x v$) and define

$$f_r(s) = \left(\int_s^\infty x^{p'} V(x)^{-p'-1} v(x)dx\right)^{1/p} \chi_{(0,r)}(s), \quad r > 0 \text{ fixed.}$$

Then $f_r\downarrow$ and substituting into (1.12) we obtain

$$C\left(\int_0^r v(x) \left[\int_x^r y^{p'} V(y)^{-p'-1} v(y)dy\right] dx\right)^{1/p} = C\left(\left[\int_0^r y^{p'} V(y)^{-p'-1} v(y)\left(\int_0^y v(x)dx\right) dy\right)^{1/p}$$

$$= C\Big(\int_0^r y^{p'} V(y)^{-p'} v(y)dy\Big)^{1/p} \geq \Big(\int_r^\infty u(x)x^{-q} \Big|\int_0^r e^{-y/x} \Big(\int_y^r t^{p'} V(t)^{-p'-1} v(t)dt\Big)^{1/p} dy\Big|^q dx\Big)^{1/q}$$

$$\geq e^{-1} \Big[\int_r^\infty u(x)x^{-q}\Big|\int_0^r\Big(\int_y^r t^{p'} V(t)^{-p'-1} v(t)dt\Big)^{1/p} dy\Big|^q dx\Big]^{1/q}.$$

But the inner integral is not smaller than

$$\int_0^r y^{p'/p} \Big(\int_y^r V(t)^{-p'-1} v(t)dt\Big)^{1/p} dy = p\int_0^r y^{p'-1} \int_y^r \Big[\int_s^r V^{-p'-1} v\, dx\Big]^{-1/p'} V(s)^{-p'-1} v(s)dsdy$$

$$= p\int_0^r V(s)^{-p'-1} v(s) \Big(\int_0^s y^{p'-1} dy\Big) \Big(\int_s^r V(\alpha)^{-p'-1} v(\alpha)d\alpha\Big)^{-1/p'} ds$$

$$\geq (p-1)\int_0^r s^{p'} \Big(\int_s^\infty V(\alpha)^{-p'-1} v(\alpha)d\alpha\Big)^{-1/p'} V(s)^{-p'-1} v(s)ds$$

$$= (p-1)\int_0^r s^{p'} V(s)^{-p'-1} v(s)\Big(\frac{V(s)^{-p'}}{p'}\Big)^{-1/p'} ds = (p-1)(p')^{1/p'} \int_0^r s^{p'} V(s)^{-p'} v(s)ds.$$

Substituting, we obtain

$$C\Big(\int_0^r y^{p'} V(y)^{-p'} v(y)dy\Big)^{1/p} \geq e^{-1}(p-1)(p')^{1/p'} \Big(\int_r^\infty u(x)x^{-q}dx\Big)^{1/q} \Big(\int_0^r s^{p'} v(s)\, V(s)^{-p'} ds\Big)$$

from which $A_1 < \infty$ follows.

On the other hand, if $f = \chi_{(0,r)}$ is substituted into (1.12) then

$$C\left(\int_0^r v(x)dx\right)^{1/p} \geq \left(\int_0^\infty u(x)x^{-q}\left(\int_0^r e^{-y/x}\,dy\right)^q dx\right)^{1/q}$$

$$= \left(\int_0^\infty u(x)\,[1 - e^{-r/x}]^q dx\right)^{1/q} \geq \left(1 - e^{-1}\right)\left(\int_0^r u(x)dx\right)^{1/q}$$

and this implies $A_0 < \infty$, which completes the proof.

We noted before that for $p, q < 1$ and $f \geq 0$ the inequality

$$\left(\int_0^\infty f(t)^p\, v(t)dt\right)^{1/p} \leq C\left(\int_0^\infty u(x)\left(\int_0^x f(t)dt\right)^q dx\right)^{1/q} \tag{1.13}$$

is possible ([3]). However for most other values of the indices the following lemma, which I owe to Eric Sawyer, shows that (1.13) cannot hold.

<u>Lemma 1.13.</u> If (1.13) holds for $0 < q < \infty$, $p > 1$ and all $f \geq 0$, then $v \equiv 0$ a.e. in any interval (r, ∞) such that

$$\int_r^\infty u(t)dt < \infty.$$

Conversely, for such trivial weights, (1.13) clearly holds.

Proof. Let $F = \min(1, v^{-1/p})$ and $f_{r,s}(t) = \chi_{(r,s)}(t)F(t)$, where $0 < r < s < \infty$. Let $f = f_{r,s}$ in (1.13), then

$$\left(\int_r^s F(t)^p v(t)dt\right)^{1/p} \le C\left(\int_r^\infty u(x)dx\right)^{1/q} \left(\int_r^s F(t)dt\right) \le C(s-r)\left(\int_r^\infty u(x)dx\right)^{1/q}$$

Therefore

$$\left[\frac{1}{s-r}\int_r^s [\min(v(t), 1)]dt\right]^{1/p} \le C(s-r)^{1/p'} \left(\int_r^\infty u(x)dx\right)^{1/q}$$

and as $s - r \to 0$ and (r, s) shrinks to a Lebesgue point x of $\min(v, 1)$ we obtain $\min(v(x), 1) \le 0$ if

$$\int_{x-\epsilon}^\infty u(x)dx < \infty, \quad \epsilon > 0.$$

This proves the lemma.

In light of this lemma it is perhaps surprising that for decreasing f a weighted inequality of the form (1.13) can hold.

Theorem 1.14. Let $f\downarrow$ and non-negative. If $1 \le q \le p < \infty$, then (1.13) holds if and only if for each $r > 0$

$$\left(\int_0^r v(x)dx\right)^{1/p} \le C\left(\int_0^r u(x)dx + \int_r^\infty (r/x)^q u(x)dx\right)^{1/q}.$$

Proof. For $p = q$ this result is due to Neugebauer [23] and the case $q < p$ follows from it easily.

2. <u>Weighted Fourier inequalities.</u> The first L^p–estimates involving power weights appeared in the early and mid 1930's with the work of Hardy–Littlewood–Paley–Titchmarsh and Pitt. Although many others extended and generalized these early results, it was in 1978 when Muckenhoupt ([22]) formally posed the problem of characterizing weight functions u and v for which the inequality

$$\|\hat{f}\|_{q,u} \leq C\|f\|_{p,v} \tag{2.1}$$

holds for all $f \in L^1$. This problem has been studied by various authors, including Muckenhoupt, and the general result may be formulated in the following theorem ([4]):

<u>Theorem 2.1.</u> Suppose $0 < p, q < \infty$, $p \geq 1$ and u, v are non–negative even functions, such that $u\downarrow$ and $1/v\downarrow$ on $(0, \infty)$. Then

(i) for $1 \leq p \leq q < \infty$, (2.1) is equivalent to the condition

$$\sup_{r>0} \left(\int_0^r u(t)dt\right)^{1/q} \left(\int_0^{1/r} v(t)^{1-p'}dt\right)^{1/p'} < \infty \tag{2.2}$$

(ii) for $0 < q < p < \infty$, $p \geq 1$, the two conditions

$$\int_0^\infty \left[\left(\int_0^{1/x} u(t)dt\right)^{1/q} \left(\int_0^x v(t)^{1-p'}dt\right)^{1/q'}\right]^r v(x)^{1-p'}dx < \infty$$

and

$$\int_0^\infty \left[\left(\int_{1/x}^\infty t^{-q/2}\, u(t)dt\right)^{1/q} \left(\int_x^\infty t^{-p'/2} v(t)^{1-p'} dt\right)^{1/q'}\right]^r x^{-p'/2} v(x)^{1-p'} dx < \infty,$$

where $1/r = 1/q - 1/p$, are sufficient for (2.1).

Remark 2.2.

(i) The condition that u and v are even, nor the monotonicity conditions imposed on u and v are required in the sufficiency part of Theorem 2.1. One requires only that in (2.2) and in the integral conditions of (ii) that u and $1/v$ are replaced by u^* respectively $(1/v)^*$, their decreasing rearrangements.

(ii) If $0 < q \leq p < 1$, weighted Fourier transform inequalities are still possible – however then the domain space of the operator must be replaced by the boundary values of functions in weighted (atomic) Hardy spaces (c.f. [11], [12]).

In the study of weighted norm inequalities for singular integral operators, the A_p–weights are the most effective weights and have been studied intensely during the last twenty years ([11], [34]). Recall that a non-negative locally integrable function w on $\mathbb{R}^n$ belongs to the A_p–weight class, $1 < p < \infty$, if there is a constant $C > 0$, such that for all n–balls $B \subset \mathbb{R}^n$ with volume $|B|$

$$\left(\frac{1}{|B|}\int_B w(x)dx\right)^{1/p} \left(\frac{1}{|B|}\int_B w(x)^{1-p'} dx\right)^{1/p'} \leq C < \infty.$$

Similarly we say $w \in A_p^*$, $1 < p < \infty$, if

$$\sup_{I \subset J} \left(\frac{1}{|I|} \int_I w(x)dx\right)^{1/p} \left(\frac{1}{|I|} \int_I w(x)^{1-p'}dx\right)^{1/p'} < \infty,$$

where J denotes the collection of intervals $I = [a_1, b_1] \times \dots \times [a_n, b_n]$ in $\mathbb{R}^n$ and $|I|$ its Lebesgue measure.

It is obvious that $A_p = A_p^*$, if $n = 1$, but in general $A_p^* \subset A_p$. In fact if $w(x) = |x|^\alpha$, then $w \in A_p$, if and only if $-n < \alpha < n(p-1)$, and if

$$w(x) = \left(\sum_{i=1}^{n} |x_i|\right)^\alpha$$

then $w \in A_p^*$ implies $-1 < \alpha < p - 1$. As mentioned, these weight classes are typical Hilbert transform – singular integral criteria, so it may be somewhat surprising that they are also important in establishing weighted Fourier transforms norm inequalities.

The next part of this section describes Fourier inequalities with A_p–weights, while in the second, Fourier inequalities of functions in certain moment subspaces are discussed.

2.1. Fourier transform inequalities with A_p–weights. In the sequel we require the following lemma:

Lemma 2.3.

(a) If $w \in A_p$, $1 < p < \infty$, then

$$\int_{|x| \geq r} |x|^{-np} w(x)dx \leq C\, r^{-np} \int_{|x| \geq r} w(x)dx \tag{2.3}$$

(b) If w is radial and as radial function increasing, then $w \in A_p$ if and only if $w^{\varepsilon} \in A_{1+\varepsilon(p-1)}$, $\varepsilon > 0$.

The first part of the lemma is well known and may be found in [11] or [34]. Indeed for $n = 1$ this is essentially (1.10) of Theorem 1.9. That $w \in A_p$ implies $w^{\varepsilon} = A_{1+\varepsilon(p-1)}$ is Hölders inequality for $0 < \varepsilon < 1$. The details for $\varepsilon \geq 1$ and the converse may be found in [15].

The lemma permits us now to prove the following weighted Paley–Titchmarsh theorem:

Theorem 2.4. Suppose w is radial and as radial function increasing. Let $1 < p \leq q \leq p'$, then

$$\left[\int_{\mathbb{R}^n} |\hat{f}(x)|^q |x|^{n(q/p'-1)} w(1/|x|)^{q/p} dx\right]^{1/q} \leq C\left[\int_{\mathbb{R}^n} |f(x)|^p w(x) dx\right]^{1/p}$$

if and only if $w \in A_p$.

For $n = 1$ the proof of this result may be found in [5] and the general case (with different proof) in [15]. Note that for $q = p'$ this is a weighted Hansdorff–Young inequality.

We sketch the sufficiency part of the proof. If $p = 2$, then by Lemma 2.3 (a) it follows that (the n–dimensional analogue of) (2.2) holds with $p = q = 2$ and $v = w$ and $u(x) = w(1/|x|)$. Therefore by Theorem 2.1

$$\int_{\mathbb{R}^n} |\hat{f}(x)|^2 w(1/|x|) dx \leq C \int_{\mathbb{R}^n} |f(x)|^2 w(x) dx. \tag{2.4}$$

Now for any $u \geq 0$ define

$$a(x) = \int_0^{|x|} u(1/t)t^{n-1}dt \quad \text{and} \quad \mu(E) = \int_E a(x)^{-2}u(1/|x|)dx, \ E \subset \mathbb{R}^n\backslash\{0\}.$$

then

$$\mu(\{x \in \mathbb{R}^n : |a(x)\hat{f}(x)| > \lambda\}) \le \mu(\{x \in \mathbb{R}^n : |a(x)| \|f\|_1 > \lambda\})$$

$$= \int_{\{x : a(x) > \lambda/\|f\|_1\}} a(x)^{-2}u(1/|x|)dx = \int_\Sigma d\sigma \int_{\{t>0 : a(t)>\lambda/\|f\|_1\}} t^{n-1}a(t)^{-2}u(1/t)dt$$

$$= C \int_{a^{-1}(\lambda/\|f\|_1)}^{\infty} a^{-2}(t)d\,a(t) = C\|f\|_1/\lambda,$$

where $a^{-1}(y) = \inf\{x > 0 : a(x) > y\}$ so that $a(a^{-1}(h)) = y$. Now let $u = w^{p'-1}$ and replace f by f/u in this estimate, then

$$\mu(\{x \in \mathbb{R}^n : |a(x)(f/u)\hat{\ }(x)| > \lambda\}) \le C\|f\|_{1,1/u}/\lambda$$

which is a weighted "weak" (1,1) estimate. But Lemma 2.3(b) shows that $w \in A_p$, if and only if $u = w^{p'-1} \in A_{1+(p'-1)(p-1)} = A_2$, so that by (2.4) with f replaced by f/u

$$\int_{\mathbb{R}^n} |a(x)(f/u)\hat{\ }(x)|^2 d\mu(x) \le C\|f\|_{2,1/u}^2.$$

This inequality together with the "weak" type (1,1) above yields (via the Marcinkiewicz interpolation theorem)

$$\int_{\mathbb{R}^n} |a(x)(f/u)\hat{\ }(x)|^p d\mu(x) \le C\|f\|_{p,1/u}^p \qquad 1 < p < 2$$

and with f/u replaced by f

$$\int_{\mathbb{R}^n} |\hat{f}(x)|^p \, a(x)^{p-2} u(1/|x|)dx \le C \int_{\mathbb{R}^n} |f(x)|^p \, u(x)^{p-1} dx . \tag{2.5}$$

But since Lemma 2.3(a) implies that

$$a(x) = \int_{1/|x|}^{\infty} t^{-n-1} u(t)dt = C \int_{|y|>1/|x|} |y|^{-2n} u(y)dy \le C|x|^{2n} \int_{|y|\le 1/|x|} u(y)dy$$

$$\le C|x|^n u(1/|x|),$$

substitution into the left side of (2.5) (note $p - 2 < 0$) yields

$$\int_{\mathbb{R}^n} |\hat{f}(x)|^p |x|^{n(p-2)} u(1/|x|)^{p-1} dx \le C \int_{\mathbb{R}^n} |f(x)|^p u(x)^{p-1} dx.$$

But $u = w^{p'-1}$ so that Theorem 2.4 holds with $q = p$. The general case follows similarly.

The monotonicity conditions imposed on the weights both in Theorem 2.1 and Theorem 2.4 are highly undesirable. In [15] the condition of radial monotonicity was replaced by insisting on appropriate monotonicity condition in each variable separately. In this way one obtains the following weighted Hausdorff–Young inequality:

$$\Big(\int_{\mathbb{R}^n} |\hat{f}(x)|^{p'} u(1/x)dx\Big)^{1/p'} \le C\Big(\int_{\mathbb{R}^n} |f(x)|^p u(x)^{p-1} dx\Big)^{1/p}, \quad 1 < p \le 2,$$

holds if and only if $u \in A_2^*$.

A detailed study of weighted Fourier estimates with weights of this type was made by Bloom, Jurkat and Sampson [7]. One of their results is the following:

Theorem 2.5. Suppose $1 < p \le q < \infty$ and w and $1/v$ are even and decreasing in $(0, \infty)$ in each of their variables separately. If

$$u(x) = |x_1 x_2 \dots x_n|^{q-2} w(1/x)$$

is increasing in each x_i, then

$$\Big(\int_{\mathbb{R}^n} |\hat{f}(x)|^q \, w(x)dx\Big)^{1/q} \le C\Big(\int_{\mathbb{R}^n} |f(x)|^p v(x)dx\Big)^{1/p}$$

if and only if

$$\Big(\int_{\mathbb{R}^n} |(P_n f)(x)|^q u(x)dx\Big)^{1/q} \le C\Big(\int_{\mathbb{R}^n} |f(x)|^p v(x)dx\Big)^{1/p}$$

where

$$(P_n f)(x) = \frac{1}{|x_1 x_2 \cdots x_n|} \int_{|t_1| \le |x_1|} \dots \int_{|t_1| \le |x_n|} f(t)dt$$

is the n-dimensional averaging operator.

We point out here, however, that weight characterizations for $P_n : L_v^p \to L_u^q$ are only known for $n = 1$ and $n = 2$. The problem for $n > 2$ is still open.

Although it is possible to weaken the monotonicity conditions on weights in these last theorems somewhat ([4]) its total deletion – and hence a complete characterization of weights – have not been proved. There are also several abstractions of Theorem 2.1 in the case $1 < p \leq q < \infty$ to locally compact connected Abelian groups and certain locally compact totally disconnected groups. However, either the weight conditions given are only sufficient, or additional restrictions on the weights – corresponding to monotonicity – must be obtained to prove characterizations. (c.f. [17], [24], [25]).

If instead of imposing monotonicity conditions on the weight functions one restricts the functions f, then much more may be said. In fact in that case we have the following characterization:

Theorem 2.6. Let f be a non-negative even function in L^p_v, $v \geq 0$, $1 \leq p < \infty$, such that f is decreasing in $(0, \infty)$ and $f(\infty) = 0$. Then

$$\int_{-\infty}^{\infty} |\hat{f}(1/x)/x|^p v(x)dx \leq C \int_{-\infty}^{\infty} |f(x)|^p v(x)dx,$$

if and only if (2.3) holds with $n = 1$ and $w = v$.

It is easy to see that $v(x) = |x|^{-\epsilon}$, $0 < \epsilon < 1$ satisfies (2.3) with $n = p = 1$ and $w = v$.

2.2. Fourier transform inequalities of functions with vanishing moments. The last theorem suggests that if restrictions are imposed on the functions f rather than the weights, then one might expect also that a larger class of weights can be generated for which weighted Fourier norm inequalities hold. For example, if $\hat{f}(0) = 0$, then

$$|\hat{f}(x)| = \left|\int_{-\infty}^{\infty} (e^{-ixt} - 1)f(t)dt\right| = \left|\int_{|xt|\leq 1} + \int_{|xt|>1}\right|$$

$$\leq 2\int_{|t|\leq 1/|x|} |xt||f(t)|dt + 2\int_{|t|>1/|x|} |f(t)|dt = 2\left\{\int_{|y|\geq|x|} |y^{-3}f(1/y)|dy + \int_{|y|\leq|x|} |y^{-2}f(1/y)|dy\right\}.$$

But these integrals are essentially the Hardy operator and its dual of $|y^{-3}f(1/y)|$ and $|y^{-2}f(1/y)|$ respectively. Thereforc, applying the weighted form of the Hardy operator and its dual one is led to the following result of Sadosky and Wheeden [26]:

Theorem 2.7. If $1 < p < \infty$ and $w \in A_p$, then

$$\int_{-\infty}^{\infty} |\hat{f}(x)|^p w(1/x) \frac{dx}{|x|^2} \leq C \int_{-\infty}^{\infty} |f(x)|^p |x|^p w(x)dx \tag{2.6}$$

for all f satisfying $\hat{f}(0) = 0$.

Compare this with Theorem 2.4 taking $q = p$ and $n = 1$.

If $w(x) = 1/|x|$, then $w \notin A_p$, $p > 1$ and one might ask if (2.6) with this weight and $\hat{f}(0) = 0$ is satisfied? The answer is no, and in fact, no norm inequality of the form

$$\left(\int_{-\infty}^{\infty} |\hat{f}(x)|^q u(x)dx\right)^{1/q} \leq C \left(\int_{-\infty}^{\infty} |f(x)|^p |x|^{p-1}dx\right)^{1/p},$$

$1 < p < \infty$, $0 < q \leq \infty$, $u \equiv 0$, can hold for f satisfying $\hat{f}(0) = 0$ ([26]).

The idea to consider Schwartz functions f satisfying

$$\int_{-\infty}^{\infty} f(x)x^j dx = 0, \quad j = 0,1,2,\dots ;$$

and whose Fourier transforms have compact support away from the origin has been exploited in [6] and [26] to prove Fourier norm inequalities for substantially larger weight classes. These subspaces are known to be dense in weighted L^p spaces ([6], [21]) and the Fourier transform has a natural extension in these spaces. It is also clear that corresponding results carry over to mixed indices.

To apply Fourier inequalities to obtain uncertainty relations of the Heisenberg–Weyl type, or Fourier restriction theorems it is desirable to replace the weights in the range space by general measures. This in turn requires corresponding measure estimates for the Hardy operator and its dual. Such results – indeed characterizations – were proved by Sinnamon [29], and the corresponding measure weighted Fourier inequalities on moment subspaces were studied in [6]. We state here only the one dimensional case:

Theorem 2.8. Suppose

(i) $v \in L^1_{loc}(\mathbb{R})$, $v > 0$ a.e. and μ a positive measure.

(ii) $1 < p \le q < \infty$, $v^{1-p'} \in L^1_{loc}(\mathbb{R}\setminus[-y,y])$ for each $y > 0$.

(iii) $$B_1 \equiv \sup_{y>0} \left\{\int_{|t|<y} |t|^q d\mu(t)\right\}^{1/q} \left\{\int_{|t|<1/y} |t|^{p'} v(t)^{1-p'} dt\right\}^{1/p'}$$

and

$$B_2 \equiv \sup_{y>0} \left\{\int_{|t|>y} d\mu(t)\right\}^{1/q} \left\{\int_{|t|>1/y} v(t)^{1-p'} dt\right\}^{1/p'}$$

are both finite.

(iv) $f \in L^p_v(\mathbb{R}) \cap L^1(\mathbb{R})$ and $\hat{f}(0) = 0$ with supp f compact.

Then there is a constant $C > 0$ (depending on B_1 and B_2) such that

$$\|\hat{f}\|_{q,\mu} \leq C\|f\|_{p,v}.$$

For example if $v(t) = |t|^{1+\varepsilon}$, $0 < \varepsilon < 2$, $\mu = \Sigma' |n|^{-1-\varepsilon} \delta_n$, where δ is the Dirac distribution and $p = q = 2$, then all conditions of this theorem are satisfied and one obtains

$$\Sigma' \frac{|\hat{f}(n)|}{|n|^{1+\varepsilon}} \leq 2^5(\pi B_1 + B_2)^2 \int_{\mathbb{R}} |f(t)|^2 |t|^{1+\varepsilon} dt.$$

(For the details see [6, Ex. 2.5(c)].)

Some progress has been made to characterize the Fourier transform in weighted Lorentz spaces although a complete solution seems not to be available. We conclude this section with a characterization of a special Fourier transform in a very special setting due to Braverman.

Let $(\Omega, \mathcal{F}, P)$ be a probability space and $X(w)$ a random variable. In this context, the Calderon $L^{p,q}$ spaces are then defined by $X \in L^{p,q}$, if and only if

$$\|X\|^*_{p,q} = \left(\int_0^\infty (P[|X| > x])^{q/p} dx^q\right)^{1/q} < \infty,$$

where $0 < p, q < \infty$. Let $F(y) = P[X > y]$ and the characteristic function (Fourier transform) be defined by

$$\hat{F}(y) = \int_{-\infty}^{\infty} e^{ity} d\, F(y).$$

The result alluded to is then the following:

Theorem 2.9. If $1 < p < 2$, $1 \leq q < \infty$, then

$$\left\{\int_0^\infty \left[1 - \mathrm{Re}\left(t \int_0^{1/t} \hat{F}(y)dy\right)\right]^{q/p} dt^q\right\}^{1/q} \approx \|X\|^*_{p,q} .$$

3. Some applications. In 1974, Zygmund ([35]) considered the two dimensional Fourier transform

$$\hat{f}(x) = \int_{\mathbb{R}^2} e^{-i\, x\cdot y} f(y)dy \quad x \in \mathbb{R}^2$$

and estimated the restriction of the Fourier transform on circles $|x| = \rho$, in terms of the L^p-norm of f. His result states that if $f \in L^p(\mathbb{R}^2)$, $1 \leq p < 4/3$, then $\hat{f}(x)$ exists a.e. on $|x| = \rho$ and

$$\left(\int_{|x|=\rho} |\hat{f}(x)|^q ds\right)^{1/q} \leq A_p \rho^{1/p'} \|f\|_p \tag{3.1}$$

for $q = p'/3$. Moreover, the result is sharp in the sense that if $p = 4/3$ then inequality (3.1) fails.

At almost the same time Sjölin [30] provided an extension theorem – that is, a dual Fourier estimate of measures carried by smooth curves in $\mathbb{R}^2$, which, when restricted to curves of constant curvature (circles) implies Zygmund's results. Since then, a vast literature on the subject evolved with many significant applications.

The proof of the result of Zygmund (and also Sjölin's) utilizes among other things, a duality argument, the Hausdorff–Young inequality and an L^p–estimate of the Riemann–Liouville fractional integral operator. But all these components permit generalizations and weighted extensions. It is not surprising therefore, that these restriction–extension theorems have weighted extensions in $\mathbb{R}^2$. These generalizations, proved in [9] have the form

$$\|\hat{f}\,|K|^{\delta}\|_{L^r(\gamma,ds)} \leq A\|f\|_{L^p_v(\mathbb{R}^2)} \tag{3.2}$$

where γ is a smooth plane curve in $\mathbb{R}^2$ of curvature K and arclength measure ds.

A very specific case of the generalizations of Zygmund's estimate (3.1) is given here ([9, Corollary 2, with $\lambda = 0$]).

<u>Proposition 3.1.</u> If $4/3 < q < \infty$, $1 < p < 6q/(3q+2)$ and

$$\max[0,\ 2(1/p' - 1/(3q))] \leq \alpha < 2/p',$$

then for $1 \leq r \leq q$,

$$\left(\int_{|x|=\rho} |\hat{f}(x)|^r ds\right)^{1/r} \leq C\rho^{1/r+\alpha-2/p'}\|f\|_{p,\alpha}\,.$$

In particular, if $q = p'/3$, then $1 < p < 4/3$ and then taking $\alpha = 0$ in Proposition 3.1, we obtain Zygmund's result.

A special case of the generalization of Sjölin's theorem has the same form. In fact under the hypotheses Proposition 3.1 one obtains (3.2) with $v(x) = |x|^{\alpha}$, $\delta > 1/(3q)$ and $\gamma(t) = (\gamma_1(t), \gamma_2(t))$, $t \in [ab]$. Moreover (3.2) holds also if $\delta = 1/(3q)$, provided $K(t) \geq 0$. (c.f [9, Cor. 1]).

It should be noted that these results are strictly two dimensional since the weights in the appropriate Fourier inequalities which lead to these results are especially adapted to the geometry of the curve. Higher dimensional weighted spherical restriction theorems can be obtained directly from the n–dimensional measure weighted Fourier inequalities. That is the n–dimensional version of Theorem 2.8.

Theorem 3.2. ([6, Theorem 5.3]) Let $v \in L^1_{loc}(\mathbb{R}^n)$, $v > 0$, a.e., radial and $1 < p \leq q < \infty$. Assume that $v^{1-p'} \in L^1_{loc}(\mathbb{R}^n \backslash B(0,y))$ for each $y > 0$ and

$$C(p,q,n,\rho) \equiv A(n,p,q)\rho^{\frac{n-1}{q}} \left\{\rho\left(\int_0^{1/(\rho\pi)} r^{n-1+p'} v(r)^{1-p'} dr\right)^{1/p'} + \left(\int_{1/(\rho\pi)}^{\infty} r^{n-1} v(r)^{1-p'} dr\right)^{1/p'}\right\}, \rho > 0$$

Then for all $f \in M_0(n) \cap L^p_v(\mathbb{R}^n)$

$$\left(\int_{\Sigma_{n-1}(\rho)} |\hat{f}(x)|^q d\sigma_{n-1}\right)^{1/q} \leq C(p,q,n,\rho)\|f\|_{p,v}.$$

We conclude with an example of the uncertainty inequality. Given $(x_0, y_0) \in \mathbb{R}^2$, then the classical uncertainty inequality states that for all $f \in \mathscr{S}(\mathbb{R})$

$$\int_{\mathbb{R}} |f(x)|^2 dx \leq 4\pi\left(\int_{\mathbb{R}} |(x - x_0)f(x)|^2 dx\right)^{1/2} \left(\int_{\mathbb{R}} |(y - y_0)\hat{f}(y)|^2 dy\right)^{1/2}$$

with equality if $f(x) = c\, e^{-s(x-x_0)^2} e^{2\pi i x y_0}$, $s > 0$, c a complex constant.

There are numerous variants and generalizations of this inequality with applications in science and engineering. It is not difficult to see that the change $f(x) \to f(x + x_0)e^{-2\pi ixy_0}$ shows that one may take without loss of generality $x_0 = y_0 = 0$ in this inequality.

Recall from Section 1, if

$$(Qf)(x) = \int_1^\infty f(xt)\frac{dt}{t} \quad x \in \mathbb{R}^n$$

and $(Q^*f)(x) = -x\cdot\nabla f(x)$, $x \in \mathbb{R}^n$, then $QQ^*f(x) = f(x)$. Instead of applying Theorem 1.2(i) with ρ replaced by p' and $q = p'$, $v(x) = u(x) = |x|^{-p'}$ we prove the result directly in the following lemma:

<u>Lemma 3.3.</u> If $f \in C_0^1(\mathbb{R}^n)$, $1 < p' < n$, then

$$\left(\int_{\mathbb{R}^n} \left|\frac{f(x)}{x}\right|^{p'} dx\right)^{1/p'} \le \frac{p'}{n - p'} \left(\int_{\mathbb{R}^n} \left|\frac{x\cdot\nabla f}{x}\right|^{p'} dx\right)^{1/p'}$$

<u>Proof.</u>

$$\left[\int_{\mathbb{R}^n} \left|\frac{f(x)}{x}\right|^{p'} dx\right]^{1/p'} = \left[\int_{\mathbb{R}^n} \left|\frac{QQ^*f(x)}{x}\right|^{p'} dx\right]^{1/p'} = \left[\int_{\mathbb{R}^n} |x|^{-p'} \left|\int_1^\infty (Q^*f)(xs)\frac{ds}{s}\right|^{p'} dx\right]^{1/p'}$$

$$= \left\{\int_\Sigma \int_0^\infty t^{n-p'-1} \left|\int_1^\infty (Q^*f)(t\sigma s)\frac{ds}{s}\right|^{p'} dt d\sigma\right\}^{1/p'} = \left\{\int_\Sigma \int_0^\infty t^{n-p'-1} \left|\int_t^\infty (Q^*f)(y\sigma)\frac{dy}{y}\right|^{p'} dt d\sigma\right\}^{1/p'}$$

$$\le \frac{p'}{n - p'} \left\{\int_\Sigma \int_0^\infty |Q^*(t\sigma)|^{p'} t^{n-p'-1} dt d\sigma\right\}^{1/p'} = \frac{p'}{n - p'} \left(\int_{\mathbb{R}^n} \left|\frac{x\cdot\vec{\nabla}f(x)}{x}\right|^{p'} dx\right)^{1/p'}$$

where the last inequality is the classical conjugate Hardy inequality.

<u>Theorem 3.4.</u> If $f \in C_0^1(\mathbb{R}^n)$, $1 < p \le 2$ and $n > p'$, then

$$\int_{\mathbb{R}^n} |f(x)|^2 dx \le C\Big(\int_{\mathbb{R}^n} |xf(x)|^p dx\Big)^{1/p} \Big(\int_{\mathbb{R}^n} |y\hat{f}(y)|^p dy\Big)^{1/p},$$

where

<u>Proof.</u> Hölder's inequality and Lemma 3.3 yield

$$\int_{\mathbb{R}} |f(x)|^2 dx \le \Big(\int_{\mathbb{R}^n} |xf(x)|^p dx\Big)^{1/p} \Big(\int_{\mathbb{R}^n} |\tfrac{f(x)}{x}|^{p'} dx\Big)^{1/p'}$$

$$\le \frac{p'}{n-p'} \Big(\int_{\mathbb{R}^n} |xf(x)|^p dx\Big)^{1/p} \Big(\int_{\mathbb{R}^n} |\tfrac{x\cdot\vec{\nabla} f(x)}{x}|^{p'} dx\Big)^{1/p'}$$

By Minkowski's inequality, and writing $\frac{\partial f}{\partial x_i} = \check{h}$, $i = 1,2,\dots,n$; it follows that the right integral product is

$$\Big(\int_{\mathbb{R}^n} \Big|\sum_{i=1}^n \frac{x_i}{|x|} \frac{\partial f}{\partial x_i}(x)\Big|^{p'} dx\Big)^{1/p'} \le \sum_{i=1}^n \Big(\int_{\mathbb{R}^n} \Big|\frac{x_i}{|x|}\, \check{h}\,(x)\Big|^{p'} dx\Big)^{1/p'}$$

$$= \sum_{i=1}^n \Big(\int_{\mathbb{R}^n} |(R_i h)\check{}\,(x)|^{p'} dx\Big)^{1/p'} \le \sum_{i=1}^n \Big(\int_{\mathbb{R}^n} |(R_i h)(x)|^p dx\Big)^{1/p}$$

$$\le C \sum_{i=1}^n \Big(\int_{\mathbb{R}^n} |h(x)|^p dx\Big)^{1/p} \le n^{1/p'} C\Big[\sum_{i=1}^n \int_{\mathbb{R}^n} |h(x)|^p dx\Big]^{1/p}$$

$$= Cn^{1/p'}\Big[\int_{\mathbb{R}^n} \sum_{i=1}^{n} \Big|\Big(\frac{\partial f}{\partial x_i}\Big)^{\wedge}(x)\Big|^p dx\Big]^{1/p} = Cn^{1/p'}\Big[\int_{\mathbb{R}^n} \Sigma |y_i \hat{f}(y)|^p dy\Big]^{1/p}$$

$$\leq cn^{1/2}\Big[\int_{\mathbb{R}^n} |\hat{f}(y)y|^p dy\Big]^{1/p},$$

where we applied the Hausdorff–Young inequality, the L^p–boundedness of the Riesz transform R and Hölder's inequality twice. Substituting we get the result.

It is clear that on using Theorem 1.2(i) and the well known weighted L^p–boundedness of the Riesz transform ([11][34]) a corresponding weighted inequality can be obtained . This unweighted form is however curious since the classical $n = 1$ – case does not follow from it.

For additional recent application of weighted Fourier inequalities we refer to [4] [15] were generalizations of the Paley–Wiener theorems were given and Laplace representations in weighted Bergman spaces were established.

References

[1] K.F. Andersen and H.P. Heinig, Weighted norm inequalities for certain integral operators, SIAM J. Math. Anal. 14, (1983), 834–844.

[2] M.A. Ariño and B. Muckenhoupt, Maximal functions on classical Lorentz spaces and Hardy's inequality with weights for non–increasing functions, (Preprint).

[3] P.R. Beesack and H.P. Heinig, Hardy's inequalities with indices less than 1; Proc. Amer. Math. Soc. 83(3)(1981) 532–536.

[4] J.J. Benedetto, H.P. Heinig and R. Johnson, Weighted Hardy spaces and the Laplace transform II, Math. Nachr. 132, (1987), 29–55.

[5] ————; Fourier inequalities with A_p–weights, Proc. Conf. Oberwolfach 1986, General Inequalities 5, Int. Ser. Num. Math. 80 Birkhäuser, Basel (1987), 217–232.

[6] J.J. Benedetto and H.P. Heinig, Fourier transform inequalities with measure weights, Advances in Math. (to appear).

[7] S. Bloom, W.B. Jurkat and G. Sampson, Two weighted (L^p, L^q) estimates for the Fourier transform, (Preprint).

[8] M. Sh. Braverman, On a class of operators, (Preprint).

[9] C. Carton–Lebrun and H.P. Heinig, Weighted extensions and restriction theorems for the Fourier transform, (Preprint).

[10] J.A. Cochran and C.S. Lee, Inequalities related to Hardy's and Heinig's, Math. Proc., Cambr. Phil. Soc. 96 (1984), 1–7.

[11] G. Garcia–Cuerva and J.L. Rubio de Francia, Weighted norm inequalities and related topics, North Holland, Amsterdam 1985.

[12] H.P. Heinig, Fourier operators on weighted Hardy spaces, Math. Proc. Cambr. Phi. Soc. 101 (1987) 103–121.

[13] ———; Some extensions of Hardy's inequality, SIAM J. Math. Anal. 6 (1975) 698–713.

[14] ———; Weighted norm inequalities for certain integral operators II, Proc. A.M.S. 95, (1985), 387–395.

[15] H.P. Heinig and G.J.Sinnamon, Fourier inequalities and integral representations of functions in weighted Bergman spaces over tube domains, Indiana Univ. Math. J. 38 (3), (1989), 603–628.

[16] E. Hernandez, Factorization and extrapolation of pairs of weights, (Preprint).

[17] J.A. Hogan, Weighted norm inequalities for the Fourier transform on connected locally compact groups, Pac. J. Math. 131 (2), (1988), 277–290.

[18] A. Kufner and B. Opic, Hardy's inequalities, generalizations and applications, Longman–Pittman 1990 (to appear).

[19] E.R. Love,Some inequalitics for geometric means; ISNM 80, General Inequalities 5; W. Walter ed., Birkhäuser 1987, 83–93

[20] F.J. Martin–Reyes and E.Sawyer, Weighted Inequalities for Riemann–Liouville fractional integrals of order one and greater (to appear).

[21] B. Muckenhoupt, R. Wheeden and W.S. Young, L^2–multipliers with power weights, Advances in Math. 49 (1983), 170–216.

[22] B. Muckenhoupt, Weighted norm inequalities for classical operators, Proc. Symp, Pure Math. 35, AMS Providence, R.I. (1979), 69–83.

[23] C.J. Neugebauer, Weighted norm inequalities for averaging operator of monotone functions (Preprint).

[24] C.W. Onneweer, Weak L^p–spaces and weighted norm inequalities for the Fourier transform on locally compact Vilenkin groups, Proc. Anal. Conf. Singapore (1986), 191–201, North Holland Math. Studies 150, Amsterdam, N.Y. 1988.

[25] T.S. Quek, Weighted norm inequalities for the Fourier transform on certain totally disconnected groups, Proc. A.M.S. 101 (1), (1987), 113–121.

[26] C. Sadosky and R.L. Wheeden, Some weighted norm inequalities for the Fourier transform of functions with vanishing moments, Trans. Amer. Math. Soc. 300, (1987) (2), 521–533.

[27] E. Sawyer, Boundedness of classical operators are classical Lorentz spaces, (Preprint).

[28] G. Sinnamon, A weighted gradient inequality, Proc. Roy. Soc. Edinburgh, 111 A (1989), 329–335.

[29] ———; Operators on Lebesgue spaces with general measures, Ph.D. thesis, McMaster Univ. 1987.

[30] P. Sjölin, Fourier multipliers and estimates of the Fourier transform of measures carried by smooth curves in $\mathbb{R}^2$. Studia Math. 51, (1974), 169–182.

[31] V.D. Stepanov, Two weighted estimates for the Riemann–Liouville integrals, Česk. Akad. Věd; Mat. Ustav. Preprint 339 (1988) 1–28.

[32] ———; The two weighted Hardy's inequality for non–increasing functions, (Preprint).

[33] C.R. Stromberg, Introduction to classical real analysis, Wadsworth, Belmont, Calif. 1981.

[34] A. Torchinsky, Real variable methods in harmonic analysis; Acad. Press, N.Y. 1986.

[35] A. Zygmund, On Fourier coefficients and transforms of functions of two variables, Studia Math. 50 (1974) 189–210.

Department of Mathematics and Statistics
McMaster University
Hamilton, Ontario, Canada L8S 4K1

RW/HH1/HH5/F/SC
May 9, 1990

WEIGHTED ESTIMATES FOR CLASSICAL INTEGRAL OPERATORS

Vakhtang Kokilashvili, Tbilisi

This paper deals with the problem of a full characterization of couples of weighted function spaces for which a given integral operator is bounded from one of these spaces into the other. According to coincidence and non–coincidence of the weights involved these problems are called one weight and two weight problems, resp.

Today the weighted inequalities and their applications to various mathematical problems are very rapidly developing part of the harmonic analysis. In last years, solutions to many serious problems in the theory of functions turned out to be closely linked with the weight theory.

Particularly, extensive and intensive study of integral operators in weighted function spaces began in the early 70's after the paper [1] by B. Muckenhoupt who showed that the weighted inequalities for the classical maximal operators are true iff the weight function satisfies the so called A_p condition. Later, solutions were given to one weight problems for Hilbert transform, fractional maximal functions, fractional integrals, and other classical integral operators [2], [3].

The two weight problem in Lebesgue spaces for the classical maximal functions was solved by E. Sawyer [4]. The same author solved in [5] the problem of a full charecterization of pairs of weight functions in the two weight weak type inequality, and another equivalent condition, perhaps more easily verifiable, was given by M. Gabidzashvili [6]. Quite recently, E. Sawyer [7] has found a characterization of pairs of weights, guaranteeing validity of the two weight strong type inequality for fractional integrals. It turned out that, using the results of [6] and [7], the solution to the last problem can be given in a more visible form.

In this lecture we present solutions of some weight (and unweight) problems for classical integral operators, in particular, for maximal functions, potentials, Riesz transforms, and for their generalizations in the spaces of the homogeneous type. These results have been obtained recently by participants of the seminar on weighted function spaces and integral operators in the Tbilisi Institute of Mathematics of the Georgian Academy of Sciences and some of them also in collaboration with our Czech colleagues from the Mathematical Institute of the Czechoslovak Academy of Sciences in Prague.

1. Fractional integrals in $\mathbf{R^n}$

First of all we will discuss some unweighted problems for the potential type integral operator

$$I_\gamma(f d\mu) = \int_{R^n} \frac{f(y)d\mu}{|x-y|^{\alpha-\gamma}}, \quad 0 < \gamma < \alpha \leq n,$$

where μ is a Borel measure.

The question is as follows: What is the full description of such measures μ for which the operator I_γ acts continuously from $L^p(R^n, d\mu)$ into $L^q(R^n, d\mu)$ when $1 < q \leq \dfrac{\alpha p}{\alpha - p\gamma}$, $1 < p < \dfrac{\alpha}{\gamma}$?

For the measure μ we introduce the following function

$$\Omega(x) = \sup_{r>0} \frac{\mu B(x,r)}{r^\alpha}.$$

PROPOSITION 1.1. *Let us suppose that the function $\Omega(x)$ is finite almost everywhere on a μ-measurable set E. Then for arbitrary f from $L^p(R^n, d\mu)$ with $1 < p < \dfrac{\alpha}{\gamma}$, the function $I_\gamma(f d\mu)$ is finite almost everywhere (in μ sense) on E.*

THEOREM 1.1 [9]. *Suppose that $0 < \gamma < \alpha \leq n$, $1 < p < \dfrac{\alpha}{\gamma}$, $\dfrac{1}{q} = \dfrac{1}{p} - \dfrac{\gamma}{\alpha}$. Then there exists a positive constant c_1 such that for every F from $L^p(R^n, d\mu)$ we have*

$$\left(\int_{R^n} (I_\gamma(f d\mu))^q \, (\Omega(x))^{(\gamma/\alpha - 1)q} \, d\mu \right)^{1/q} \leq c_1 \left(\int_{R^n} |f(x)|^p \, d\mu \right)^{1/q}$$

This theorem and some other consideration yield the following

THEOREM 1.2 [9]. *If $\dfrac{1}{q} = \dfrac{1}{p} - \dfrac{\gamma}{\alpha}$, then the following conditions are equivalent:*

i) $I_\gamma(f d\mu)$ acts continuously from $L^p(R^n, d\mu)$ into $L^q(R^n, d\mu)$ when $1 < p < \dfrac{\alpha}{\gamma}$.

ii) I_γ is of the weak type $(1, \dfrac{\alpha}{\alpha - \gamma})$ (with respect to the measure μ)

iii) There is a positive constant c_2 such that

$$\mu B(x,r) \leq c_2 \, r^\alpha$$

for every $x \in R$, $r > 0$.

Here $B(x,r)$ denotes a ball in R^n centered at x and with the radius r.

In general we have the following

Theorem 1.3 [9]. *Suppose that* $1 < q \le \dfrac{\alpha p}{\alpha - p\gamma}$, $1 < p \le \dfrac{\alpha}{\gamma}$. *Then the operator* I_γ *acts continuously from* $L^p(R^n, d\mu)$ *into* $L^q(R^n, d\mu)$ *if and only if there exists a positive constant* c_3 *such that*

$$\mu B(x,r) \;\le\; c_3 r^{((\alpha-\gamma)pq)/(pq-q+p)}\,.$$

Theorem 1.3 contains the well–known Sobolev embedding theorem [11]. Let E be a Borel subset of R^n with the condition that its Hausdorff α–measure $\mathcal{H}_\alpha E$ is positive.

We have the following

Theorem 1.4 [9]. *Suppose that* $1 < p < \alpha\gamma^{-1}$, $1 < q \le \dfrac{\alpha p}{\alpha - \gamma p}$. *Then the inequality*

$$\left(\int\limits_E \left(\int\limits_E \frac{f(y)\,d\mathcal{H}_\alpha}{|x-y|^{\alpha-\gamma}}\right)^q d\mathcal{H}_\alpha\right)^{1/q} \le c\left(\int\limits_E |f(x)|^p\,d\mathcal{H}_\alpha\right)^{1/p}$$

holds for every $f \in L^p(E, \mathcal{H}_\alpha)$ *with a constant* $c > 0$ *independent of* f *if and only if*

$$\mathcal{H}_\alpha(E \cap B(x,r)) \;\le\; c_1 r^{((\alpha-\gamma)pq)/(pq+p-q)}\,,$$

where c *does not depend on* x *and* r.

Now suppose that Γ is a rectifiable curve in the complex plane (with finite or infinite length). We shall consider the potential type operators

$$\mathcal{K}_\gamma f(t) = \int\limits_\Gamma \frac{f(\tau)}{|t-\tau|^{1-\gamma}}\,d\tau\,, \quad 0 < \gamma < 1\,.$$

Theorem 1.5 [8]. *If* $1 < p < \gamma^{-1}$, $1 < q \le \dfrac{p}{1-p\gamma}$, *then the following two statements are equivalent:*

i) K_γ *acts continuously from* $L^p(\Gamma)$ *into* $L^q(\Gamma)$.

ii) The curve Γ *satisfies the condition*

$$|\Gamma \cap B(z,r)| \le c\,r^{((1-\gamma)pq)/(pq-q+p)}\,,$$

where c *does not depend on* $z \in \Gamma$ *and* $r > 0$.

Note that a generalization of these results for another variant of the fractional order integral in spaces of the homogeneous type was proposed in [12].

Now we consider weight problems for integral operators with positive kernels in weighted Lorentz space. This is a set of measurable functions $f : R^n \to R^1$ for which the following quasinorm

$$\|f\|_{L^{ps}_w} = \begin{cases} \left(p\int\limits_0^\infty (w\{y \in R^n : |f(x)| > \lambda\})^{s/p}\,\lambda^{s-1}\,d\lambda\right)^{1/s} & \text{for } 1 \le p < \infty,\ 1 \le s < \infty\,, \\ \sup\limits_\lambda \lambda\,(w\{x \in R^n : |f(x)| > \lambda\})^{1/p} & \text{for } 1 \le p < \infty,\ s = \infty\,, \end{cases}$$

is finite.

Further, we shall say that a function $w : R \to R$ is weight function, if w is local integrable and positive almost everywhere. For the weight function w and a Borel set E we denote

$$wE = \int_E w(x)\,dx\,.$$

Let us consider the integral operator

$$\mathcal{K}f(x) = \int_{R^n} K(x,y)f(y)\,dy\,,$$

where $K : R^n \times R^n \to R^1$ is positive and $f \geq 0$.

THEOREM 1.6 [14]. *Let $1 \leq s \leq p < q < \infty$, $w : R^n \to R^1$ be a weight function, let $\psi : R^n \to R^1$ be a nonnegative locally integrable function, and ν a Borel measure. If*

$$\sup_{\substack{x\in R^n\\ r>0}} (\nu B(x,2r))^{1/q}\,\|\chi_{R^n\setminus B(x,r)} w^{-1}\psi K(x,.)\|_{L_w^{p's'}} < \infty\,, \tag{1.1}$$

then there exists a positive constant c_4 such that for any $\lambda > 0$ and an arbitrary nonnegative f from L_w^{ps} we have

$$\nu\{x \in R^n : \mathcal{K}(f\psi)(x) > \lambda\} \;\leq\; c_4\,\lambda^{-q}\,\|f\|^q_{L_w^{ps}}\,.$$

In [17] a condition for pairs of weight functions can be found which is necessary and sufficient for the operator $\mathcal{K}$ to be the weak type mapping from one weighted Lorentz space into another one in the case that the kernel K is an anisotropic radial decreasing function and $1 < p \leq q < \infty$.

Let us consider the generalized potential operator

$$T_\gamma(\phi,t) = \int_{R^n} \frac{\phi(y)}{(|x-y|+t)^{n-\gamma}}\,dy\,, \qquad 0<\gamma<n\,, \quad t>0.$$

We suppose that β is a measure on $R^{n+1}_+ = R^n \times [0,+\infty)$, $\widehat{B}(x,r) = B(x,r) \times [0,2r)$.

THEOREM 1.7 [15]. *Let $1 \leq s \leq p < q < \infty$. Then the following statements are equivalent:*

i) There is a positive constant c_5 such that the inequality

$$\beta\{(x,t) \in R^{n+1}_+ : T_\gamma(f\psi)(x,t) > \lambda\} \;\leq\; c\lambda^{-q}\|f\|^q_{L_w^{ps}}\,,$$

holds for every $\lambda > 0$ and an arbitrary function f from L_w^{ps}

ii) $\displaystyle\sup_{\substack{x\in R^n\\ r>0,t>0}} (\beta\widehat{B}(x,r))^{1/q}\|\chi_{R^n\setminus B(x,r)}\psi(|x-.|+t)^{\gamma-n}w^{-1}\|_{L_w^{p's'}} < \infty.$

From the above results and solution of the two weight problem for the Riesz potential which was recently given by E. Sawyer [7] we are able to give a necessary and sufficient condition for validity of the two weight strong type inequality which is more convenient from the point of view of the verification.

Theorem 1.8. *Let $1 < p < q < \infty$. Then the inequality*

$$\left(\int_{R^n} (I_\gamma(f)(x))^q v(x)\,dx\right)^{1/q} \le c\left(\int_{R^n} |f(x)|^p w(x)\,dx\right)^{1/q}$$

with a constant $c > 0$ independent of $f \in L^p_w(R^n)$ holds if and only if

$$\sup_{\substack{x\in R^n\\ r>0}} (vB(x,r))^{1/q} \|\chi_{R^n\setminus B(x,r)} w^{-1} |x - .|^{\gamma-n}\|_{L^{p'}_w} < \infty$$

and

$$\sup_{\substack{x\in R^n\\ r>0}} (w^{1-p'}(B(x,r)))^{1/p'} \|\chi_{R^n\setminus B(x,r)} |x - .|^{\gamma-n}\|_{L^q_v} < \infty.$$

2. Maximal functions and Riesz transforms

We start this section with an unweighted problem for the Hardy–Littlewood maximal function and the Riesz transforms. For a locally integrable function $f : R^n \to R^1$, the Hardy-Littlewood maximal function is defined by

$$Mf(x) = \sup \frac{1}{|B|} \int_B |f(y)|\,dy\,,$$

where sup is taken over all balls B containing the point x.

For a measurable function $f : R^n \to R^1$ satisfying the condition

$$\int_{R^n} \frac{|f(x)|\,dx}{(1+|x|)^n} < \infty,$$

the Riesz transforms $R_j f$ $(j = 1, 2, \ldots, n)$ are defined as follows:

$$R_j f(x) = \lim_{\varepsilon\to 0} c_n \int_{|y|>\varepsilon} \frac{y_j}{|y|^{n+1}} f(x-y)\,dy\,,$$

where $y = (y_1, \ldots, y_n)$ and $c_n = \pi^{-(n+1)/2}\Gamma((n+1)/2)$.

We shall say that a function $\phi : R^1 \to R^1$ belongs to the class Φ, if ϕ is an even, nonnegative, nondecreasing function on $(0, \infty)$ such that $\phi(0+) = 0$, $\lim_{t\to\infty} \phi(t) = \infty$. By $\phi(L)$ we denote the set of all measurable functions f for which

$$\int_{R^n} (\phi \circ f)(x)\,dx < \infty\,.$$

Definition. A nonnegative function $\phi : R^1 \to R^1$ is called *quasiconvex* if there exists a convex function ψ and a constant $c > 1$ such that

$$\psi(t) \le \phi(t) \le c\,\psi(ct) \qquad \text{for all } t \in R^1.$$

Proposition 2.1. *If $\phi \in \Phi$, then the inequality*

$$\phi(\lambda)\, m\{x \in R^n : Mf(x) > \lambda\} \leq c \int_{R^n} (\phi \circ (cf))(x)\, dx$$

holds for every $\lambda > 0$ and $f \in \phi(L)$ with a constant c independent of λ and f if and only if the function ϕ is quasiconvex.

Proposition 2.2. *Let $\phi \in \Phi$. Then the following two conditions are equivalent:*

i) There is a positive constant c such that

$$\phi(\lambda)\, m\{x \in R^n : R_j f(x) > \lambda\} \leq c \int_{R^n} (\phi \circ (cf))(x)\, dx\,.$$

ii) ϕ is quasiconvex and satisfies the Δ_2 condition.

Theorem 2.1 [18]. *The following statements are equivalent:*

i) There exists a positive constant c_1 such that

$$\int_{R^n} (\phi \circ Mf)(x)\, dx \leq c_1 \int_{R^n} (\phi \circ (c_1 f))(x)\, dx$$

for every f such that $c_1 f \in \phi(L)$.

ii) ϕ^α is quasiconvex for some $\alpha \in \phi(L)$.

iii) There exists a positive constant c_2 such that

$$\int_0^\sigma \frac{\phi(s)}{s^2}\, ds \leq c_2 \frac{\phi(c_2\sigma)}{\sigma}, \qquad 0 < \sigma < \infty.$$

If ϕ is a Young function, then it can be shown that all the above statements are equivalent to the Δ_2 condition for the function complementary to function ϕ.

For the Riesz transforms the following theorem is true.

Theorem 2.2. *For the validity of the inequalities*

$$\int_{R^n} (\phi \circ R_j f)(x)\, dx \leq c \int_{R^n} (\phi \circ (cf))(x)\, dx \qquad (j = 1, 2, \ldots, n)$$

it is necessary and sufficient that $\phi \in \Delta_2$ and ϕ^α be quasiconvex for some α, $0 < \alpha < 1$.

For the Hilbert transform and the conjugate function the analogous problem was solved in [21], [22], [23] in a different way. Further we discuss the following question: What is the relation between the classes of functions ϕ for which either the inequality for the scalar maximal function (the Riesz transform) holds and the classes of those functions ϕ for which the corresponding inequality for the vector–valued maximal function (the Riesz transform) is true?

THEOREM 2.3 [19]. *Suppose that $1 < \theta < \infty$. Then the following conditions are equivalent:*

i) The inequality

$$\phi(\lambda)\, m\{x \in R^n :. \Big(\sum_{j=1}^{\infty}(Mf_j(x))^\theta\Big)^{1/\theta} > \lambda\} \le c_1 \int_{R^n} (\phi\circ f)(x)\, dx$$

is valid for some positive c_1 independent of $f = (f_1, \cdots, f_n, \cdots)$ and $\lambda > 0$.

ii) The inequality

$$\phi(\lambda)\, m\{x \in R^n : \Big(\sum_{j=1}^{\infty}|R_k f_j(x)|^\theta\Big)^{1/\theta} > \lambda\} \le c_2 \int_{R^n} \phi\circ\Big(\sum_{j=1}^{\infty}\|f_j(x)\|^\theta\Big)^{1/\theta} dx$$

is valid for some positive c_2 independent of $f = (f_1, \ldots, f_n, \ldots)$ and $\lambda > 0$.

iii) ϕ is quasiconvex and $\phi \in \Delta_2$.

THEOREM 2.4 [19]. *Let us suppose that $\phi \in \Phi$ and $1 < \theta < \infty$. Then the following statements are equivalent:*

i) There is a positive constant c_1 such that for each $f = (f_1, \ldots, f_n, \ldots)$ the inequality

$$\int_{R^n} \phi\circ\Big(\sum_{j=1}^{\infty}(Mf_j)^\theta\Big)^{1/\theta}(x)\, dx \le c_1 \int_{R^n} \phi\circ\Big(\sum_{j=1}^{\infty}|f_j|^\theta\Big)^{1/\theta}(x)\, dx$$

holds.

ii) There is a positive constant c_2 such that

$$\int_{R^n} \phi\circ\Big(\sum_{j=1}^{\infty}|R_k f_j|^\theta\Big)^{1/\theta}(x)\, dx \le c_2 \int_{R^n} \phi\circ\Big(\sum_{j=1}^{\infty}|f_j|^\theta\Big)^{1/\theta}(x)\, dx\,.$$

iii) $\phi \in \Delta_2$ and ϕ^α is quasi-convex for some $\alpha \in (0,1)$.

In the sequel, we present some general inequalities with Carleson measures for the maximal operator

$$\widetilde{M}f(x,t) = \sup |Q|^{-1} \int_Q |f(y)|\, dy$$

where the supremum is taken over all cubes $Q \subset R^n$ containing x and whose sides are of length at least $2^{-1}t$.

Let ϱ be a positive measure on $R^{n+1}_+ = R^n \times [0,\infty)$. L. Carleson [24] proved that $\widetilde{M}$ is bounded from $L^p(R^n)$ into $L^p(R^{n+1}_+, \varrho)$ and from $L^1(R^n)$ into $L^1_*(R^{n+1}, \varrho)$ if and only iff there exists a constant $c > 0$ such that

$$\text{(C)} \qquad \varrho\widehat{B}(x,r) \le cr^n\,,$$

where $\widehat{B}(x,r) = B(x,r)\times[0,2r)$. A measure ϱ on R^{n+1}_+ satisfying (C) will be called the Carleson measure.

We generalize this result for the functions from the Orlicz–Morrey spaces.

Theorem 2.5 [25]. *If $\phi \in \Phi$ and ϕ^α is quasi-convex for some $\alpha \in (0,1)$, $0 \le \lambda < n$, then the inequality*

$$\sup_{x\in R^n, r>0} r^{-\lambda} \int_{\widehat{B}(x,r)} (\phi o \widetilde{M} f)(y,\tau)\, d\varrho \le c_1 \sup_{x\in R^n, r>0} r^{-\lambda} \int_{B(x,r)} (\phi o c_1 f)(y)\, dy$$

holds if and only if ϱ is a Carleson measure.

The case $\lambda = 0$ will be reformulated in the next theorem together with a weak type inequality. We shall consider Orlicz spaces in their standard form.

Theorem 2.6 [25]. *Let ϱ be a positive measure on R_+^{n+1}, $\phi \in \Phi$ and suppose that ϕ^α is convex for some $\alpha \in (0,1)$. Then the following statements are equivalent:*

i) $\phi(\lambda)\, \varrho\left\{(x,t) \in R_+^{n+1} : \widetilde{M} f(x,t) > \lambda\right\} \le c_1 \int_{R^n} (\phi o f)(x)\, dx$.

ii) $\int_{R^n} (\phi o \widetilde{M} f)(x,t)\, d\varrho \le c_2 \int_{R^n} (\phi o f)(x)\, dx$.

iii) ϱ is a Carleson measure.

iv) There exists $c > 0$ such that

$$\phi\left(\frac{\varepsilon |E|}{|B|}\right) \varrho(\widehat{B}) \le c\, \phi(\varepsilon)\, |E|$$

for every ball B and measurable $E \subset B$, and every positive ε.

If, moreover, ϕ is a Young function, then any of the above conditions implies

v) $\|\widetilde{M} f\|_{L_{\phi,\varrho}} \le c \|f\|_{L_\phi}$

Additionally, if ϕ is a Young function satisfying the Δ_2 condition, then (v) implies (i)–(iv).

For the Riesz transforms we have

Theorem 2.7 [25], [26]. *Let ϕ and ψ be a couple of complementary Young functions, both satisfying the Δ_2 condition. Suppose w is a weight function. Then the statements below are equivalent:*

i) $\int_{R^n} (\phi o R_j f)(x) w(x)\, dx \le c_1 \int_{R^n} (\phi o f)(x) w(x)\, dx$

ii) $\|R_j f\|_{L_\phi(\varepsilon w)} \le c_2 \|f\|_{L_\phi(\varepsilon w)}, \quad \varepsilon > 0$

iii) $\phi(\lambda) \int_{\{x : |R_j f(x)| > \lambda\}} w(x)\, dx \le c_3 \int_{R^n} (\phi o f)(x)\, w(x)\, dx, \quad j = 1,2,\ldots,n$.

iv) $w \in A_{i(\phi)}$ (the Muckenhoupt class) where $i(\phi)$ is the lower index of ϕ.

The proof of theorem 2.7 rests on the pioneering paper of R.Kerman and A. Torchinsky [27]. The generalization of results from [27] for anisotropic fractional maximal functions and Riesz potentials can be found in [28], [29].

Now we discuss the weight problem for the fractional maximal function.

$$M_\gamma f(x) \;=\; \sup_{Q\ni x} |Q|^{1-\gamma/n} \int\limits_Q |f(y)|\,dy\,, \quad 0<\gamma<n\,.$$

THEOREM 2.8 [31]. *Let $1 \le s \le p < q < \infty$. Then the inequality*

$$\|M_\gamma(f)\|_{L_v^{q\infty}} \le c_1 \|f\|_{L_w^{ps}}$$

is valid for some positive c independent of f if and only if

$$(vB)^{1/q} \|\chi_{B(x,r)} w^{-1}\|_{L_w^{p',s'}} \;\le\; c_2\,|B|^{1-\gamma/n}$$

for c_2 independent of B.

THEOREM 2.9 [31]. *Let $1 < p < \dfrac{n}{\gamma}$, $\dfrac{1}{q} = \dfrac{1}{p} - \dfrac{n}{\gamma}$, $1 < s < \infty$. Then the following estimates are equivalent:*

i) There is a constant $c > 0$ such that for each $f \in L_w^{ps}$,

$$\|M_\gamma(f w^{\gamma/n})\|_{L_w^{qs}} \le c_1 \|f\|_{L_w^{ps}}\,.$$

ii) There exists a constant $c_2 > 0$ such that

$$\|T_\gamma(f w^{\gamma/n})\|_{L_w^{qs}} \le c_2 \|f\|_{L_w^{ps}}\,.$$

iii) $w \in A_s$ where $s = 1 + \dfrac{q}{p'}$.

The case of the Hardy–Littlewood maximal function was discussed earlier by H. M. Chung, R. Hunt, D. Kurtz [32].

Now we return to the Riesz transforms and present a complete description of the class of functions w for which the Riesz transforms map $L\log^+ L(w)$ into L_w^1 or weak L^1. These problems for maximal functions were solved earlier by A. Carbery, S.-Y. A. Chang, J. Garnett [33]. For the Riesz transforms we have

THEOREM 2.10 [34]. *The following statements are equivalent:*

i) There exists a constant $c_1 > 0$ such that for any measurable function f supported in a cube Q we have the inequality

$$\int\limits_Q |R_j f(x)| w(x)\,dx \;\le\; c_1 \left(\int\limits_Q w(x)\,dx \;+\; \int\limits_Q |f(x)| \log^+ |f(x)| w(x)\,dx \right)$$

where c_1 is independent of Q and f.

ii) There exist constants $\varepsilon > 0$ and $c_2 > 0$ such that for any cube Q and measurable subset $E \subset Q$,

$$\frac{1}{|Q|} \int\limits_Q \exp\left(\varepsilon \frac{|R_j(\chi_E w)(x)|}{w(x)}\right) dx \leq c_2 , \qquad j = 1, 2, \ldots, n .$$

Theorem 2.11 [34]. *Let $0 < \alpha < \infty$. Then the following conditions are equivalent:*

i) There exists a constant $c_3 > 0$ such that for any $\lambda > 0$ and any measurable function f we have the estimate

$$w\{x \in R^n : |R_j f(x)| > \lambda\} \leq c_3 \int\limits_{R^n} \frac{|f(x)|}{\lambda} \left(1 + \log^+ \frac{|f(x)|}{\lambda}\right)^\alpha w(x)\, dx \quad (j = 1, 2, \ldots, n).$$

ii) There exists a constant $c_4 > 0$ such that for any $\lambda > 0$ and any measurable function f we have

$$w\{x \in R^n : Mf(x) > \lambda\} \leq c_4 \int\limits_{R^n} \frac{|f(x)|}{\lambda} \left(1 + \log^+ \frac{|f(x)|}{\lambda}\right)^\alpha w(x)\, dx .$$

iii) There exists a constant $\varepsilon > 0$ such that

$$\sup \frac{1}{|Q|} \int\limits_Q \exp\left(\varepsilon \frac{1}{|Q| w(x)} \int\limits_Q w(y)\, dy\right)^{1/\alpha} dx < \infty .$$

where the supremum is taken over all cubes Q.

Now we consider the strong maximal function

$$M_s f(x) = \sup_{x \in J} \frac{1}{|J|} \int\limits_J |f(y)|\, dy,$$

where the supremum is taken over all rectangles with faces which are parallel to coordinate axes and contain the point x.

A collection of rectangles $\{R_j\}$ is said to satisfy *the weak overlapping condition*, if

$$\left| R_j \setminus \bigcup_{i=1}^{j-1} R_i \right| > \frac{1}{2} |R_j|$$

for each $j = 1, 2, \ldots$.

The following theorem provides a full description of the class of weight functions w for which the weak type weighted inequality for the strong maximal function is true.

Theorem 2.12 [35]. *The weighted inequality*

$$w\{x \in R^n : M_s f(x) > \lambda\} \le c \int\limits_{R^n} \frac{|f(x)|}{\lambda} \left(\log^+ \frac{|f(x)|}{\lambda}\right)^{n-1} w(x)\, dx$$

holds if and only if there exists a positive constants ε *and* c_1 *such that for every collection* $\{R_j\}$ *of rectangles satisfying the weak overlapping condition the inequality*

$$\int\limits_{\bigcup_j R_j} \exp\left\{\varepsilon \sum_j \frac{w(R_j)}{|R_j| w(x)} \chi_{R_j}(x)\right\}^{1/(n-1)} w(x)\, dx \le c\, w\Big(\bigcup_j R_j\Big)$$

is fulfilled.

In the paper [16] it was shown that if $w \in A_1$ defined by rectangles, then the above weak type inequality holds, but there are examples of weight functions w which satisfy the above condition, but do not belong to A_1.

3. Maximal functions and potential type integrals in spaces of the homogeneous type

Let (X, ρ, μ) be the space of the homogeneous type, i.e. the space with a measure μ and equipped with a quasimetric ρ; the latter being a mapping $\rho : X \times X \to R^1_+$ such that

1) $\rho(x,y) = \rho(y,x)$ for every $x, y \in X$;

2) $\rho(x,y) = 0$ if and only if $x = y$;

3) $\rho(x,y) \le \eta(\rho(x,z) + \rho(z,y))$ for every $x, y, z \in X$

where the constant $\eta > 0$ is independent of x, y, and z.

We shall assume that all balls $B(x,r) = \{y \in X : \rho(x,y) < r\}$ are μ–measurable and that there is a constant $c > 0$ such that for every $x \in X$ and $r > 0$,

$$0 < \mu B(x, 2r) \le c\, \mu B(x,r) < \infty.$$

For a locally integrable function f on X we define the following maximal function

$$\widetilde{M}_\gamma f(x,t) = \sup (\mu B)^{\gamma-1} \int\limits_B |f(y)|\, d\mu, \qquad 0 \le \gamma < 1, \quad t > 0,$$

where the supremum is taken over all balls B containing the point x and with radius greater than $t/2$.

In what follows we also consider integrals of the potential type

$$T_\gamma f(x,t) = \int\limits_X \frac{f(y)}{(\mu B(x, \varrho(x,y) + t))^{1-\gamma}}, \qquad 0 < \gamma < 1, \quad t \ge 0.$$

The functions M_γ and T_γ generalize the Hardy-Littlewood maximal function and the Riesz potential.

Let $\widehat{B}$ denote the cylinder $B \times [0, 2r)$, and let w be a weight function on X, i.e. w is locally integrable on X and positive almost everywhere in the μ-measure sense. Further, β will be a measure defined on the product of σ-algebras generated by balls from X and intervals from $[0, \infty)$.

We shall discuss the problem of a full description of pairs (w, β) for which weighted Lorentz-norm weak type inequalities and Lebesgue-norm strong type inequalities hold for operators $\widetilde{M}_\gamma$ and T_γ. Results about these operators in the Morrey spaces will be given, too. Below, we denote by $L^p(X, wd\mu)$ the Lebesgue space and by $L^{ps}(X, d\mu)$ the weighted Lorentz space defined for μ-measurable functions acting from X into R^1.

THEOREM 3.1 [41]. *Let $1 \le s \le p \le q < \infty$, $0 \le \gamma < 1$. Then the following statements are equivalent:*

i) There is a constant $c_1 > 0$ such that for any function f from $L^{ps}(X, wd\mu)$ and $\tau > 0$,

$$\beta\left\{(x,t) \in X \times [0,\infty) : \widetilde{M}_\gamma f(x,t) > \tau\right\} \le c_1 \tau^{-q} \|f\|^q_{L^{ps}(X,\, d\mu)}$$

ii) There is a constant $c_2 > 0$ such that for any ball B from X,

$$(\beta \widehat{B})^{1/q} \|\chi_B w^{-1}\|_{L^{p's'}(X,\, d\mu)} \le c_2 (\mu B)^{1-\gamma}.$$

Note that Theorem 3.1 was established in [38] for the case $p = 1$.

Now we consider the two-weight weak type problem for the operator T_γ.

THEOREM 3.2 [41]. *Let $1 \le s \le p < q < \infty$. Then the following two conditions are equivalent:*

i) There is a positive constant c such that for any $f \in L^{ps}(X, wd\mu)$ the inequality

$$\beta\{(x,t) \in X \times [0,\infty) : |T_\gamma f(x,t)| > \tau\} \le c\tau^{-q} \|f\|^q_{L^{ps}(X,\, wd\mu)}$$

holds.

ii) $\sup \beta\left(\widehat{B}(a, N(2r+t))\right)^{1/q} \|\chi_{R^N \setminus B(a,r)} w^{-1} (\mu B(a, \varrho(a,.)+1))^{s-1}\|_{L^{p's'}(X,\, wd\mu)} < c_1$, *with a positive constant c_1 is independent of $a \in X$, $r > 0$ and $t > 0$, $N = \eta + 4\eta^2$.*

This theorem was proved earlier in [43] in the case when $d\beta = wd\mu \otimes \delta_0$ where δ_0 is Dirac measure supported at the origin and $p = s$. In the paper [42] there was obtained the solution to the one weight problem for the operator $T_\gamma f(x,0) = T_\gamma f(x)$.

Theorem 3.3 [43]. *Let $1 < p < \frac{1}{\gamma}$, $\frac{1}{q} = \frac{1}{p} - \gamma$. Then the necessary and sufficient condition for the validity of the inequality*

$$\left(\int_X |T_\gamma(fw^\gamma)(x)|^q\, w(x)d\mu\right)^{1/q} \leq c \left(\int_X |f(x)|^p w(x)d\mu\right)^{1/p}$$

with some positive constant c independent of f is

$$\sup \left(\frac{1}{\mu B}\int_B w(x)d\mu\right)\left(\frac{1}{\mu B}\int_B w^{-1/(s-1)}(x)d\mu\right)^{s-1} < \infty\,,$$

where $s = 1 + \frac{q}{p'}$ and the supremum is taken over all balls in X.

The next theorem gives us the necessary and sufficient condition for the validity of two weight strong inequality for the operator $\widetilde{M}_\gamma$.

Theorem 3.4 [44]. *Let $1 < p \leq q < \infty$, $0 \leq \gamma < 1$ and let the measure $\beta = w^{-1/(p-1)}d\mu$ satisfy the doubling condition, i.e.*

$$\int_{B(a,2r)} w^{-1/(p-1)}(x)\, d\mu \leq c \int_{B(a,r)} w^{-1/(p-1)}(x)\, d\mu$$

for every ball $B(a,r)$. Then the following conditions are equivalent:

i) The inequality

$$\left(\int_{X\times[0,\infty)} \left(\widetilde{M}_\gamma f(x,t)\right)^q d\beta\right)^{1/q} \leq c_1 \left(\int_X |f(x)|^p\, w(x)\, d\mu\right)^{1/p}$$

holds for every function $f \in L^p(X,d\mu)$ with a constant c_1 independent of f.

ii) There exists a constant c_2 such that

$$\left(\int_{\widehat{B}} \left(\widetilde{M}_\gamma(\chi_B w^{-1/(p-1)})(x,t)\right)^q d\beta\right)^{1/q} \leq c_2 \left(\int_B w^{-1/(p-1)}(x)\, d\mu\right)^{1/p}.$$

Considering the case $w \equiv 1$ in Theorem 3.4 we have

Theorem 3.5 [42]. *If $1 < p < q < \infty$ and $0 < \gamma < 1$, then the following statements are equivalent:*

i) The inequality

$$\left(\int_{X\times[0,\infty)} \left(\widetilde{M}_\gamma f(x,t)\right)^q d\beta\right)^{1/q} \leq c_1 \left(\int_X |f(x)|^p\, d\mu\right)^{1/p}$$

holds for every function $f \in L^p(X, d\mu)$.

ii) The inequality

$$\left(\int\limits_{X\times[0,\infty)} |T_\gamma f(x,t)|^q \, d\beta \right)^{1/q} \leq c_2 \left(\int\limits_X |f(x)|^p \, d\mu \right)^{1/p}$$

holds for any function $f \in L^p(X, d\mu)$.

iii) It is

$$\beta\widehat{B} \leq c_3 \, (\mu B)^{q(1/p-\gamma)}$$

for every ball B *from* X.

The constants c_1, c_2, c_3 *are independent of* f *and* B.

The equivalence of ii) and iii) in the case $d\beta = w d\mu \otimes \delta_0$ has been proved earlier in [45].

Let us mention several other results. In [46] the necessary and sufficient condition on the weight function was found which ensures continuity of the fractional maximal function with respect to the basis of the convex comparable sets in the weighted Lebesgue spaces. Under an additional condition on this basis a complete description of pairs of weight functions is given such that the fractional maximal function is continuous from one weighted Lebesgue space to the other.

Further, in [43] the Koosis problems were solved for the fractional maximal functions and fractional integrals defined on the homogeneous type spaces. In [47] there was obtained a complete description of pairs (v, w) of weights for which the potential operator in the space of the homogeneous type for a critical index acts continuously from the weighted Lebesgue space L^p_w into the space BMO_v — the weighted space of functions with bounded mean oscillation. In this case the Koosis type problem have been also solved.

REFERENCES

[1] B. Muckenhoupt: Weighted norm inequalities for the Hardy maximal function. Trans. Amer. Math. Soc. 165(1972), 207-226.

[2] R. Hunt, B. Muckenhoupt, R. Wheeden: Weighted norm inequalities for the conjugate function and Hilbert transform. Trans. Amer. Math. Soc. 176(1973), 227-251.

[3] B. Muckenhoupt, R. Wheeden: Weighted norm inequalities for fractional integrals. Trans. Amer. Math. Soc. 132(1974), 261-274.

[4] E. Sawyer: A characterization of a two-weight norm inequalities for maximal operators. Studia Math. 75(1982), 1-11.

[5] E. Sawyer: A two weight type inequalities for fractional integrals. Trans. Amer. Math. Soc. 281,1(1981), 339-345.

[6] M. Gabidzashvili: Weighted inequalities for anisotropic potentials. Trudy Tbiliss. Mat. Inst. Razmadze Akad. Nauk. Gruzin. SSR 82(1988), 23-36.

[7] E. Sawyer: A characterization of two weight norm inequalities for fractional and Poisson integrals. Trans. Amer. Math. Soc. 308(1988), 533-545.

[8] V. Kokilashvili: On the fractional integrals on curves (Russian). Dokl. Akad. Nauk SSSR 305,1(1989), 33-34. Engl.translation in Soviet Math. Dokl. 39,2(1989), 251-252.

[9] V. Kokilashvili: Potentials on thin sets. Dokl. Akad. Nauk SSSR, 308,5(1989), 1042-1044.

[10] V. Kokilashvili: Fractional integrals on curves. Trudy Tbiliss. Mat. Inst. Razmadze Akad. Nauk. Gruzin. SSR 95(1989), 56-70.

[11] S. L. Sobolev: On a theorem in functional analysis (Russian). Mat. Sb. 4,86(1938), 471-497. Engl. translation in Amer. Math. Soc. Transl. 2,34(1963), 39-68.

[12] V. Kokilashvili, A.Kufner: Fractional integrals on spaces of homogeneous type. Comm. Math. Univ. Carolinae 30,3(1989), 511-523.

[13] V. Kokilashvili: Anisotropic potentials on thin sets. Bull. Georgian Acad. Sci. (1989).

[14] V. Kokilashvili, M. Gabidzashvili: Two weight weak type inequalities for fractional type integrals. Preprint #45, Math. Inst. Czech. Acad. Sci. (1989), 1-11.

[15] M. Gabidzashvili, I. Genebashvili, V. Kokilashvili: Two weight weak type inequalities

for generalised potentials. Trudy Math. Inst. Acad. Sci. SSSR. To appear.

[16] V. Kokilashvili: Two weight problem for integrals with positive kernel. Trudy Tbiliss. Mat. Inst. Rasmadse Akad. Nauk. Grusin. SSR. To appear.

[17] V. Kokilashvili, J. Rákosník: A two weight weak inequality for potential type operators. To appear.

[18] A. Gogatishvili, V. Kokilashvili, M.Krbec: Maximal functions in $\phi(L)$ classes. Dokl. Akad. Nauk SSSR. To appear.

[19] A. Gogatishvili: Riess transform and maximal functions in $\phi(L)$ classes. Bull. Georgian Acad. Sci. 137,3(1990), 489–492

[20] D. Gallardo: Orlicz spaces for which the Hardy-Littlewood maximal operator is bounded. Publ. Math. 32(1988), 261-266.

[21] R. Ryan: Conjugate functions in Orlicz spaces. Pacific J.Math. 13(1963), 1371-1377.

[22] O.Cereteli: On integrability of conjugate functions. Trudy Tbiliss. Mat. Inst. Razmadze Akad. Nauk. Grusin. SSR 43(1973), 149-167.

[23] P. Oswald: Fourier series and conjugate functions in $\phi(L)$. Anal. Math. 8,4(1982) 287-303.

[24] L. Carleson: Interpolation by bounded analytic functions and the corona problem. Annals of Math. 76(1962), 547-559.

[25] V. Kokilashvili, M. Krbec: Maximal functions in $\phi(L)$ classes and Carleson measures. Bull. Georgian Acad. Sci. 137,2(1990), 269-271.

[26] V. Kokilashvili: Singular integrals in the Orlicz weighted classes. Trudy Tbiliss. Mat. Inst. Rasmadse Akad. Nauk. Grusin. SSR 89 (1988), 42-50.

[27] R. Kerman, A. Torchinsky: Integral inequalities with weights for the Hardy maximal functions. Studia Math. 71,3 (1981/1982), 277-284.

[28] V. Kokilashvili, M. Krbec: Weighted inequalities for Riesz potentials and fractional maximal functions in Orlicz spaces. (Russian.) Dokl. Akad. Nauk SSSR 283(1985), 280-283. (Engl. translation in Soviet Math. Dokl. 32(1985), 70-73.)

[29] V. Kokilashvili, M. Krbec: On boundedness of anisotropic fractional maximal functions

and potentials in weighted Orlicz spaces (Russian). Trudy Tbiliss. Mat. Inst. Razmadze Akad. Nauk. Gruzin. SSR 82(1986), 106–115.

[30] V. Kokilashvili: Maximal functions and singular integrals in weighted functional spaces. Trudy Tbiliss. Mat. Inst. Razmadze Akad. Nauk. Gruzin. SSR 80(1985), 1-113.

[31] V. Kokilashvili: Weighted inequalities for maximal functions and fractional integrals in Lorentz spaces. Math. Nachr. 133(1987), 33–42.

[32] H. M. Chung, R. Hunt, D. Kurtz: The Hardy–Littlewood maximal function in $L(p,q)$ spaces with weights. Indiana Univ. Math. J. 31(1982), 109–120.

[33] A. Carbery, S.-Y. A. Chang, J. Garnett: Weights and $L \log L$. Pacific J. Math. 120,1(1985), 33–45.

[34] A. Gogatishvili: Riesz transforms and maximal functions in Morrey–Campanato type spaces and in weighted Zygmund spaces. Bull. Georgian Acad. Sci. To appear.

[35] A. Gogatishvili: Weak type weight inequality for strong maximal functions. Bull. Georgian Acad. Sci. To appear.

[36] R. Bagby, D. S. Kurtz: $L \log L$ spaces and weights for the strong maximal function. J. Analyse Math. 44(1984/1985), 22–31.

[37] A. Calderón: Inequalities for the maximal function relative to a metric. Studia Math. 57(1976), 297–306.

[38] R. Macías, C. Segovia: A well-behaved quasi-distance for spaces of homogeneous type. Trab. Mat. Inst. Argent. Mat. 32(1981), 1-18.

[39] F. Ruiz, J. Torrea: Vector-valued Calderon-Zygmund theory and Carleson measures on spaces of homogeneous nature. Studia Math. 88(1987), 221-243.

[40] F. Ruiz: A unified approach to Carleson measures and A_p weights. Pacific J. Math. 117,2 (1985), 397-404.

[41] I. Genebashvili: Two weight norm inequalities for fractional maximal functions in Lorentz spaces (Russian). Bull. Georgian Acad. Sci. 136(1989), 21–24.

[42] I. Genebashvili: Carleson measures and potentials defined on the spaces of homogeneous type. Bull. Georgian Acad. Sci. 135,3(1989), 505–508.

[43] M. Gabidzashvili: Weighted norm inequalities for potential type integrals and fractional maximal functions in homogeneous spaces. Trudy. Conf. Molod. Tech. MGU (1985), 15–18.

[44] I. Genebashvili: Two weight problem for maximal functions defined on the spaces of homogeneous type. Bull. Georgian Acad. Sci. 135,2, Part 2 (1989), 53–56.

[45] M. Gabidzashvili: Weighted inequalities for potentials in the homogeneous type spaces. Trudy Tbiliss. Mat. Inst. Razmadze Akad. Nauk. Gruzin. SSR 95 (1990), 3–9.

[46] A. Gogatishvili: Maximal functions with respect to the basis of the convex comparable sets in weighted Lebesgue space. Trudy Tbiliss. Mat. Inst. Razmadze Akad. Nauk. Gruzin. SSR 95(1989), 20–37.

[47] A. Gogatishvili: Weighted norm inequalities for potentials in spaces of homogeneous type. Bull. Georgian Acad. Sci. 129,3(1988), 493–495.

[48] E. Sawyer, R. Wheeden: Weighted inequalities for fractional integrals on Euclidean and homogeneous spaces. Preprint, 1989.

Mappings of Monotone Type: Theory and Applications

VESA MUSTONEN, OULU

1. INTRODUCTION

Let X be a real reflexive Banach space and let X^* stand for its dual space with respect to the continuous pairing $\cdot \cdot$. A mapping T from its domain $D(T)$ in X to X^* is said to be monotone if

$$T(u) - T(v)u - v \geq 0 \qquad \text{for all } u, v \in D(T).$$

Mappings from a subset of X to X^* are rather natural framework for the approach to deal with the problems of calculus of variations. Indeed, if ϕ is a convex functional from a subset G of X to , then its derivative ϕ' is a mapping from G to X^* and in fact, ϕ is convex if and only if ϕ' is monotone. The basic result for monotone mappings obtained by Minty in early sixties says that every monotone hemicontinuous coercive mapping form X to X^* is surjective. This theorem has many extensions to more general classes of mappings of monotone type as was shown by Browder, Višik, Brezis, Leray and Lions, Hess, Gossez and others in the following two decades. The principal tool in proving such extension was a Galerkin method which uses approximations in finite dimensional spaces.

On the other hand, the degree of mapping has been one of the most important tools in nonlinear functional analysis applied to obtain existence and multiplicity results for solutions of functional equations. The classical degree for continuous mappings from a bounded open subset G of ^{N} to ^{N} was constructed by Brouwer in 1912. In the celebrated paper by Leray and Schauder in 1934 the degree was extended for mappings in infinite dimensional Banach spaces of the form $F = I + C$ where I is the identity map and C is compact. Since then a number of further generalizations have been introduced. Important recent contributions are due to Browder in the framework of his larger program of studying the mapping degree for general classes of mappings with domains in one Banach space X and ranges in another Banach space Y (**[Bro 3,4,5]**).

The present notes are concerned with the construction of approximative degree theories for some classes of mappings of monotone type and applications to nonlinear partial differential equations. We shall start in Section 2 by giving the definitions and basic properties of various classes of mappings of monotone type relevant for our applications. In Section 3 we survey the construction of the degree for mappings of class (S_+) and for quasimonotone mappings. This approach was introduced in (**[BM 1]**, **[Be]**) and it is based on the Leray-Schauder theory rather than on Brouwer degree and Galerkin procedure used in the previous approach by Browder (**[Bro 4,5]**). Section 4 is devoted to some general theorems obtained by homotopy argument whenever a classical degree theory is available. The results are applied in Section 5 to the study of nonlinear elliptic partial differential operators in divergence form

$$A(u) = \sum_{|\alpha| \leq m} (-1)^{|\alpha|} D^\alpha A_\alpha(x, u, \dots, D^m u).$$

In Section 6 we consider mappings defined in a Hilbert space H having a representation of the form

$$F = Qg + Pf,$$

where $H = M \oplus M^\perp$, Q and P are the orthogonal projections to M and $M^\perp$, respectively, $g = I + C$ is a Leray-Schauder map and f is a mapping of class (S_+). We show that the degree for such mappings can be constructed as an extension of the Leray-Schauder degree. The results of Section 6 are applied to the study of existence and multiplicity of periodic solutions for semilinear wave equations of the form

$$\begin{cases} \frac{\partial^2 u}{\partial t^2} - \frac{\partial^2 u}{\partial x^2} - g(x,t,u) = h & \text{in } (0,\pi) \times (0, 2\pi) \\ u(0,\cdot) = u(\pi,\cdot) = 0 \\ u(\cdot, t+2\pi) = u(\cdot, t) \end{cases}$$

as we shall show later in Section 7. These two sections survey the results of recent papers ([BM 2,3,4]). Finally, in Section 8 we deal with the extension of the degree for mappings of the form $F = T + S$, where T is a maximal monotone map from a subset $D(T)$ of X to X^* and S is of class (S_+). The case where $F = L + S$, L is a linear densely defined maximal monotone map and S is of class (S_+) or pseudomonotone with respect to $D(L)$, is discussed in Section 9. This makes possible to apply the results to the study of nonlinear parabolic operators of the form

$$\frac{\partial u}{\partial t} + \sum_{|\alpha| \le m} (-1)^{|\alpha|} D^\alpha A_\alpha(x, t, u, \dots, D^m u).$$

The material of these two sections is mainly based on Browder's paper ([Bro 5]) and on the ideas indicated therein.

2. Mappings of monotone type

Let X be a real reflexive Banach space. We can assume without loss of generality that X and its dual space X^* are locally uniformly convex (see [De]). We shall be dealing with mappings T acting from a subset $D(T)$ of X into X^*. The norm convergence in X and in X^* is denoted by $\to$ and the weak convergence by $\rightharpoonup$, respectively. We start with the following concepts of continuity. T is called *hemicontinuous* if $T(u + t_n v) \rightharpoonup T(u)$ as $t_n \to 0+$, and *demicontinuous* if $u_n \to u$ implies $T(u_n) \rightharpoonup T(u)$.

A mapping $T: D(T) \to X^*$ is called

- *monotone* (we denote $T \in (MON)$) if $\langle T(u) - T(v), u - v\rangle \ge 0$ for all $u, v \in D(T)$.
- *quasimonotone* ($T \in (QM)$) if for any sequence $\{u_n\}$ in $D(T)$ with $u_n \rightharpoonup u$, we have $\limsup \langle T(u_n), u_n - u\rangle \ge 0$.
- *pseudomonotone* ($T \in (PM)$) if for any sequence $\{u_n\}$ in $D(T)$ with $u_n \rightharpoonup u$ and $\limsup\langle T(u_n), u_n - u\rangle \le 0$, we have $\lim\langle T(u_n), u_n - u\rangle = 0$, and if $u \in D(T)$, then $T(u_n) \rightharpoonup T(u)$.
- *of class* (S_+) ($T \in (S_+)$) if for any sequence $\{u_n\}$ in $D(T)$ with $u_n \rightharpoonup u$ and $\limsup\langle T(u_n), u_n - u\rangle \le 0$, we have $u_n \to u$.
- *bounded* if it takes bounded subsets of $D(T)$ into bounded subsets of X^*.
- *locally bounded* if for each $u \in D(T)$ there exists a neighbourhood $\mathcal{U}$ of u such that $T(\mathcal{U})$ is bounded in X^*.
- *compact* ($T \in (COMP)$) if it is continuous and takes any bounded subset of $D(T)$ into a relatively compact subset of X^*.

Moreover, a monotone mapping T is *strictly monotone* if $\langle T(u) - T(v), u - v\rangle > 0$ for all $u \ne v$ in $D(T)$, and *strongly monotone* (denote $T \in (MON)_S$) if there exists a continuous strictly increasing function g from $[0,\infty)$ to $[0,\infty)$ with $g(0) = 0$ such that

$$\langle T(u) - T(v), u - v\rangle \ge g(\|u - v\|)\|u - v\| \qquad \text{for all } u, v \in D(T).$$

If we assume that all mappings are demicontinuous and defined in the whole space X, we have the following inclusions

$$
\begin{array}{ccccccc}
(MON)_S & \longrightarrow & (S_+) & \longrightarrow & (PM) & \longrightarrow & (QM) \\
 & & & & \uparrow & & \uparrow \\
 & & & & (MON) & & (COMP).
\end{array}
$$

For the sequel it will be important to observe that if $S \in (S_+)$ and $T \in (QM)$ are arbitrary demicontinuous mappings, then $T + S \in (S_+)$. Otherwise, all the classes defined above have at least a conical structure.

It is well known that the conditions

$$\|\mathcal{J}(u)\| = \|u\|, \quad \langle \mathcal{J}(u), u\rangle = \|u\|^2 \qquad \text{for all } u \in X,$$

determine a unique mapping $\mathcal{J}$ from X to X^*, which is called the *duality map*. In our case it is bijective bicontinuous strictly monotone and of class (S_+).

The concept of a monotone map has a natural generalization for multivalued maps. A subset $A \subset X \times X^*$ is called *monotone* if for each pair $[u_1, w_1]$, $[u_2, w_2] \in A$ we have $\langle w_1 - w_2, u_1 - u_2\rangle \geq 0$. Let T be a *multi* from $D(T) \subset X$ to 2^{X^*}, the subsets of X^*, where $D(T) = \{u \in X \mid T(u) \neq \emptyset\}$. Then T is called *monotone* if its graph

$$G(T) = \{[u, w] \subset X \times X^* \mid u \in D(T),\ w \in T(u)\}$$

is a monotone set in $X \times X^*$. A monotone multi T is said to be *maximal monotone* (denote $T \in (MM)$) if $G(T)$ is not a proper subset of any monotone set in $X \times X^*$, or equivalently, if $[u_0, w_0] \in X \times X^*$ and $\langle w_0 - w, u_0 - u\rangle \geq 0$ for all $[u, w] \in G(T)$, then $[u_0, w_0] \in G(T)$, i.e., $u_0 \in D(T)$ and $w_0 \in T(u_0)$. Any monotone multi T has the inverse T^{-1} from $\mathcal{R}(T) = \bigcup\{T(u) \mid u \in D(T)\}$ to 2^X defined by

$$T^{-1}(w) = \{u \in D(T) \mid w \in T(u)\},$$

which is also monotone. Moreover, T is maximal monotone if and only if T^{-1} is maximal monotone. Using the duality map $\mathcal{J}$ one can show that T is maximal monotone if and only if $\mathcal{R}(T + \lambda\mathcal{J}) = X^*$ for every $\lambda > 0$. The results above remainnaturally true if T is a single valued mapping. In particular, any single valued hemicontinuous monotone mapping from X to X^* is also maximal monotone.

Finally, if L is a linear monotone multi from $D(L)$ to 2^{X^*}, then its graph $G(L)$ is a subspace of $X \times X^*$. If $D(L)$ is dense in X, then L is necessarily single valued. We can also define the adjoint of L as a map $L^*\colon D(L^*) \to 2^{X^*}$ by

$$w \in L^*u \quad \text{if and only if} \quad \langle w, v\rangle = \langle Lv, u\rangle \quad \text{for all } v \in D(L).$$

A necessary and sufficient condition for the maximal monotony of L is that $G(L)$ is closed and L^* is monotone. For more details and proofs on multis we refer to **[De]**.

3. The degree for mappings of class (S_+), (PM) and (QM)

The topological degree of mappings is one of the most effective tools for studying the existence and multiplicity of solutions of nonlinear equations. In our discussion the basic concept to start with will be the degree for mappings in infinite-dimensional Banach spaces of the form $F = I+C$ with C compact, which we shall call the LS-degree according to Leray and Schauder. For later reference we recall however the following more general definition of classical topological degree.

DEFINITION 3.1. *Let X and Y be topological spaces and $\mathcal{O}$ a family of open subsets of X. With each $G \in \mathcal{O}$ we associate a class $\mathcal{F}_G$ of mappings $F: \overline{G} \to Y$ and with each triplet (F, G, y) with $F \in \mathcal{F}_G$, $G \in \mathcal{O}$ and $y \in Y$, $y \notin F(\partial G)$ we associate an integer $d(F, G, y)$. The function d is said to be a classical topological degree if the following conditions are satisfied:*

(a) *If $d(F, G, y) \neq 0$, then there exists $u \in G$ such that $F(u) = y$.*

(b) *If $G_0, G \in \mathcal{O}$, $G_0 \subset G$ and $F \in \mathcal{F}_G$, then the restriction $F|_{\overline{G_0}} \in \mathcal{F}_{G_0}$. If $G_1, G_2 \in \mathcal{O}$ are disjoint subsets of $G \in \mathcal{O}$ such that $y \notin F(\overline{G} \setminus (G_1 \cup G_2))$, then*

$$d(F, G, y) = d(F, G_1, y) + d(F, G_2, y).$$

(c) *For an admissible class of homotopies $\mathcal{H}_G = \{F_t \in \mathcal{F}_G \mid 0 \leq t \leq 1\}$, $G \in \mathcal{O}$, and for a continuous curve $\{y(t) \mid 0 \leq t \leq 1\}$ in Y with $y(t) \notin F_t(\partial G)$ for all $t \in [0,1]$ we have $d(F_t, G, y(t))$ is constant for all $t \in [0,1]$.*

(d) *There exists a normalizing map $j: X \to Y$ such that $j|_{\overline{G}} \in \mathcal{F}_G$ for each $G \in \mathcal{O}$, and $d(j, G, y) = 1$ whenever $y \in j(G)$.*

The unique LS-degree function d_{LS} is obtained by choosing $X = Y$ a real Banach space, $\mathcal{O}$ all open bounded subsets of X, $\mathcal{F}_G$ all continuous mappings from $\overline{G}$ to X of the form $F = I + C$, C compact. The admissible class of homotopies $\mathcal{H}_G$ consists of *LS-homotopies* $F_t = I + C_t$, where $(t, u) \to C_t(u)$ is compact from $[0,1] \times \overline{G}$ to X. Obviously the normalizing map for LS-degree is I.

Our task now will be to introduce approximative procedures which extend the LS-degree to further classes of mappings of monotone type. From now on we assume that X is a real reflexive separable Banach space and that X and X^* are locally uniformly convex.

In virtue of the embedding theorem by Browder and Ton **[BT]** there exists a separable Hilbert space H and a linear compact injection $\Psi: H \to X$ such that $\Psi(H)$ is dense in X. We define a further map $\hat{\Psi}: X^* \to H$ by

$$(\hat{\Psi}(w), v) = \langle w, \Psi(v) \rangle, \quad v \in H, \; w \in X^*,$$

where $(\cdot,\cdot)$ stands for the inner product in H. It is obvious that $\hat{\Psi}$ is also linear compact injection. Let G be an open bounded subset in X. We denote

$$\mathcal{F}_G(S_+) = \{F: \overline{G} \to X^* \mid F \in (S_+), \text{ bounded and demicontinuous}\}$$

and

$$\mathcal{H}_G(S_+) = \{F_t: \overline{G} \to X^*, \; 0 \leq t \leq 1 \mid F_t \text{ bounded homotopy of class } (S_+)\},$$

where F_t is said to be a *bounded homotopy of class* (S_+) if it is uniformly bounded in $t \in [0,1]$ and if for any sequences $\{u_n\}$ in $\overline{G}$, $\{t_n\}$ in $[0,1]$ with $u_n \rightharpoonup u$ in X and $t_n \to t$ such that $\limsup \langle F_{t_n}(u_n), u_n - u \rangle \leq 0$, we have $u_n \to u$ in X and $F_{t_n}(u_n) \rightharpoonup F_t(u)$ in X^*. With each $F \in \mathcal{F}_G(S_+)$ we can now associate a family of mappings $\{F_\epsilon \mid \epsilon > 0\}$ defined by

$$F_\epsilon(u) = u + \frac{1}{\epsilon} \Psi \hat{\Psi} F(u), \qquad u \in \overline{G}. \tag{3.1}$$

For any fixed $\epsilon > 0$, F_ϵ maps $\overline{G}$ into X and has the form $I + C_\epsilon$ where $C_\epsilon = \frac{1}{\epsilon} \Psi \hat{\Psi} F$ is compact. Hence the LS-degree is defined for the triplets (F_ϵ, G, y) whenever $y \notin F_\epsilon(\partial G)$. We have the following basic

LEMMA 3.1. *Let $F \in \mathcal{F}_G(S_+)$, $A \subset \overline{G}$ a closed subset and $0 \notin F(A)$. Then there exists $\hat{\epsilon} > 0$ such that $0 \notin F_\epsilon(A)$ for all $0 < \epsilon < \hat{\epsilon}$. Moreover, if $0 \notin F(\partial G)$, there exists $\epsilon_0 > 0$ such that $d_{LS}(F_\epsilon, G, 0)$ is constant for all $0 < \epsilon < \epsilon_0$.*

PROOF: If the first assertion were false, there would exist sequences $\{\epsilon_n\}$ and $\{u_n\} \subset A$ such that $\epsilon_n \to 0+$ and $F_{\epsilon_n}(u_n) = 0$. Choosing subsequences, if necessary, (we will not change notations) we may assume that $u_n \rightharpoonup u$ in X and $F(u_n) \rightharpoonup w$ in X^*. Since $\Psi\mathring{\Psi}$ is linear and compact, $\Psi\mathring{\Psi}F(u_n) \to \Psi\mathring{\Psi}w$ in X. On the other hand, $\Psi\mathring{\Psi}F(u_n) = -\epsilon_n u_n \to 0$ implying that $w = 0$. Hence $F(u_n) \rightharpoonup 0$ and

$$\begin{aligned}\limsup \langle F(u_n), u_n - u\rangle &= \limsup \langle F(u_n), u_n\rangle \\ &= \limsup \langle F(u_n), -\frac{1}{\epsilon_n}\Psi\mathring{\Psi}F(u_n)\rangle \\ &= \limsup \left\{-\frac{1}{\epsilon_n}\|\mathring{\Psi}F(u_n)\|_H^2\right\} \le 0.\end{aligned}$$

Since $F \in (S_+)$, we have $u_n \to u$ with $u \in A$. By demicontinuity $F(u_n) \rightharpoonup F(u)$ implying a contradiction $F(u) = 0$. Hence we may conclude the existence of $\hat{\epsilon} > 0$ such that $F_\epsilon(u) \neq 0$ for all $u \in A$ and $0 < \epsilon < \hat{\epsilon}$.

For the second assertion we assume that $\epsilon_0 = \hat{\epsilon}$ for $A = \partial G$, $0 < \epsilon_1 < \epsilon_2 < \epsilon_0$ and consider the following LS-homotopy

$$tF_{\epsilon_1} + (1-t)F_{\epsilon_2} = I + \left(\frac{t}{\epsilon_1} + \frac{1-t}{\epsilon_2}\right)\Psi\mathring{\Psi}F := F_{\epsilon_t},$$

where $\epsilon_t = (\frac{t}{\epsilon_1} + \frac{1-t}{\epsilon_2})^{-1}$, $0 \le t \le 1$. Since $\epsilon_1 \le \epsilon_t \le \epsilon_2 < \epsilon_0$ for all $t \in [0,1]$ we have $F_{\epsilon_t}(u) \neq 0$ for all $u \in \partial G$ and $t \in [0,1]$. Since the LS-degree satisfies (c), we obtain

$$d_{LS}(F_{\epsilon_1}, G, 0) = d_{LS}(F_{\epsilon_2}, G, 0)$$

completing the proof. ∎

In view of Lemma 3.1 it is relevant to define

$$d_{S_+}(F, G, 0) = d_{LS}(F_\epsilon, G, 0) \qquad \text{where } 0 < \epsilon < \epsilon_0. \tag{3.2}$$

Moreover, for any $y \in X^*$ with $y \notin F(\partial G)$ we can define

$$d_{S_+}(F, G, y) = d_{S_+}(F - y, G, 0). \tag{3.3}$$

To convince ourselves that we have obtained a classical topological degree function d_{S_+} for mappings in $\mathcal{F}_G(S_+)$ the conditions (a) to (d) have to be verified. It is obviously sufficient to deal with the case $y = 0$ or $y(t) \equiv 0$.

(a) If $0 \notin F(\overline{G})$, it follows from Lemma 3.1 that $0 \notin F_\epsilon(\overline{G})$ for all $0 < \epsilon < \hat{\epsilon}$. Hence $d_{LS}(F_\epsilon, G, 0) = 0$ for all $0 < \epsilon < \hat{\epsilon}$ implying $d_{S_+}(F, G, 0) = 0$. Therefore $d_{S_+}(F, G, 0) \neq 0$ implies $0 \in F(G)$.

(b) If G_1 and G_2 are open disjoint subsets of G and $0 \notin F(\overline{G} \setminus (G_1 \cup G_2))$, we can apply again Lemma 3.1 with $A = \overline{G} \setminus (G_1 \cup G_2)$ and use the property (b) for the LS-degree to derive (b) for d_{S_+}.

(c) If $F_t \in \mathcal{H}_G(S_+)$ we can extend Lemma 3.1 for homotopies in the obvious way. The property (c) follows then from the corresponding property for LS-homotopies.

(d) To show that $\mathcal{J}$ plays the role of normalizing map we consider the affine LS-homotopy $(1-t)I + t\mathcal{J}_\epsilon$. Since $\mathcal{J}(u) = 0$ if and only if $u = 0$, and since

$$\langle \mathcal{J}(u), (1-t)u + t\mathcal{J}_\epsilon(u)\rangle = \|u\|^2 + \frac{t}{\epsilon}\|\mathring{\Psi}\mathcal{J}(u)\|_H^2 > 0$$

for all $u \neq 0$ and $0 \le t \le 1$, we obtain

$$d_{S_+}(\mathcal{J}, G, 0) = \lim_{\epsilon \to 0+} d_{LS}(\mathcal{J}_\epsilon, G, 0) = d_{LS}(I, G, 0) = 1$$

whenever $0 \in \mathcal{J}(G)$.

Using the fact that LS-degree is unique it is not hard to show that also S_+-degree is unique (see [Be]). Thus we can conclude

Theorem 3.1. *Let X be a separable reflexive Banach space, G an open bounded subset in X and $\mathcal{F}_G(S_+)$ the class of admissible mappings. Then there exists exactly one degree function d_{S_+}, the S_+-degree, satisfying the properties* (a) *to* (d) *with respect to $\mathcal{H}_G(S_+)$ and normalising map $\mathcal{J}$.*

Remark 3.1: Originally Browder [Bro 4] constructed the S_+-degree by using Galerkin approximations, for which in each finite dimensional subspace the Brouwer degree is defined. His approach works also when X is not separable.

Remark 3.2: The S_+-degree can be constructed to all demicontinuous mappings of class (S_+), not only for bounded ones as above. It is essential that a demicontinuous (S_+)-mapping $F: X \to X^*$ is proper on bounded sets, i.e., for any bounded closed set $A \subset X$ and for any compact set $K \subset X^*$, $F^{-1}(K) \cap A$ is compact. For the detailed discussion in this direction we refer to [Be].

Remark 3.3: The S_+-degree can be extended for quasimonotone mappings, i.e., to the class $\mathcal{F}_G(QM)$ by using the fact that $F + \epsilon\mathcal{J} \in (S_+)$ whenever $F \in (QM)$ and $\epsilon > 0$. However, we face here the difficulty that the image $F(A)$ of a closed subset $A \subset \overline{G}$ is no more closed. Therefore the QM-degree obtained through approximations

$$d_{QM}(F, G, y) = \lim_{\epsilon \to 0+} d_{S_+}(F + \epsilon\mathcal{J}, G, y) \tag{3.4}$$

is not a classical degree in the sense of Definition 3.1. For instance we have:

(a)′ If $d_{QM}(F, G, y) \neq 0$, then $y \in \overline{F(G)}$.

In fact, for the definition (3.4) we assume $y \notin \overline{F(\partial G)}$. For more details on weak degree theories we refer to [Bro 4,5], [Be].

Remark 3.4: Since $(S_+) \subset (PM) \subset (QM)$, the QM-degree is defined for all mappings $F \in \mathcal{F}_G(PM)$. We shall see from applications that many results obtained for (S_+)-mappings hold true also for pseudomonotone mappings. This is based on the fact that for each $F \in \mathcal{F}_G(PM)$ the set $F(A)$ is closed whenever $A \subset \overline{G}$ is weakly closed. Indeed, if $\{w_n\} \subset F(A)$ with $w_n \to w$, then $w_n = F(u_n)$ for some $\{u_n\} \subset A$. Since G is bounded, $u_n \rightharpoonup u$ for some $u \in A$, at least for a subsequence. Thus

$$\limsup \langle F(u_n), u_n - u \rangle = 0$$

implying $F(u_n) \rightharpoonup F(u)$ and hence $w = F(u) \in F(A)$. In particular, if we assume that G is convex, then $\overline{G}$ is weakly closed implying that $F(\overline{G})$ is closed. Consequently, for $F \in \mathcal{F}_G(PM)$ and G convex, we can conclude instead of (a)′, that

$$d_{QM}(F, G, y) \neq 0 \quad \text{implies} \quad y \in F(\overline{G}).$$

4. Some applications of continuation method

We shall describe some standard results obtained by a continuation method (homotopy argument) when a classical degree theory is available. We use here the S_+-degree but analogous results hold for other degree functions which will be introduced in our further discussion.

Let X be a real reflexive Banach space, G an open bounded subset in X and $F \in \mathcal{F}_G(S_+)$. If $y \in X^* \setminus F(\partial G)$, a sufficient condition for the solvability of the equation

$$F(u) = y$$

is that $d_{S_+}(F, G, y) \neq 0$. In many cases this can be shown by using the homotopy argument (c) for a suitable homotopy involving F and some *reference map* $T \in \mathcal{F}_G(S_+)$, i.e., an injection satisfying $d(T, G, w) \neq 0$ whenever $w \in T(G)$. Then the property (c) for the (S_+)-degree yields

Theorem 4.1. *Let G be an open bounded subset in X, $T \in \mathcal{F}_G(S_+)$ a reference map and $F \in \mathcal{F}_G(S_+)$. If for a given $y \in X^*$ there exists $w \in T(G)$ such that*

$$tF(u) + (1-t)T(u) \neq ty + (1-t)w \qquad \text{for all } u \in \partial G,\ 0 \leq t \leq 1, \tag{4.1}$$

then $d_{S_+}(F, G, y) \neq 0$ and the equation $F(u) = y$ admits a solution u in G.

The obvious reference map for the (S_+)-degree is the duality map $\mathcal{J}$. Bearing in mind the properties of $\mathcal{J}$ we can rewrite condition (4.1) to obtain from Theorem 4.1 the following

Theorem 4.2. *Let G be an open bounded subset in X and $F \in \mathcal{F}_G(S_+)$. If there exists $\bar{u} \in G$ such that*

$$\langle F(u) - y, u - \bar{u}\rangle > \|F(u) - y\|\|u - \bar{u}\| \qquad \text{for all } u \in \partial G, \tag{4.2}$$

then $d_{S_+}(F, G, y) = 1$ and the equation $F(u) = y$ admits a solution u in G.

For mappings F which satisfy some coercivity conditions we can derive surjectivity results

Theorem 4.3. *Let $F \in \mathcal{F}_X(S_+)$ satisfy the conditions*

(i) *if $y \in X^*$ and $F(u_n) \to y$ in X^*, then $\{u_n\}$ is bounded in X*

(ii) *there exists $R > 0$ such that*

$$\frac{\langle F(u), u\rangle}{\|u\|} + \|F(u)\| > 0 \quad \text{for all } u \in X \text{ with } \|u\| \geq R.$$

Then $F(X) = X^*$.

Proof: Let $y \in X^*$ be given. Then there exist $k > 0$ and $R' \geq R$ such that

$$\|F(u) - ty\| \geq k \qquad \text{for all } t \in [0,1] \text{ and } \|u\| \geq R'. \tag{4.3}$$

Indeed, if we assume the contrary, we find sequences $\{u_n\}$ in X and $\{t_n\}$ in $[0,1]$ such that $\|u_n\| \to \infty$, $\|F(u_n) - t_n y\| \to 0$ and $t_n \to t$. Hence $F(u_n) \to ty$ and by (i) we get a contradiction. From (4.3) we then obtain by the property (c)

$$d_{S_+}(F, B_{R'}, y) = d_{S_+}(F, B_{R'}, 0), \tag{4.4}$$

where $B_{R'} = \{u \in X \mid \|u\| < R'\}$. In view of (ii) we have $\langle F(u), u\rangle > -\|F(u)\|\|u\|$ for all $\|u\| = R'$. By Theorem 4.2 we get $d_{S_+}(F, B_{R'}, 0) = 1$ implying by (4.4) that $y \in F(B_{R'})$. ∎

Remark 4.1: It is useful to observe that (i) is met if

$$\|F(u_n)\| \to \infty \qquad \text{whenever } \|u_n\| \to \infty \text{ in } X. \tag{i$_1$}$$

Moreover, if F satisfies the condition

$$\frac{\langle F(u), u\rangle}{\|u\|} + \|F(u)\| \to \infty \quad \text{as } \|u\| \to \infty, \tag{i$_2$}$$

then clearly both (i) and (ii) are met. Finally, the strong coercivity condition

$$\frac{\langle F(u), u\rangle}{\|u\|} \to \infty \quad \text{as } \|u\| \to \infty \tag{i$_3$}$$

implies (i)$_2$.

In view of the Remarks 3.3 and 3.4 we can expect that Theorems 4.1 to 4.3 have generalizations for mappings F in $\mathcal{F}_G(QM)$ and $\mathcal{F}_G(PM)$. Indeed, the following modifications are easily obtained (see [**BM 1**], [**Be**]).

THEOREM 4.4. *Let G be an open bounded subset in X and $F \in \mathcal{F}_G(QM)$. If there exists $\bar{u} \in G$ such that the condition* (4.2) *holds, then the equation $F(u) = y$ is almost solvable in the sense that $y \in \overline{F(\overline{G})}$. In particular, if G is convex and $F \in \mathcal{F}_G(PM)$, then $y \in F(\overline{G})$, i.e., the equation $F(u) = y$ admits a solution u in $\overline{G}$.*

THEOREM 4.5. *Let $F \in \mathcal{F}_X(QM)$ satisfy the conditions* (i) *and* (ii), *then $\overline{F(X)} = X^*$. In particular, if $F \in \mathcal{F}_X(PM)$, then $F(X) = X^*$.*

REMARK 4.2: For monotone demicontinuous mappings $F\colon X \to X^*$ one can show that $F(X) = X^*$ if and only if the condition (i) is satisfied.

For odd mappings we can obtain generalizations of Borsuk's theorem. If G is an open bounded symmetric set in X containing the origin and $F \in \mathcal{F}_G(S_+)$ with $F(-u) = -F(u)$ for all $u \in \partial G$, then also F_ϵ given by (3.1) is odd on ∂G for any $\epsilon > 0$. Hence Borsuk's theorem for LS-mappings ([**De**]) implies the existence of $u_\epsilon \in \overline{G}$ such that

$$F_\epsilon(u_\epsilon) = 0 \qquad \text{for any } \epsilon > 0.$$

Moreover, if $0 \notin F_\epsilon(\partial G)$, then $d_{LS}(F_\epsilon, G, 0)$ is odd. As in the proof of Lemma 3.1 we conclude $0 \in \overline{F(\overline{G})}$. For the case $0 \notin F(\partial G)$ we have also that $d_{S_+}(F, G, 0)$ is odd. If we assume that $F \in \mathcal{F}_G(QM)$ and F is odd on ∂G, we can consider its S_+-approximation $F + \epsilon\mathcal{J}$, which is also odd on ∂G for every $\epsilon > 0$. Thus Borsuk's theorem extends to the class $\mathcal{F}_G(QM)$ in an obvious way. In particular, we have the following standard surjectivity theorem, which can be proved as Theorem 4.3.

THEOREM 4.6. *Let $F \in \mathcal{F}_X(QM)$ satisfy the condition* (i) *and the condition*

(iii) *there exists $R > 0$ such that $F(-u) = -F(u)$ for all $\|u\| \geq R$.*

Then $\overline{F(X)} = X^*$. In particular, if $F \in \mathcal{F}_X(PM)$, then $F(X) = X^*$.

5. APPLICATIONS TO NONLINEAR ELLIPTIC PROBLEMS

Since the contributions by Minty, Browder, Višik, Brezis, Leray and Lions in the sixties the theory of mappings of monotone type has become a standard frame work (see [**PS**], [**FK**]) for the study of boundary value problems for nonlinear elliptic partial differential operators in divergence form

$$Au(x) = \sum_{|\alpha| \leq m} (-1)^{|\alpha|} D^\alpha A_\alpha(x, u(x), Du(x), \ldots, D^m u(x)), \qquad x \in \Omega, \tag{5.1}$$

where Ω is an open subset in $\mathbf{R}^N$ $(N \geq 2)$ and $m \geq 1$. The coefficients A_α are functions of the point $x \in \Omega$ and of $\xi = (\eta, \zeta) \in \mathbf{R}^{N_0}$ with $\eta = \{\eta_\beta \mid |\beta| \leq m-1\} \in \mathbf{R}^{N_1}$, $\zeta = \{\zeta_\beta \mid |\beta| = m\} \in \mathbf{R}^{N_2}$ and $N_1 + N_2 = N_0$. We assume that each $A_\alpha(x, \xi)$ is a *Carathéodory function*, i.e., measurable in x for fixed $\xi = (\eta, \zeta) \in \mathbf{R}^{N_0}$ and continuous in ξ for allmost all $x \in \Omega$. We shall assume here, for simplicity, that Ω is a bounded subset of $\mathbf{R}^N$. Then the familiar growth condition

($\mathbf{A_1}$) There exist $p > 1$, $c_1 > 0$ and $k_1 \in L^{p'}(\Omega)$, $p' = \frac{p}{p-1}$, such that

$$|A_\alpha(x, \xi)| \leq c_1(|\zeta|^{p-1} + |\eta|^{p-1} + k_1(x))$$

for all $x \in \Omega$, $\xi = (\eta, \zeta) \in \mathbf{R}^{N_0}$, $|\alpha| \leq m$, implies that the operator (5.1) gives rise to a bounded continuous mapping T from any Sobolev space X with $W_0^{m,p}(\Omega) \subset X \subset W^{m,p}(\Omega)$ into its dual space X^* by the rule

$$(5.2) \qquad \langle T(u), v\rangle = \int_\Omega \sum_{|\alpha|\leq m} A_\alpha(x, \xi(u))D^\alpha v, \qquad u, v \in X.$$

In the sequel we shall consider the case $X = W_0^{m,p}(\Omega)$. Then for a given $h \in L^{p'}(\Omega) \subset X^*$ the solutions $u \in X$ of the equation

$$(5.3) \qquad T(u) = h$$

are called *weak solutions* of the Dirichlet boundary value problem

$$(5.4) \qquad \begin{cases} Au(x) = h(x) & \text{in } \Omega \\ D^\alpha u(x) = 0 & \text{on } \partial\Omega \text{ for all } |\alpha| \leq m-1. \end{cases}$$

Hence we are in a position to apply the results of Section 4 to (5.4) as soon as T belongs to one of the classes of mappings of monotone type. A condition which obviously ensures that T is monotone is

$$(A_2)_M \qquad \sum_{|\alpha|\leq m} \{A_\alpha(x,\xi) - A_\alpha(x,\xi^*)\}(\xi_\alpha - \xi_\alpha^*) \geq 0 \qquad \text{for all } x \in \Omega,\ \xi, \xi^* \in \mathbf{R}^{N_0}$$

A condition implying, that T is pseudomonotone, is the classical Leray-Lions condition [LL] (cf. [LM])

$$(A_2)_S \qquad \sum_{|\alpha|=m} \{A_\alpha(x,\eta,\zeta) - A_\alpha(x,\eta,\zeta^*)\}(\zeta_\alpha - \zeta_\alpha^*) > 0 \qquad \text{for all } x \in \Omega,\ \eta \in \mathbf{R}^{N_1} \text{ and } \zeta \neq \zeta^* \text{ in } \mathbf{R}^{N_2}$$

A remarkable feature on the condition $(A_2)_S$ is that monotonicity is assigned only to the top order part

$$(5.5) \qquad A_1u(x) = \sum_{|\alpha|=m} (-1)^{|\alpha|} D^\alpha A_\alpha(x, u(x), Du(x), \ldots, D^m u(x))$$

and the lower order part

$$(5.6) \qquad A_2u(x) = \sum_{|\alpha|\leq m-1} (-1)^{|\alpha|} D^\alpha A_\alpha(x, u(x), Du(x), \ldots, D^m u(x))$$

is to obey the growth condition (A_1), only. It is interesting to observe that if we define the corresponding mappings T_1 and $T_2 \colon X \to X^*$ by

$$\langle T_1(u), v\rangle = \int_\Omega \sum_{|\alpha|=m} A_\alpha(x, \eta(u), \zeta(u))D^\alpha v, \qquad u, v \in X$$

and

$$\langle T_2(u), v\rangle = \int_\Omega \sum_{|\alpha|\leq m-1} A_\alpha(x, \eta(u), \zeta(u))D^\alpha v, \qquad u, v \in X,$$

then of course (A_1) and $(A_2)_S$ imply $T_1 \in (PM)$ and (A_1) implies that $T_2 \in (QM)$. Actually, by the Sobolev embedding theorem $\lim\langle T_2(u_n), u_n - u\rangle = 0$ for any sequence $\{u_n\} \subset X$ with

$u_n \rightharpoonup u$. Moreover, T_1 remains in the class (PM) although we weaken the strict monotonicity condition $(\mathrm{A}_2)_\mathrm{S}$ to

$$(\mathrm{A}_2) \qquad \sum_{|\alpha|=m} \{A_\alpha(x,\eta,\zeta) - A_\alpha(x,\eta,\zeta^*)\}(\zeta_\alpha - \zeta_\alpha^*) \geq 0.$$

If we assume only the conditions (A_1) and (A_2), then $T = T_1 + T_2$ is in the class (QM) but no more necessarily in the class (PM), unless T_2 is weakly continuous which is related to linear behaviour of A_2 with respect to highest order derivatives. For a more detailed discussion with some further refinements we refer to **[GM]**.

In order to apply the existence and surjectivity results of Section 4 to the problem (5.4) a further condition is needed. If the operator A satisfies the condition

(A_3) There exists $c_2 > 0$ and $k_2 \subset L^1(\Omega)$ such that

$$\sum_{|\alpha|\leq m} A_\alpha(x,\xi)\xi_\alpha \geq c_2|\zeta|^p - k_2(x)$$

for all $x \in \Omega$ and $\xi = (\eta,\zeta) \in \mathbf{R}^{N_0}$.

then T satisfies the strong coercivity condition $(\mathrm{i})_3$, since by the Poincare inequality in $X = W_0^{m,p}(\Omega)$,

$$\frac{\langle T(u),u\rangle}{\|u\|} \geq c_2\frac{\|D^m u\|^p_{L^p(\Omega)}}{\|u\|} + \frac{\|k_2\|_{L^1(\Omega)}}{\|u\|} \to \infty$$

as $\|u\| \to \infty$. Consequently, we get from Theorem 4.3 and 4.5 the following

COROLLARY 5.1. *Let Ω be a bounded open subset in $\mathbf{R}^N$ and assume that the operator A satisfies the conditions* (A_1) *and* (A_3).

(a) *If A satisfies one of the conditions* $(\mathrm{A}_2)_\mathrm{M}$ or $(\mathrm{A}_2)_\mathrm{S}$, *then T is surjective and the problem* (5.4) *admits a weak solution u in $W_0^{m,p}(\Omega)$ for any given $h \in L^{p'}(\Omega)$.*
(b) *If A satisfies* (A_2), *then $T(W_0^{m,p}(\Omega))$ is dense in $W^{-m,p'}(\Omega)$ and the problem* (5.4) *is almost solvable for any given $h \in L^{p'}(\Omega)$.*
(c) *If A satisfies* (A_2) *and T_2 is weakly continuous, then T is surjective and the problem* (5.4) *admits a weak solution u in $W_0^{m,p}(\Omega)$ for any given $h \in L^{p'}(\Omega)$.*

REMARK 5.1: It is not difficult to see that in fact the conditions (A_1), $(\mathrm{A}_2)_\mathrm{S}$ and (A_3) imply that T belongs to the class (S_+). If we assume that the problem (5.4) has variational structure, i.e., there exists a real valued functional $f\colon X \to \mathbf{R}$ such that $f' = T$, and if T happens to be quasimonotone and $f(u) \to \infty$ as $\|u\| \to \infty$, then the result by Hess **[He]** tells us that f is weakly lower semicontinuous and has a minimum. This means that for such operators the assertion in the case (b) will be the same as in the cases (a) and (c).

6. THE DEGREE FOR A CLASS OF MAPPINGS IN A HILBERT SPACE

In this section we shall consider mappings of monotone type acting in a real separable Hilbert space H. The definitions of Section 2 are to be understood with respect to the inner product $(\cdot,\cdot)$ of H. Hence (LS) is a subclass of (S_+). We assume that $H = M \oplus M^\perp$ where both M and $M^\perp$ are infinite-dimensional. Let Q and P denote the orthogonal projections to M and $M^\perp$, respectively. If G is an open bounded subset of H, we denote

$$\mathcal{F}_G(LS;S_+) = \{F\colon \overline{G} \to H \mid F = Qg + Pf \text{ for some bounded demicontinuous } g \in (LS) \text{ and } f \in (S_+)\},$$

the class of admissible mappings and

$$\mathcal{H}_G(LS;S_+) = \{F_t \mid 0 \le t \le 1,\ F_t = Qg_t + Pf_t\},$$

the class of admissible homotopies, where g_t is a LS-homotopy and f_t a S_+-homotopy. For each $F \in \mathcal{F}_G(LS;S_+)$ we will find LS-approximations by a partial Galerkin procedure. Indeed, let $\{N_n\}$ be a sequence of finite-dimensional subspaces of $M^\perp$ such that $N_n \subset N_{n+1}$ for each n and $\cup N_n$ is dense in $M^\perp$. Denoting by P_n the orthogonal projection to N_n we associate with $F = Q(I+C) + Pf$ a sequence $\{F_n\}$ where

$$F_n = I + QC + P_n f - P_n, \qquad n \in \mathbb{N}. \tag{6.1}$$

Since $QC + P_n f - P_n$ is compact, $F_n \in (LS)$ for each $n \in \mathbb{N}$. Similarly, each $F_t \in \mathcal{H}_G(LS;S_+)$ admits an approximation

$$(F_t)_n = I + QC_t + P_n f_t - P_n, \qquad n \in \mathbb{N}. \tag{6.2}$$

For each $y \in H$ we also need the approximation

$$y_n = Qy + P_n y, \qquad n \in \mathbb{N}. \tag{6.3}$$

Then we have

Lemma 6.1. *Let $F \in \mathcal{F}_G(LS;S_+)$ and $y \in H$ with $y \notin F(\partial G)$. Then there exists $n_0 \in \mathbb{N}$ such that $y_n \notin F_n(\partial G)$ for all $n \ge n_0$. Moreover, there exists $n_1 > n_0$ such that $d_{LS}(F_n, G, y_n)$ is constant for all $n \ge n_1$*

Proof: Since $(F-y)_n = F_n - y_n$, we may assume that $y = 0$. If the first assertion were false, then there would exist sequences $\{u_k\} \subset \partial G$ and $\{n_k\} \subset \mathbb{N}$ such that $F_{n_k}(u_k) = 0$ and $n_k \to \infty$. Separating the components in M and $M^\perp$ we have

$$Qu_k + QC(u_k) = 0 \quad \text{and} \quad Pu_k + P_{n_k} f(u_k) - P_{n_k} u_k = 0. \tag{6.4}$$

Taking a subsequence we have $u_k \rightharpoonup u$, $C(u_k) \to z$ and $f(u_k) \rightharpoonup w$ in H. Hence $Qu_k \to Qu$ and $Pu_k = P_{n_k} u_k$ for all k implying that $P_{n_k} f(u_k) = 0$ for all k. By (6.4) we obtain $P_{n_k} f(u_k) \rightharpoonup Pw = 0$. Therefore $(f(u_k), Pu) \to 0$ and

$$\begin{aligned}
\limsup (f(u_k), u_k - u) &= \limsup (f(u_k), Pu_k - Pu + Qu_k - Qu) \\
&= \limsup (f(u_k), P_{n_k} u_k) \\
&= \limsup (P_{n_k} f(u_k), u_k) \\
&= 0.
\end{aligned}$$

Since $f \in (S_+)$, $u_k \to u$ with $u \in \partial G$ and $F_{n_k}(u_k) \rightharpoonup F(u) = 0$, a contradiction. Hence there exists $n_0 \in \mathbb{N}$ such that the LS-degree $d_{LS}(F_n, G, 0)$ is defined for all $n \ge n_0$.

Assume that the second assertion is false. Then there exists a sequence $\{n_k\} \subset \mathbb{N}$, $n_k \ge n_0$, $n_k \to \infty$ such that

$$d_{LS}(F_{n_k}, G, 0) \ne d_{LS}(F_{n_{k+1}}, G, 0)$$

for all $k = 1, 2, \dots$. In view of the property (c) for the LS-degree we find sequences $\{u_k\} \subset \partial G$ and $\{t_k\} \subset (0,1)$ such that

$$(1 - t_k) F_{n_k}(u_k) + t_k F_{n_{k+1}}(u_k) = 0. \tag{6.5}$$

Bearing in mind that $F = Q(I+C) + Pf$ we get from (6.5)

$$u_k + QC(u_k) + (1-t_k)P_{n_k}(f(u_k) - u_k) + t_k P_{n_{k+1}}(f(u_k) - u_k) = 0.$$

Separating the components in M and $M^\perp$ we have

$$\text{(6.6)} \qquad \begin{cases} Qu_k + QC(u_k) = 0 \\ Pu_k + (1-t_k)P_{n_k}(f(u_k) - u_k) + t_k P_{n_{k+1}}(f(u_k) - u_k) = 0. \end{cases}$$

We can assume again that $u_k \rightharpoonup u$, $C(u_k) \to z$ and $f(u_k) \rightharpoonup w$ in H. As in the former proof we get $Qu_k \to Qu$, $Pu_k = P_{n_{k+1}}u_k$, $P_{n_k}f(u_k) = 0$ and $Pw = 0$. Moreover (6.6) also implies that

$$P_{n_{k+1}}f(u_k) = -\frac{(1-t_k)}{t_k}(P_{n_{k+1}}u_k - P_{n_k}u_k).$$

Hence

$$\begin{aligned} \limsup(f(u_k), u_k - u) &= \limsup(f(u_k), P_{n_{k+1}}u_k) \\ &= \limsup\left\{-\frac{(1-t_k)}{t_k}\|P_{n_{k+1}}u_k - P_{n_k}u_k\|^2\right\} \le 0. \end{aligned}$$

By the (S_+)-property of f we obtain $u_k \to u$ with $u \in \partial G$ and $F(u) = 0$, a contradiction. ∎

REMARK 6.1: The first part of Lemma 6.1 extends for admissible homotopies $F_t \in \mathcal{H}_G(LS; S_+)$ and continuous curves $\{y(t) \mid 0 \le t \le 1\}$ in H.

By virtue of Lemma 6.1 it is relevant to define

$$\text{(6.7)} \qquad d_H(F, G, y) = \lim d_{LS}(F_n, G, y_n)$$

for any given $F \in \mathcal{F}_G(LS; S_+)$ and $y \in H$ with $y \notin F(\partial G)$. It is not hard to verify that the function d_H defined by (6.7) satisfies the properties (a) to (d) of the classical topological degree with respect to $\mathcal{H}_G(LS; S_+)$ and the identity map I as the normalizing map. We indicate here how to verify (c), for example. Let $F_t \in \mathcal{H}_G(LS; S_+)$ and let $\{y(t) \mid 0 \le t \le 1\}$ be a continuous curve in H such that $y(t) \notin F_t(\partial G)$ for all $t \in [0,1]$. To show that $d_H(F_t, G, y(t))$ is constant for all $t \in [0,1]$ we first use the extension of Lemma 6.1 as mentioned in Remark 6.1. Hence there exists $n_0 \in \mathbb{N}$ such that $(y(t))_n \notin (F_t)_n(\partial G)$ for all $t \in [0,1]$ and $n \ge n_0$. For any pair $t_1, t_2 \in [0,1]$ we can then apply the definition 6.7 to find $k \ge n_0$ such that

$$d_H(F_{t_i}, G, y(t_i)) = d_{LS}((F_{t_i})_k, G, (y(t_i))_k)$$

for each $i = 1, 2$. Using the property (c) for the LS-degree the assertion follows for d_H.

For (d) it is sufficient to note that $I_n = I$ for all $n \in \mathbb{N}$ and $y_n \in G$ for all large n whenever $y \in G$. Therefore

$$d_H(I, G, y) = \lim d_{LS}(I, G, y) = 1$$

for all $y \in G$.

Finally the uniqueness of the degree d_H can also be shown (see [**BM 2**]) and hence we have

THEOREM 6.1. *Let H be a real separable Hilbert space, G a bounded open subset in H and $\mathcal{F}_G(LS; S_+)$ the class of admissible mappings. Then there exists exactly one degree function d_H satisfying the properties* (a) *to* (d) *with respect to $\mathcal{H}_G(LS; S_+)$ and normalizing map I.*

REMARK 6.2: A weak degree theory can be established here (cf. Remark 3.3) for mappings $\mathcal{F}_G(LS; QM)$ by using the approximations $f_\epsilon = f + \epsilon I$ for $f \in (QM)$.

It is now obvious from Section 4 how the existence and surjectivity results can be derived for admissible mappings F.

THEOREM 6.2. *Let G be an open bounded subset in H, $T \in \mathcal{F}_G(LS;S_+)$ a reference map and either $F \in \mathcal{F}_G(LS;S_+)$ or $F \in \mathcal{F}_G(LS;PM)$, G convex,and $F = Q(I + C_1 + C_2 f) + Pf$ with C_1, C_2 compact linear operators.*

If for a given $y \in H$ there exists $w \in T(G)$ such that

$$tF(u) + (1-t)T(u) \neq (1-t)w + ty \qquad \text{for all } u \in \partial G,\ t \in [0,1]. \tag{6.8}$$

then the equation $F(u) = y$ admits a solution u in G.

PROOF: Since the case $F \in \mathcal{F}_G(LS;S_+)$ is obvious by Theorem 4.1, we consider the latter case. It is sufficient to show that $F(\overline{G})$ is closed. In order to show this, let $\{w_n\}$ be a sequence in $F(\overline{G})$ with $w_n \to w$. Then $w_n = F(u_n)$ for some $\{u_n\} \subset \overline{G}$ and since G is bounded and convex, $u_n \rightharpoonup u$ with $u \in \overline{G}$ for a subsequence. Using the representation of F we have

$$Qw_n = Qu_n + QC_1(u_n) + QC_2(f(u_n)) \quad \text{and} \quad Pw_n = Pf(u_n).$$

Since f is bounded and C_1, C_2 are compact linear operators, we can choose a further subsequence of $\{u_n\}$ such that $Qu_n \to Qu$ and $Pf(u_n) \to Pw$. Therefore

$$\begin{aligned}\limsup(f(u_n), u_n - u) &= \limsup\{(Pf(u_n), u_n - u) + (f(u_n), Qu_n - Qu)\}\\ &= 0.\end{aligned}$$

Using the fact that $f \in (PM)$ we have $f(u_n) \rightharpoonup f(u)$ implying $F(u_n) = Qu_n + QC_1(u_n) + QC_2(f(u_n)) + Pf(u_n) \rightharpoonup F(u)$. Hence $w = F(u)$ with $u \in \overline{G}$ completing the proof. ∎

We are now in a position to derive surjectivity results from Theorem 6.2 along the lines of Theorems 4.3 to 4.6. For our applications we shall need the following

THEOREM 6.3. *Let $F \in \mathcal{F}_H(LS;S_+)$ or $F \in \mathcal{F}_H(LS;PM)$ with a representation $F = Q(I + C_1 + C_2 f) + Pf$ where C_1, C_2 are compact linear operators. Assume that F satisfies the condition*

(i) *if $y \in H$ and $F(u_n) \to y$, then $\{u_n\}$ is bounded,*

and one of the conditions

(ii) *there exists $R > 0$ such that $\dfrac{(F(u),u)}{\|u\|} + \|F(u)\| > 0$ for all $\|u\| \geq R$,*

(iii) *there exists $R > 0$ such that $F(-u) = -F(u)$ for all $\|u\| \geq R$.*

Then $F(H) = H$.

7. APPLICATIONS TO SEMILINEAR WAVE EQUATIONS

We consider semilinear wave equation of the form

$$\begin{cases} u_{tt} - u_{xx} - g(x,t,u) = h & \text{in } (0,\pi)\times \mathbf{R}\\ u(0,\cdot) = u(\pi,\cdot) = 0 & \\ u(\cdot,t+2\pi) = u(\cdot,t), & t \in \mathbf{R}\end{cases} \tag{7.1}$$

where the function $(x,t,s) \to g(x,t,s)$ from $\Omega \times \mathbf{R}$ to $\mathbf{R}$ is measurable in $w = (x,t) \in \Omega = (0,\pi)\times(0,2\pi)$ for each $s \in \mathbf{R}$ and continuous in s for almost all $w \in \Omega$ and h is a given function in $L^2(\Omega)$. Moreover we assume that there are positive constants α and β that

$$\begin{cases} \alpha \leq \dfrac{g(\cdot,\cdot,s)}{s} \leq \beta & \text{for all } s \neq 0\\ g(x,\cdot,s) \text{ is } 2\pi\text{-periodic and } g(\cdot,\cdot,0) = 0 & \\ g(x,t,\cdot) \text{ is nondecreasing.} & \end{cases} \tag{7.2}$$

Then it is clear that g gives rise to a Nemytskii operator N in $L^2(\Omega)$ by

$$N(u) = g(\cdot, \cdot, u), \quad u \in L^2(\Omega).$$

Moreover, N is continuous bounded and monotone. We denote in the sequel the Hilbert space $L^2(\Omega)$ by H. Let C^2 stand for twice continuously differentiable functions $v\colon [0,\pi] \times \mathbf{R} \to \mathbf{R}$ such that $v(0,\cdot) = v(\pi,\cdot) = 0$ and $v(x,t)$ is 2π-periodic in t. Then a function $u \in H$ is said to be ***a weak solution*** of (7.1) if

$$(u, v_{tt} - v_{xx}) - (N(u), v) = (h, v) \qquad \text{for all } v \in C^2.$$

Define

$$\phi_{lm} = \begin{cases} \frac{\sqrt{2}}{\pi}\sin(lx)\sin(mt), & l, m \in \mathbf{Z}_+ \\ \frac{1}{\pi}\sin(lx), & l \in \mathbf{Z}_+,\ m = 0 \\ \frac{\sqrt{2}}{\pi}\sin(lx)\cos(mt), & l, -m \in \mathbf{Z}_+. \end{cases}$$

Then the set $\{\phi_{lm} \mid l \in \mathbf{Z}_+,\ m \in \mathbf{Z}\}$ forms an orthonormal basis in H and each $u \in H$ has a representation

$$u = \sum_{l=1}^{\infty} \sum_{m=-\infty}^{+\infty} a_{lm}\phi_{lm}.$$

The wave operator has an abstract realization in H defined by

$$Lu = \sum_{l=1}^{\infty} \sum_{m=-\infty}^{+\infty} (l^2 - m^2) a_{lm}\phi_{lm}, \tag{7.3}$$

from $D(L) = \{u \in H \mid \sum_{l,m} |(l^2 - m^2)|^2 |a_{lm}|^2 < \infty\}$ to H. It can be shown that $u \in H$ is a weak solution of (7.1) if and only if

$$Lu - N(u) = h \qquad \text{with } u \in D(L). \tag{7.4}$$

The operator L is linear, densely defined, self adjoint and closed with closed range. In particular $H = \operatorname{Im} L \oplus \operatorname{Ker} L$ and L has a pure point spectrum of eigenvalues

$$\sigma(L) = \{\lambda_{lm} = l^2 - m^2 \mid l \in \mathbf{Z}_+,\ m \in \mathbf{Z}\}$$

with corresponding eigenvectors ϕ_{lm}. Here $\operatorname{Ker} L$ is infinite-dimensional, but all non-zero eigenvalues have finite multiplicity. If we denote by L_0 the restriction of L into $\operatorname{Im} L \cap D(L)$, then its inverse $L_0^{-1}\colon \operatorname{Im} L \to \operatorname{Im} L \cap D(L)$ is well-defined linear compact operator. To see the connection between the equation (7.4) and the class $\mathcal{F}_G(LS; S_+)$ we consider the mapping

$$F = Q(I - L_0^{-1}QN) + PN, \tag{7.5}$$

where Q and P are the orthogonal projections to $\operatorname{Im} L$ and $\operatorname{Ker} L$, respectively. If we assume that N is strongly monotone, i.e.,

$$\frac{g(\cdot,\cdot,s) - g(\cdot,\cdot,t)}{s - t} \geq \mu, \qquad s \neq t$$

for some constant $\mu > 0$, then obviously $F \in \mathcal{F}_G(LS; S_+)$. Otherwise we only have $F \in \mathcal{F}_G(LS; PM)$, where F admits a representation $F = Q(I + C_2 N) + P$ with $C_2 = -L_0^{-1}Q$ a compact linear operator as required in Theorems 6.2 and 6.3. Hence the results of Section 6 are available for the equation (7.4) as soon as we observe that $(L-P)(L_0^{-1}Q-P)=Q+P=I$ implying the equivalence of the equations

$$\text{(7.6)} \qquad Lu - N(u) = h \qquad \text{with } u \in D(L)\cap \overline{G}$$

$$\text{(7.7)} \qquad F(u) = y \qquad \text{with } u \in \overline{G},\ y=(L_0^{-1}Q-P)h,$$

where F is given by (7.5).

The obvious reference map in the class $\mathcal{F}_G(LS;S_+)$ is the identity map I corresponding to $L-P$ in the setting of (7.6). If we assume that G is an open bounded (convex in the case $N\in(PM)$) subset in H with $0\in G$, the condition (6.8) of Theorem 6.2 with $w=0$ and $T=I$ becomes

$$\text{(7.8)} \qquad Lu - tN(u) - (1-t)Pu \neq th \qquad \text{for all } u\in\partial G,\ 0\le t\le 1.$$

We get the following well known result ([**BN**], [**Ma**])

COROLLARY 7.1. *If g satisfies (7.2) with $0<\alpha<\beta<1$, then the equation (7.1) admits a weak solution u in $D(L)$ for any given $h\in H$.*

PROOF: Let $h\in H$ be given. It suffices to show that there exists R such that (7.8) holds for $\|u\|=R$, $0\le t\le 1$. We argue by contradiction. Assume there exist sequences $\{u_n\}\subset D(L)$, $\{t_n\}\subset[0,1]$ with $\|u_n\|\to\infty$ and $t_n\to t$ such that

$$\text{(7.9)} \qquad Lu_n - t_nN(u_n) - (1-t_n)Pu_n = t_nh, \qquad n\in\mathbb{N}.$$

Hence $\|Lu_n\| = t_n\|Qh+QN(u_n)\|\le t_n(\|h\|+\|N(u_n)\|)$. By (7.3) we have $(Lu_n,u_n)\le\|Lu_n\|^2$ and by (7.2) we get $\frac{1}{\beta}\|N(u_n)\|^2\le(N(u_n),u_n)$ and $\|u_n\|\le\frac{1}{\alpha}\|N(u_n)\|$. Using also the fact that $0\le t_n^2\le t_n\le 1$ we obtain from (7.9)

$$\begin{aligned} t_n^2\|N(u_n)\|^2 + 2t_n\|h\|\|N(u_n)\| + \|h\|^2 &\ge (L(u_n),u_n)\\ &= t_n(N(u_n),u_n) + (1-t_n)(Pu_n,u_n) + t_n(h,u_n)\\ &\ge \frac{1}{\beta}t_n^2\|N(u_n)\|^2 + (1-t_n)\|Pu_n\|^2 - t_n\|h\|\frac{1}{\alpha}\|N(u_n)\|. \end{aligned}$$

Since $0<\alpha<\beta<1$, we can conclude that $\{t_n\|N(u_n)\|\}$ and $\{(1-t_n)\|Pu_n\|^2\}$ remain bounded. In the case that $t>0$ we obtain a contradiction from $\|u_n\|\le\frac{1}{\alpha}\|N(u_n)\|$. If $t=0$, then $\{\|Pu_n\|\}$ is bounded and by (7.9) also $\{\|Lu_n\|\}$ is bounded. Since $\|Qu_n\|\le\|L_0^{-1}\|\|Lu_n\|$, a contradiction follows from $\|u_n\|\le\|Qu_n\|+\|Pu_n\|$. ∎

The result of Corollary 7.1 can be easily extended to cover the non-resonance cases where we assume that $[\alpha,\beta]\cap\sigma(L)=\emptyset$. This is shown by replacing in (7.8) the reference map $L-P$ by $A_1=L-\tau I$, where $\tau=\frac{1}{2}(\alpha+\beta)$, for example. The corresponding reference map in the setting of the equation is $F_1=Q(I-\tau L_0^{-1}Q)+\tau P$.

COROLLARY 7.2. *If g satisfies (7.2) with $0<\alpha<\beta$ and $[\alpha,\beta]\cap\sigma(L)=\emptyset$, then the equation (7.1) admits a weak solution u in $D(L)$ for any given $h\in H$.*

The resonance problem around an eigenvalue $\lambda\in\sigma(L)$, $\lambda>0$, can be tackled by choosing the reference map $A_2=-\lambda I+P_\lambda$, where P_λ stands for the orthogonal projection into $\mathrm{Ker}(L-\lambda I)$. For the results in this direction we refer to [**BM 2**], where the condition (7.2) is also given in a more general form. Similar results by different methods can be found in [**BN**], [**Ma**], [**MW**], for example.

Next we indicate briefly how the problem of multiple periodic solutions for semilinear wave equations can be dealt with using the framework of Section 6. We are interested in the case where the nonlinearity g interacts on the spectrum of L. For convenience we consider an autonomous wave equation of the form

$$(7.10)\qquad \begin{cases} u_{tt} - u_{xx} = g(u) & \text{in } (0,\pi)\times \mathbf{R} \\ u(0,\cdot) = u(\pi,\cdot) = 0 & \\ u(\cdot, t+2\pi) = u(\cdot, t), & t \in \mathbf{R}, \end{cases}$$

where g is continuous non-decreasing with $g(0) = 0$ and satisfies the condition

$$(7.11)\qquad 0 < \tilde{\alpha} = \inf_{s\neq 0} \frac{g(s)}{s} < \sup_{s\neq 0} \frac{g(s)}{s} = \tilde{\beta} < \infty$$

where $(\tilde{\alpha},\tilde{\beta}) \cap \sigma(L) \neq \emptyset$. We also assume that

$$(7.12)\qquad \lambda_{l-1} < \tilde{\alpha} \leq h^*(\infty) < \lambda_l \leq \cdots \leq \lambda_m < h_*(0) \leq \tilde{\beta} < \lambda_{m+1},$$

where $h^*(\infty) = \limsup_{|s|\to\infty} \frac{g(s)}{s}$ and $h_*(0) = \liminf_{|s|\to 0} \frac{g(s)}{s}$. The condition (7.12) means that $h(s) = \frac{g(s)}{s}$ crosses finitely many positive eigenvalues when s goes from 0 to ∞. The idea is to use the homotopy argument similar to (7.8) with a suitable reference map $F_0 = L - A$ with $A \in (S_+)$ and find open bounded convex sets G in H such that for some $w \in F_0(G \cap D(L))$,

$$(7.13)\qquad Lu - tN(u) - (1-t)A(u) \neq (1-t)w \qquad \text{for all } u \in \partial G \cap D(L),\ 0 \leq t \leq 1.$$

Then we can conclude that the equation (7.10) admits at least one solution u in $G \cap D(L)$. Since (7.10) obviously has the trivial solution, we must require that $0 \notin \overline{G}$.

To deal with the simplest possible case we assume that only one simple eigenvalue $\lambda_m = \bar{\lambda}$ is crossed. (In fact, for the wave operator such eigenvalues are $\lambda_1 = 1$ and $\lambda_3 = 4$, only.) Let $\bar{\phi}$ be the eigenvector associated with $\bar{\lambda}$. We denote $H_2 = \text{sp}\{\bar{\phi}\}$, $H_1 = \overline{\text{sp}}\{\phi_{lm} \mid l^2 - m^2 < \bar{\lambda}\}$ and $H_3 = \overline{\text{sp}}\{\phi_{lm} \mid l^2 - m^2 > \bar{\lambda}\}$. Then $H = H_1 \oplus H_2 \oplus H_3$ and each $u \in H$ has a representation $u = P_1u + P_2u + P_3u$ where P_i is the orthogonal projection to H_i, $i = 1,2,3$. A suitable homotopy will be $F_0 = L - \bar{\lambda}I + P_2$, i.e., $A = \bar{\lambda}I - P_2$ in (7.13). Thus we are led to find a priori estimates for the solutions of the equation

$$(7.14)\qquad Lu - tN(u) - (1-t)(\bar{\lambda}u - P_2u) = (1-t)w, \qquad u \in D(L),\ 0 \leq t \leq 1,$$

where $w = \bar{\phi}$ or $w = -\bar{\phi}$. It is shown in [**BM 4**] that using (7.11) and (7.12) such estimates can be obtained and we get the following

COROLLARY 7.3. *Let $\bar{\lambda} \in \{1,4\}$ and let g satisfy the conditions* (7.11) *and* (7.12) *with $\lambda_l = \lambda_m = \bar{\lambda}$. Then there exist constants ρ and R with $0 < \rho < 1 < R$ such that* (7.10) *admits at least one nontrivial solution in each of the sets $G_+ = \{u \in D(L) \mid \|u\| < R,\ P_2u = \theta\bar{\phi},\ \rho < \theta < R\}$ and $G_- = \{u \in D(L) \mid \|u\| < R,\ P_2u = -\theta\bar{\phi},\ \rho < \theta < R\}$.*

It turns out that $\dim H_2$ plays a crucial role when we look for sets G where to apply homotopy argument. The case where $\dim H_2 = 2$ and $\bar{\lambda} = l^2 - m^2$ for some $l, m \in \mathbf{Z}_+$ can be dealt with as shown in [**BM 3**] (cf [**Hi**]) but all other cases where $\dim H_2 \geq 3$ seem to be open.

On the other hand, we can generalize Corollary 7.3 by constructing subspaces V of H which are invariant under L and N and apply Corollary 7.3 to the reduced problem

$$L_V u = N_V(u), \qquad u \in D(L) \cap V,$$

where L_V and N_V are the restrictions of L and N into V, respectively. This idea for wave equations was used by Vejvoda [**Ve**] and Coron [**Co**]. For more general results we refer to [**BM 3,4**].

8. The degree for perturbations of maximal monotone mappings

We shall assume again that X is a real reflexive Banach space and that X and X^* are locally uniformly convex. We consider mappings $F = T+S$ where T is a maximal monotone multi (or a single valued map) from $D(T) \subset X$ to 2^{X^*} with $0 \in T(0)$, and S is a bounded demicontinuous mapping from X to X^* of class (S_+). With T we associate the family of generalized Yosida approximations

$$T_\epsilon(u) = (T^{-1} + \epsilon\mathcal{J}^{-1})^{-1}(u), \qquad u \in X,\ \epsilon > 0.$$

It is well-known (see **[De]**, **[Bro 5]**) that for each $\epsilon > 0$, T_ϵ is a bounded continuous maximal monotone single valued mapping from all of X to X^* with $T_\epsilon(0) = 0$. Moreover, $T_\epsilon(u) \rightharpoonup T^0(u)$ on $D(T)$ as $\epsilon \to 0+$, where $T^0(u)$ denotes the unique element of the set $T(u)$ having minimal norm, i.e., $\|T^0(u)\| = \text{dist}(0, T(u))$. On the other hand, $\|T_\epsilon(u)\| \to \infty$ as $\epsilon \to 0+$, if $u \notin \overline{D(T)}$.

We shall sketch the extension of the degree function for the mappings $F = T+S$ by using approximations $F_\epsilon = T_\epsilon + S$ in the class (S_+). Let G be an open bounded set in X. We denote

$$\begin{aligned}\mathcal{F}_G(MM; S_+) = \{F = T + S \mid T\colon X \supset D(T) \to 2^{X^*} &\text{ maximal monotone}\\ \text{with } 0 \in T(0) \text{ and } S\colon \overline{G} \to X^* &\text{ bounded demicontinuous}\\ \text{of class } (S_+)\}.&\end{aligned}$$

The obvious idea now is to show that we can define the degree function d_{MM} for all admissible triplets (F, G, y), $F \in \mathcal{F}_G(MM; S_+)$, $y \notin F(\partial G)$, as the limit

$$d_{MM}(F, G, y) = \lim_{\epsilon \to 0+} d_{S_+}(F_\epsilon, G, y). \tag{8.1}$$

To this end we start with the following (cf. **[Bro 5]**)

Lemma 8.1. *Let T be a maximal monotone multi: $X \supset D(T) \to 2^{X^*}$ with $0 \in T(0)$ and let $\{u_n\}$, $\{\epsilon_n\}$, $\{\delta_n\}$, $\{s_n\}$ be sequences such that $u_n \rightharpoonup u$ in X, $\epsilon_n \to 0+$, $\delta_n \to 0+$, $0 \le s_n \le 1$. Let $v_n = T_{\epsilon_n}(u_n)$, $z_n = T_{\delta_n}(u_n)$ and $w_n = (1-s_n)v_n + s_n z_n$ and assume that $w_n \rightharpoonup w$ in X^* and $\limsup\langle w_n, u_n\rangle \le \langle w, u\rangle$. Then $u \in D(T)$, $w \in T(u)$ and $\lim\langle w_n, u_n\rangle = \langle w, u\rangle$.*

Proof: Since $v_n = T_{\epsilon_n}(u_n)$ and $z_n = T_{\delta_n}(u_n)$, $v_n \in T(u_n - \epsilon_n\mathcal{J}^{-1}(v_n))$ and $z_n \in T(u_n - \delta_n\mathcal{J}^{-1}(z_n))$. Since T is monotone and $0 \in T(0)$, we have

$$\langle v_n, u_n - \epsilon_n\mathcal{J}^{-1}(v_n)\rangle \ge 0 \quad\text{and}\quad \langle z_n, u_n - \delta_n\mathcal{J}^{-1}(z_n)\rangle \ge 0.$$

Therefore

$$\epsilon_n\langle v_n, \mathcal{J}^{-1}(v_n)\rangle = \epsilon_n\|v_n\|^2 \le \langle v_n, u_n\rangle$$

and similarly

$$\delta_n\|z_n\|^2 \le \langle z_n, u_n\rangle.$$

Multiplying these inequalities by $(1-s_n)$ and s_n, respectively, and adding we get

$$(1-s_n)\epsilon_n\|v_n\|^2 + s_n\delta_n\|z_n\|^2 \le \langle w_n, u_n\rangle \qquad \text{for all } n \in \mathbf{N}.$$

Since $\{\langle w_n, u_n\rangle\}$ remains bounded, we conclude that $(1-s_n)\epsilon_n\|v_n\| \to 0$ and $s_n\delta_n\|z_n\| \to 0$. Let $[x, y] \in G(T)$ be arbitrary. By the monotony of T we have

$$\langle v_n - y, u_n - \epsilon_n\mathcal{J}^{-1}(v_n) - x\rangle \ge 0 \quad\text{and}\quad \langle z_n - y, u_n - \delta_n\mathcal{J}^{-1}(z_n) - x\rangle \ge 0.$$

Hence

$$\langle v_n - y, u_n - x\rangle \ge -\epsilon_n\|y\|\|v_n\| \quad\text{and}\quad \langle z_n - y, u_n - x\rangle \ge -\delta_n\|y\|\|z_n\|.$$

By multiplying and adding as above again we get

$$\langle w_n - y, u_n - x\rangle \geq -\|y\| \{(1-s_n)\epsilon_n\|v_n\| + s_n\delta_n\|z_n\|\}$$

implying

$$\liminf\langle w_n - y, u_n - x\rangle \geq 0.$$

Bearing in mind the assumption we obtain

$$\langle w, u\rangle \geq \limsup\langle w_n, u_n\rangle \geq \liminf\langle w_n, u_n\rangle \geq \langle w, x\rangle + \langle y, u\rangle - \langle y, x\rangle.$$

Hence $\langle w-y, u-x\rangle \geq 0$ and since $T \in (MM)$ we conclude $u \in D(T)$ and $w \in T(u)$. Substituting $[u, w]$ for $[x, y]$ above we also get $\lim\langle w_n, u_n\rangle = \langle w, u\rangle$. ∎

Now we are in the position to continue in the familiar way.

Lemma 8.2. *Let $F \in \mathcal{F}_G(MM, S_+)$ and $y \in X^*$ with $y \notin F(\partial G)$. Then there exists $\hat{\epsilon} > 0$ such that $y \notin F_\epsilon(\partial G)$ for all ϵ with $0 < \epsilon < \hat{\epsilon}$. Moreover, there exists $\epsilon_0 < \hat{\epsilon}$ such that $d_{LS}(F_\epsilon, G, y)$ is constant for all ϵ with $0 < \epsilon < \epsilon_0$.*

Proof: Assume that the first assertion is false. Then there exist sequences $\{u_n\} \subset \partial G$ and $\{\epsilon_n\}$ such that $u_n \rightharpoonup u$, $\epsilon_n \to 0+$ and

$$F_{\epsilon_n}(u_n) = T_{\epsilon_n}(u_n) + S(u_n) = y \qquad \text{for all } n \in \mathbb{N}.$$

Since S is bounded, we can assume that $v_n = T_{\epsilon_n}(u_n) = y - S(u_n) \rightharpoonup v$. Since $S \in (S_+) \subset (QM)$ we also have

$$\limsup\langle v_n, u_n - u\rangle = \limsup\langle y - S(u_n), u_n - u\rangle \leq 0$$

and we can employ Lemma 8.1 with the specialization $s_n = 0$ for all $n \in \mathbb{N}$. Consequently $u \in D(T)$, $v \in T(u)$ and $\langle v_n, u_n\rangle \to \langle v, u\rangle$ implying that

$$\limsup\langle S(u_n), u_n - u\rangle = 0.$$

Using the fact that $S \in (S_+)$ we get $u_n \to u$ with $u \in \partial G$ and $S(u_n) \rightharpoonup S(u)$. Thus $v = y - S(u) \in T(u)$ and $y \in (T+S)(u)$, a contradiction.

If the second assertion were false, we could find, using the property (c) of the S_+-degree, sequences $\{u_n\} \subset \partial G$, $\{\epsilon_n\}$, $\{\delta_n\}$ and $\{s_n\}$ such that $u_n \rightharpoonup u$, $\epsilon_n \to 0+$, $\delta_n \to 0+$, $0 \leq s_n \leq 1$ and

$$(1-s_n)F_{\epsilon_n}(u_n) + s_n F_{\delta_n}(u_n) = y \qquad \text{for all } n \in \mathbb{N}.$$

Thus

$$(1-s_n)T_{\epsilon_n}(u_n) + s_n T_{\delta_n}(u_n) + S(u_n) = y \qquad \text{for all } n \in \mathbb{N}.$$

Denoting $v_n = T_{\epsilon_n}(u_n)$, $z_n = T_{\delta_n}(u_n)$ and $w_n = (1-s_n)v_n + s_n z_n$ we can assume that $w_n = y - S(u_n) \rightharpoonup w$ in X^*. Hence we can invoke Lemma 8.1 if we know that $\limsup\langle w_n, u_n\rangle \leq \langle w, u\rangle$. But this is again a consequence of the fact that $S \in (S_+)$. Thus we may conclude that $u \in D(T)$, $w \in T(u)$ and $\lim\langle w_n, u_n\rangle = \langle w, u\rangle$, which implies as above that $\limsup\langle S(u_n), u_n - u\rangle = 0$ and $u_n \to u$ with $u \in \partial G$. A contradiction is achieved by the fact that $y \in (T+S)(u)$. ∎

By Lemma 8.2 the definition (8.1) is relevant. To verify that d_{MM} is a classical degree function we need the class of admissible homotopies. We shall call a homotopy F_t, $0 \leq t \leq 1$, permissible if $F_t = T_t + S_t$, where S_t is a bounded homotopy of class (S_+) and T_t is a *permissible homotopy of maximal monotone mappings* in the sense that $T_t \in (MM)$ for each $t \in [0,1]$ and the mapping $t \to (T_t + \mathcal{J})^{-1}(v)$ is continuous from $[0,1]$ to the strong topology of X. In fact, it is shown by Browder ([Bro 5]) that an equivalent condition for T_t is the following generalized pseudomonotonicity condition: if $t_n \to t$ in $[0,1]$, $[u_n, w_n] \in G(T_{t_n})$ with $u_n \rightharpoonup u$, $w_n \rightharpoonup w$ and $\limsup\langle w_n, u_n\rangle \leq \langle w, u\rangle$, then $u \in D(T_t)$, $w \in T_t(u)$ and $\limsup\langle w_n, u_n\rangle = \langle w, u\rangle$. However, for applications and even for uniqueness of the degree function (8.1) it seems adequate to deal with the class of affine homotopies of the form

$$\mathcal{H}_G(MM; S_+) = \{F_t = tF_0 + (1-t)S_1 \mid F_0 \in \mathcal{F}_G(MM; S_+),\ S_1 \in \mathcal{F}_G(S_+)\}$$

Then we have

Theorem 8.1. *Let X be a real reflexive Banach space, G an open bounded subset in X and $\mathcal{F}_G(MM;S_+)$ the class of admissible mappings. Then there exists exactly one degree function d_{MM} satisfying the properties* (a) *to* (d) *with respect to $\mathcal{H}_G(MM;S_+)$ and normalizing map $\mathcal{J}$*

For the proof we need generalized versions of Lemmas 8.1 and 8.2 for homotopies as shown by Browder [**Bro 5**]. We check only the property (a). Indeed, assume $d_{MM}(T+S,G,y) \neq 0$. Then $d_{S_+}(T_\epsilon + S, G, y)$ is constant and non-zero for ϵ small enough, and there exist sequences $\{u_n\} \subset G$ and $\{\epsilon_n\}$ such that $T_{\epsilon_n}(u_n) + S(u_n) = y$ for all $n \in \mathbb{N}$, $u_n \rightharpoonup u$ and $\epsilon_n \to 0+$. We can proceed exactly the same way as in proving Lemma 8.2 to conclude that $u_n \to u \in \overline{G}$ and $y \in T(u) + S(u)$.

Existence and surjectivity results for mappings $F = T + S \in \mathcal{F}_G(MM;S_+)$ can be derived along the lines of Section 4. For example we get from Theorem 4.3 the following

Corollary 8.1. *Let $F = T + S \in \mathcal{F}_X(MM;S_+)$ satisfy the conditions*

(i) *F^{-1} is locally bounded*

(ii) *there exists $R > 0$ such that*

$$\frac{\langle w + S(u), u\rangle}{\|u\|} + \|w + S(u)\| > 0 \qquad \text{for all } \|u\| \geq R,\ w \in T(u).$$

Then $\mathcal{R}(T+S) = X^*$.

Remark 8.1: It can be shown by different methods that for $T \in (MM)$, $\mathcal{R}(T) = X^*$ if and only if T^{-1} is locally bounded (cf. Remark 4.2). We also know that $\mathcal{R}(T + \lambda\mathcal{J}) = X^*$ for all $\lambda > 0$ if and only if $T \in (MM)$ (for more details consult [**De**]).

Remark 8.2: The degree function obtained above can be extended to the class $\mathcal{F}_G(MM;QM)$ as in the previous cases by approximations $T + S + \epsilon\mathcal{J}$, $T \in (MM)$, $S \in (QM)$, $\epsilon > 0$. On the other hand we may also use approximations $T_\epsilon + S$ for which the weak degree is already defined. To indicate the results in this direction we have

Corollary 8.2. *Let $F = T+S$, where T is maximal monotone with $0 \in T(0)$ and $S\colon X \to X^*$ is bounded demicontinuous pseudomonotone and strongly coercive. Then $\mathcal{R}(T+S) = X^*$.*

Proof: We know that T_ϵ is bounded demicontinuous monotone mapping from X to X^* for each $\epsilon > 0$. Hence $T_\epsilon \in (PM)$ and $F_\epsilon = T_\epsilon + S \in (PM)$. Since $T_\epsilon(0) = 0$, we have

$$\frac{\langle F_\epsilon(u), u\rangle}{\|u\|} = \frac{\langle T_\epsilon(u), u\rangle}{\|u\|} + \frac{\langle S(u), u\rangle}{\|u\|} \to \infty \qquad \text{as } \|u\| \to \infty.$$

By Theorem 4.5, $F_\epsilon(X) = X^*$ for each $\epsilon > 0$. Let $y \in X^*$ be arbitrary and $\{\epsilon_n\}$ a sequence with $\epsilon_n \to 0+$. For each $n \in \mathbb{N}$ there exists $u_n \in X$ such that

$$T_{\epsilon_n}(u_n) + S(u_n) = y.$$

Since S is strongly coercive, $\{u_n\}$ is bounded and we can assume that $u_n \rightharpoonup u$ and $S(u_n) \rightharpoonup h$. Denoting $v_n = T_{\epsilon_n}(u_n)$ we also have $v_n \rightharpoonup y - h = v$. Since $S \in (PM)$ we obtain $\limsup\langle v_n, u_n - u\rangle \leq 0$ and we can invoke Lemma 8.1 to derive $u \in D(T)$, $v \in T(u)$ and $\lim\langle v_n, u_n\rangle = \langle v, u\rangle$. Hence $\limsup\langle S(u_n), u_n - u\rangle = 0$ implying $S(u_n) \rightharpoonup S(u)$ and $v = y - S(u)$, i.e., $y \in (T+S)(u)$. ∎

9. Perturbations of linear maximal monotone mappings and applications to parabolic problems

The study of perturbations of maximal monotone multis is motivated for instance by the connections to generalized Hammerstein equations, variational inequalities and subdifferentials of lower semicontinuous functionals. We are interested here in the parabolic initial-boundary value problems for differential operators of the form

$$(9.1)\qquad \frac{\partial u(x,t)}{\partial t} + \sum_{|\alpha|\le m} (-1)^{|\alpha|} D^\alpha A_\alpha(x,t,u(x,t),D_x u(x,t),\ldots,D_x^m u(x,t)), \qquad (x,t) \in Q$$

where $Q = \Omega \times (0,T)$, Ω an open bounded subset in $\mathbf{R}^N$, $m \ge 1$ and the coefficients A_α are functions of $(x,t) \in Q$ and of $\xi = (\eta,\zeta) \in \mathbf{R}^{N_0}$ with $\eta \in \mathbf{R}^{N_1}$ and $\zeta \in \mathbf{R}^{N_2}$, $N_1 + N_2 = N_0$, as in Section 5. We assume also that each $A_\alpha(x,t,\xi)$ is a Carathéodory function and satisfies the polynomial growth condition (A_1) for some $c_1 > 0$, $1 < p < \infty$ and $k_1 \in L^{p'}(Q)$. The latter part of the operator (9.1) gives rise to a bounded continuous mapping $S\colon \mathcal{V} \to \mathcal{V}^*$ by the rule

$$(9.2)\qquad \langle S(u),v\rangle = \int_Q \sum_{|\alpha|\le m} A_\alpha(x,t,\xi(u))D^\alpha v, \qquad u,v \in \mathcal{V},$$

where $\mathcal{V} = L^p(0,T;X)$ with $X = W_0^{m,p}(\Omega)$, for example. If we assume, for simplicity, that $p \ge 2$ and if we give also an initail value $u(0) = 0$, then the former part of (9.1) has a realization L as a maximal monotone closed linear mapping from

$$D(L) = \left\{ v \in \mathcal{V} \mid \frac{\partial v}{\partial t} \in \mathcal{V}^*,\ v \in C(0,T;L^2(\Omega)),\ v(0) = 0 \right\}$$

to $\mathcal{V}^*$, (see [**Li**]). Hence the initial-boundary value problem for the operator (9.1) admits a weak formulation

$$(9.3)\qquad Lu + S(u) = h, \qquad u \in D(L),$$

where h is a given element in $\mathcal{V}^* = L^{p'}(0,T;X^*)$. If we knew that S is coercive and pseudomonotone we can apply the results of Section 8. Indeed, if the coefficients A_α satisfy the monotonicity condition

$$(A_2)_M\qquad \sum_{|\alpha|\le m} \{A_\alpha(x,t,\xi) - A_\alpha(x,t,\xi^*)\}(\xi_\alpha - \xi_\alpha^*) \ge 0 \quad \text{for all } (x,t) \in Q \text{ and } \xi,\xi^* \in \mathbf{R}^{N_0}$$

and the strong coercivity condition

$$(A_3)_S\qquad \sum_{|\alpha|\le m} A_\alpha(x,t,\xi)\xi_\alpha \ge c_2|\xi|^p - k_2(x,t) \qquad \text{for all } (x,t) \in Q,\ \xi \in \mathbf{R}^{N_0} \text{ with } c_2 > 0 \text{ and } k_2 \in L^1(Q),$$

then S is monotone and hence also pseudomonotone. By Corollary 8.2 the equation (9.3) has a solution for any $h \in \mathcal{V}^*$. In order to deal with more refined monotonicity conditions like

$$(A_2)_S\qquad \sum_{|\alpha|=m} \{A_\alpha(x,t,\eta,\zeta) - A_\alpha(x,t,\eta,\zeta^*)\}(\zeta_\alpha - \zeta_\alpha^*) > 0$$
$$\text{for all } (x,t) \in Q,\ \eta \in R^{N_1} \text{ and } \zeta \ne \zeta^* \in \mathbf{R}^{N_2}$$

we need a further extension of the degree for mappings of the form $F = L + S$ where L is a linear maximal monotone closed densely defined operator from $D(L)$ in $\mathcal{V}$ into $\mathcal{V}^*$ and S is *of class* (S_+) or *pseudomonotone with respect to* $D(L)$, i.e., for any sequence $\{u_n\}$ in $D(L)$ with $u_n \rightharpoonup u$ in $\mathcal{V}$, $Lu_n \rightharpoonup Lu$ in $\mathcal{V}^*$ and $\limsup\langle S(u_n), u_n - u\rangle \le 0$, we have $u_n \to u$ in $\mathcal{V}$, or $S(u_n) \rightharpoonup S(u)$ in $\mathcal{V}^*$ and $\langle S(u_n),u_n\rangle \to \langle S(u),u\rangle$, respectively.

LEMMA 9.1. *If the operator (9.1) satisfies conditions* (A_1) *and* $(A_2)_S$*, then the mapping* S *defined by (9.2) is pseudomonotone with respect to* $D(L)$*. If also the condition* $(A_3)_S$ *is satisfied, then* S *is of class* (S_+) *with respect to* $D(L)$.

PROOF: For the proof of the first part we refer to [Mu] which deals with the case where Ω may be unbounded. Assume that the conditions (A_1), $(A_2)_S$ and $(A_3)_S$ are satisfied. Hence for any sequence $\{u_n\}$ in $D(L)$ with $u_n \rightharpoonup u$ in $\mathcal{V}$, $Lu_n \rightharpoonup Lu$ in $\mathcal{V}^*$ and $\limsup\langle S(u_n), u_n - u\rangle \leq 0$ we have $S(u_n) \rightharpoonup S(u)$ in $\mathcal{V}^*$ and $\langle S(u_n), u_n\rangle \to \langle S(u), u\rangle$. Moreover we know by the proof of first part that $D^\alpha u_n \to D^\alpha u$ in $L^p(Q)$ for all $|\alpha| \leq m-1$ and $D^\alpha u_n(x,t) \to D^\alpha u(x,t)$ a.e. in Q for all $|\alpha| = m$, at least for a subsequence. Denoting $h_n(x,t) = \sum_{|\alpha|\leq m} A_\alpha(x,t,\xi(u_n(x,t)))D^\alpha u_n(x,t) + k_2(x,t)$, $h(x,t) = \sum_{|\alpha|\leq m} A_\alpha(x,t,\xi(u(x,t)))D^\alpha u(x,t) + k_2(x,t)$, we have $h_n(x,t) \geq 0$ a.e. in Q, $\|h_n\|_{L^1(Q)} \to \|h\|_{L^1(Q)}$ and $h_n(x,t) \to h(x,t)$ a.e. in Q. Hence $h_n \to h$ in $L^1(Q)$ and there exists a function $k \in L^1(Q)$ such that

$$\sum_{|\alpha|\leq m} A_\alpha(x,t,\xi(u_n(x,t)))D^\alpha u_n(x,t) \leq k(x,t) \qquad \text{a.e. in } Q$$

for some further subsequence. Using now $(A_3)_S$ we can conclude by the dominated convergence theorem that $u_n \to u$ in $\mathcal{V}$. ∎

In order to get suitable approximations for the map $F = L+S$ we denote $Y = D(L)$ equipped with the norm $\|u\|_Y = \|u\|_{\mathcal{V}} + \|Lu\|_{\mathcal{V}^*}$. Y is also a reflexive Banach space with continuous embedding $j: Y \to \mathcal{V}$. For each $\epsilon > 0$ we then define (cf. [Li, Chapitre 3])

$$a_\epsilon(u,v) = \epsilon\langle \mathcal{J}^{-1}(Lu), Lv\rangle + \langle Lu, v\rangle + \langle S(u), v\rangle, \qquad u,v \in D(L), \tag{9.4}$$

where $\mathcal{J}$ stands for the duality map: $\mathcal{V} \to \mathcal{V}^*$. It is easy to see that $v \to a_\epsilon(u,v)$ is a bounded linear functional in Y and therefore (9.4) defines a mapping $F_\epsilon: Y \to Y^*$ by

$$\langle F_\epsilon(u), v\rangle = a_\epsilon(u,v), \qquad u,v \in Y. \tag{9.5}$$

If we assume that S is of class (S_+) with respect to $D(L)$, then F_ϵ is of class (S_+) for each $\epsilon > 0$. Indeed, let $\{u_n\}$ be a sequence in Y with $u_n \rightharpoonup u$ in Y and $\limsup\langle F_\epsilon(u_n), u_n - u\rangle \leq 0$. Hence $u_n \rightharpoonup u$ in $\mathcal{V}$, $Lu_n \rightharpoonup Lu$ in $\mathcal{V}^*$ and

$$\limsup\{\epsilon\langle \mathcal{J}^{-1}(Lu_n) - \mathcal{J}^{-1}(Lu), Lu_n - Lu\rangle + \langle Lu_n - Lu, u_n - u\rangle + \langle S(u_n), u_n - u\rangle\} \leq 0. \tag{9.6}$$

Since $\mathcal{J}^{-1}$ is strictly monotone and L is monotone, we have

$$\limsup\langle S(u_n), u_n - u\rangle \leq 0$$

implying $u_n \to u$ in $\mathcal{V}$. Hence also

$$\lim\langle \mathcal{J}^{-1}(Lu_n) - \mathcal{J}^{-1}(Lu), Lu_n - Lu\rangle = 0.$$

Since $\mathcal{J}^{-1}$ has the properties of the duality map: $\mathcal{V}^* \to \mathcal{V}^{**}$, $\|Lu_n\| \to \|Lu\|$ and thus $Lu_n \to Lu$ in $\mathcal{V}^*$. Hence F_ϵ is in the class (S_+) for any $\epsilon > 0$.

Using the approximations $\{F_\epsilon \mid \epsilon < 0\}$ we can now proceed in the familiar way to obtain a degree function for mappings $F = L + S$. However, some further work is to be done due to the fact that the bounded sets in Y and $\mathcal{V}$ are not the same. A more detailed discussion on this extension will appear elsewhere. We close these notes by a remark that existence results for the equation (9.4) under the conditions (A_1), $(A_2)_S$ and $(A_3)_S$ can be obtained also directly by using approximating equations $F_\epsilon(u) = j^*h$ in Y, where $j^*: \mathcal{V}^* \to Y^*$ denotes the adjoint of j.

ACKNOWLEDGEMENT: These lecture notes were written during a visit to the University of Oklahoma. The author wishes to thank the Department of Mathematics for the hospitality.

REFERENCES

[Am] H. AMANN, *Saddle points and multiple solutions of differential equations,* Math. Z. **169** (1979), 127–166.

[AW] H. AMANN AND S. WEISS, *On the uniqueness of the topological degree,* Math. Z. **130** (1973), 39–54.

[Be] J. BERKOVITS, *On the degree theory for nonlinear mappings of monotone type,* Ann Acad. Sci. Fenn. Ser. A1, Dissertationes **58** (1986).

[BM 1] J. BERKOVITS AND V. MUSTONEN, *On the topological degree for mappings of monotone type,* Nonlinear Anal. TMA **10** (1986), 1373–1383.

[BM 2] J. BERKOVITS AND V. MUSTONEN, *An extension of Leray-Schauder degree and applications to nonlinear wave equations,* Differential and Integral Equations, to appear.

[BM 3] J. BERKOVITS AND V. MUSTONEN, *On multiple solutions for a class of semilinear wave equations,* Nonlinear Anal. TMA, to appear.

[BM 4] J. BERKOVITS AND V. MUSTONEN, *On the existence of multiple solutions for semilinear equations with monotone nonlinearities crossing finite number of eigenvalues,* to appear.

[Bre] H. BREZIS, *Équationes et inéquationes non linéaires dans les espaces vectoriels en dualité,* Ann. Inst. Fourier (Grenoble) **18** (1968), 115–175.

[BN] H. BREZIS AND L. NIRENBERG, *Characterizations of the ranges of some nonlinear operators and applications to boundary value problems,* Ann. Scuola Norm. Pisa **(4)5** (1978), 225–326.

[Bro 1] F. E. BROWDER, *Nonlinear elliptic boundary value problems,* Bull. Amer. Math. Soc. **69** (1963), 862–874.

[Bro 2] F. E. BROWDER, *Topological methods for nonlinear elliptic equations of arbitrary order,* Pacific J. Math. **17** (1966), 17–31.

[Bro 3] F. E. BROWDER, *Nonlinear equations of evolution and nonlinear operators in Banach spaces,* Proc. Sympos. Pure Math., Vol. **18** Part II, AMS, Providence, 1975.

[Bro 4] F. E. BROWDER, *Fixed point theory and nonlinear problems,* Proc. Sympos. Pure Math., Vol. **39** Part 2, AMS, Providence, 1983.

[Bro 5] F. E. BROWDER, *Degree theory for nonlinear mappings,* Proc. Sympos. Pure Math., Vol. **45** Part 1, AMS, Providence (1986), 203–226.

[BH] F. E. BROWDER AND P. HESS, *Nonlinear mappings of monotone type in Banach spaces,* J. Functional Analysis **11** (1972), 251–294.

[BT] F. E. BROWDER AND B. A. TON, *Nonlinear functional equations in Banach spaces and elliptic super-regularization,* Math. Z. **105** (1968), 177–195.

[Br] L. E. J. BROUWER, *Über Abbildungen von Mannichfaltigkeiten,* Math. Ann. **71** (1912), 97–115.

[Co] J. M. CORON, *Periodic solutions of a nonlinear wave equation without assumption of monotonicity,* Math. Ann. **262** (1983), 273–285.

[De] K. DEIMLING, *Nonlinear functional analysis,* Springer-Verlag, Berlin, 1985.

[FK] S. FUČIK AND A. KUFNER, *Nonlinear differential equations,* Elsevier, Amsterdam, 1980.

[GM] J. P. GOSSEZ AND V. MUSTONEN, *Pseudomonotonicity and the Leray-Lions condition,* in preparation.

[He] P. HESS, *A variational approach to a class of nonlinear eigenvalue problems,* Proc. Amer. Math. Soc. **29**(2) (1971), 272–276.

[Hi] N. HIRANO, *Multiple nontrivial solutions of semilinear elliptic equations,* J. Differential Equations **71** (1988), 334–347.

[LM] R. LANDES AND V. MUSTONEN, *On pseudomonotone operators and nonlinear noncoercive variational problems on unbounded domains,* Math. Ann. **248** (1980), 241–246.

[LL] J. LERAY AND J. L. LIONS, *Quelques résultats de Višik sur des problémes elliptiques non linéaires par les méthodes de Minty-Browder,* Bul. Soc. Math. France **93** (1965), 97–107.

[LS] J. LERAY AND J. SCHAUDER, *Topologie et equationes functionelles,* Ann. Sci. École Norm Sup. **51** (1934), 45–78.

[Li] J. L. LIONS, *Quelqes méthodes de resolution des problémes aux limites non linéaires,* Dunod, Gauthier-Villars, Paris, 1969.

[Ma] J. MAWHIN, *Nonlinear functional analysis and periodic solutions of semilinear wave equations,* Nonlinear Phenomena in Mathematical Sciences (Lakshmikantham, ed.), Academic Press, New York (1982), 671–681.

[MW] J. MAWHIN AND M. WILLEM, *Operators of monotone type and alternative problems with infinite dimensional kernel,* Recent Advances in Differential Equations, Academic Press, New York-London (1981), 295–307.

[Mi 1] G. J. MINTY, *Monotone (nonlinear) operators in Hilbert space,* Duke Math. Journal **29** (1962), 341–346.

[Mi 2] G. J. MINTY, *On a monotonicity method for the solution of nonlinear equations in Banach spaces,* Proc. Nat. Acad. Sc. USA, **50** (1963), 1038–1041.

[Mu] V. MUSTONEN, *On pseudomonotone operators and nonlinear parabolic initial-boundary value problems on unbounded domains,* Ann. Acad. Sci. Fenn. Ser. A1, Vol. **6** (1981), 225–232.

[PS] D. PASCALI AND S. SBURLAN, *Nonlinear mappings of monotone type,* Editura Academiei, Bukarest, 1978.

[Ve] O. VEJVODA, *Partial differential equations,* Noordhoff, Sijthoff, 1981.

[Vi] J. M. VIŠIK, *Systémes d'équations aux dérivées partielles fortement elliptiques quasi linéaires sous forme divergente,* Troudi Mosk. Mat. Obv. **12** (1963), 125–184.

GENERALIZATIONS OF SOME CLASSICAL INEQUALITIES AND THEIR APPLICATIONS

Lars Erik Persson, Luleå

0. INTRODUCTION

The main aim of this talk is to review, complement and generaliz. some inequalities I recently obtained partly together with other authors. The paper is organized in the following way: In section 1 we collect some necessary notations and definitions. We also present some of the classical inequalities we will generalize, unify and complement later on. In section 2 we prove two general "setvalued" inequalities, which in particular generalize some results recently obtained by J. Peetre and the present author [25]. The singlevalued versions of these inequalities are studied in section 3 and several applications are pointed out e.g. a recent result by J. Matkowski [22]. In section 4 we present a generalization of Hölder's inequality to the case with a family of spaces, which recently has been obtained by L. Nikolova and the present author [24]. We also include a generalized completely symmetric form of Hölder's inequality, thereby generalizing some previous results by Aczél-Beckenbach [1] and the present author [29]. In section 5 we prove a sharp generalized form of Minkowski's inequality. For the proof of this inequality we need to prove a certain generalization of Clarcson's inequality, which can be of independent interest. Section 6 is used to prove some relations between generalized versions of some classical inequalities. In particular we find that some of these inequalities, in a sense, are equivalent. In section 7 we present a recent result by T. Koski and the present author [15], where in particular the sharpest possible upper bounds for the exponential entropies $E(\alpha;f)$ are obtained for every $\alpha > 0$, $\alpha \neq 1$. In particular by letting $\alpha \to 1$ we obtain a quite new proof of the "differential entropy inequality", which is one of the corner stones in Information Theory. In section 8 we give some concluding remarks. In particular we present a new results concerning the best constant in a variant of Hardy's inequality. We also include some remarks concerning recent development of the theory and applications of generalized Gini means. In particular we point out some examples of inequalities which may be seen as limiting cases of classical inequalities.

Acknowledgements: I want to thank prof. V.I. Burenkov, Docent Ljudmila I. Nikolova and Dr. Sigrid Sjöstrand for various kinds of remarks, which have improved the final version of this paper.

1. PRELIMINARIES

In this paper we let p and q denote real numbers such that $0 < p < \infty$ and $q = p/(p-1)$. (Ω,Σ,μ) denotes a σ-finite complete measure space and $L^0(\Omega)$ is the space of all complex-valued μ-measurable functions on Ω. Moreover, X denotes a Banach function space on (Ω,Σ,μ), abbreviated BFS, i.e. $X = (X,\|\cdot\|_X)$ is a Banach subspace of $L^0(\Omega)$ satisfying that, for every $x,y \in L^0(\Omega)$, the following lattice property holds:

$$y \in X,\ |x| \leq |y|\ \mu\text{-a.e.} \Rightarrow x \in X \text{ and } \|x\|_X \leq \|y\|_X.$$

Let A be a Banach space and X a Banach function space on Ω and let $\omega = \omega(s)$, $s \in \Omega$, be a positive weight function on Ω. The spaces $X^P(A,\omega)$ consists of all strongly measurable functions (cross-sections) $x = x(s)$ satisfying

$$\|x\|_{X^P(A,\omega)} = (\|(\|(x(s)\|_A\omega(s))^P\|_X)^{1/P} < \infty,$$

see [24]. For the cases $A = R$ and $\omega = 1$ we use the abbreviated notations $X^P(\omega)$ and $X^P(A)$, respectively. In particular the spaces $X^P(R,1)$ coincides with the usual X^P-spaces. It is easy to prove that the X^P-spaces are Banach function spaces if $p \geq 1$ and at least quasi-Banach function spaces for all $p > 0$ (see e.g. [29]).

Let Γ denote a linear class of real-valued functions $x = x(t)$ defined on a non-empty set E. If $L: \Gamma \to R$, is an <u>isotone linear functional</u> ("isotone" means that for every $x,y \in L$ such that $x(t) \geq y(t)$ on E it holds that $L(x) \geq L(y)$), then the corresponding <u>generalized Gini mean</u> $G(\alpha,\beta;x)$, $-\infty < \alpha,\beta < \infty$, is defined by

$$G(\alpha,\beta;x) = G_L(\alpha,\beta;x) = \begin{cases} (L(x^\alpha)/L(x^\beta))^{1/(\alpha-\beta)} & ,\alpha \neq \beta, \\ \exp\left(\left\{\frac{d}{da}(\ln L(x^a))\right\}_{a=\alpha}\right) & ,\alpha = \beta, \end{cases}$$

whenever $x^\alpha, x^\beta \in \Gamma$ and $0 < L(x^\alpha), L(x^\beta) < \infty$. Some basic facts concerning these means can be found in [24], [25] and [30]. For the case when $L(x) = \Sigma x_k$ $(x_k \geq 0,\ k = 1,2,\ldots,n)$ these means coincide with the classical Gini means.

Let D denote an additive Abelian semigroup. Let $\mathbf{u} = (u_1, u_2, \ldots, u_n)$ and $\mathbf{v} = (v_1, v_2, \ldots, v_n) \in R^n$. We write $\mathbf{u} \leq \mathbf{v}$ if $u_1 \leq v_1, u_2 \leq v_2, \ldots, u_n \leq v_n$. We say that $f: D \to R^n$, is <u>subadditive</u> if

(1.1) $f(x + y) \leq f(x) + f(y)$

for all $x,y \in D$. If (1.1) holds in the reversed direction, then we say that f is superadditive. If equality holds in (1.1), then we say that f is affine.

$P(\Omega)$ denotes the power set of Ω, i.e. the set of subsets of Ω.

Concerning classical inequalities and its applications we refer to the books [3], [7], [8], [9], [14] and [21]. We finish this section by listing suitable integral forms of some of these inequalities.

1. *Hölder's inequality*: If $p \geq 1$, then

$$(1.2) \quad \|xy\|_{L^1(\Omega)} \leq \|x\|_{L^p(\Omega)} \|y\|_{L^q(\Omega)} .$$

If $p \leq 1$, $p \neq 0$, $xy > 0$ a.e., then (1.2) holds in the reversed direction.

2. *Minkowski's inequality*: If $p \geq 1$, then

$$(1.3) \quad \|x+y\|_{L^p(\Omega)} \leq \|x\|_{L^p(\Omega)} + \|y\|_{L^p(\Omega)} .$$

If $0 < p \leq 1$ and $x,y \geq 0$ a.e., then (1.3) holds in the reversed direction.

3. *Jensen's inequality*: Let Φ be a non-negative and convex function on $[0,\infty)$. If $\mu(\Omega) = 1$, then, for every $x \in L^1(\Omega)$,

$$(1.4) \quad \Phi\Big(\int_\Omega |x|\,d\mu\Big) \leq \int_\Omega \Phi(|x|)\,d\mu .$$

If Φ is non-negative and concave, then (1.4) holds in the reversed direction.

4. *Clarcson's inequality*: If $1 < p \leq 2$, then

$$(1.5) \quad \left(\Big\|\frac{x+y}{2}\Big\|_{L^p(\mu)}\right)^q + \left(\Big\|\frac{x-y}{2}\Big\|_{L^p(\mu)}\right)^q \leq \left(\frac{1}{2}(\|x\|_{L^p(\mu)})^p + \frac{1}{2}(\|x\|_{L^p(\mu)})^p\right)^{q-1}$$

If $p \geq 2$, then (1.5) holds in the reversed direction.

For the case $p = 2$ Clarcson's inequality reduces to the usual parallellogram law for L^p-spaces.

5. *The AG-mean inequality*: If $x = x(t) > 0$ on Ω, $0 < \mu(\Omega) < \infty$, then

$$(1.6) \quad \exp\left(\frac{1}{\mu(\Omega)}\int_\Omega \ln x\,d\mu\right) \leq \frac{1}{\mu(\Omega)}\int_\Omega x\,d\mu .$$

If $\Omega = \{1,2,\ldots n\}$, $\mu(k) = 1$ and $x(k) = x_k$, $k = 1,2,\ldots ,n$, then (1.6) reads

$$\left(\prod_1^n x_k\right)^{1/n} \leq \frac{1}{n}\sum_1^n x_k$$

which is the most well-known form of the AG-inequality. We note that (1.6) is a special case both of Jensen's inequality and of the fact that the Gini means $G(\alpha,\beta;x)$ are increasing in α (β fixed) and in β (α fixed), see [25] and our Proposition 10.

6. *Beckenbach-Dresher's inequality*: If $0 \leq \beta \leq 1 \leq \alpha$ or if $0 \leq \alpha \leq 1 \leq \beta$, $\alpha \neq \beta$, then

$$\text{(1.7)}\quad \left(\frac{\int_\Omega |x+y|^\alpha d\mu}{\int_\Omega |x+y|^\beta d\mu}\right)^{\frac{1}{\alpha-\beta}} \leq \left(\frac{\int_\Omega |x|^\alpha d\mu}{\int_\Omega |x|^\beta d\mu}\right)^{\frac{1}{\alpha-\beta}} + \left(\frac{\int_\Omega |y|^\alpha d\mu}{\int_\Omega |y|^\beta d\mu}\right)^{\frac{1}{\alpha-\beta}} .$$

If $\beta \leq 0 \leq \alpha \leq 1$ or if $\alpha \leq 0 \leq \beta \leq 1$, $\alpha \neq \beta$, then (1.7) holds in the reversed direction.

7. *Hardy's inequalities*: If $p > 1$, then

$$\text{(1.8)}\quad \int_0^\infty \left(\frac{1}{t}\int_0^t |x(u)|\,du\right)^p dt \leq \left(\frac{p}{p-1}\right)^p \int_0^\infty |x(t)|^p\,dt .$$

If $0 < p < 1$, then

$$\int_0^\infty \left(\frac{1}{t}\int_t^\infty |x(u)|\,du\right)^p dt \geq \left(\frac{p}{1-p}\right)^p \int_0^\infty |x(t)|^p d\mu .$$

The constants in these inequalities are the best possible.

Limiting cases of some of the inequalities above are pointed out in our section 8. Moreover, the following inequality may be seen as a limiting case of the inequalities we present in our section 7:

8. *The differential-entropy inequality*. Roufly speeking this inequality says that the normal distribution has the largest possible differential-entropy among all distributions with fixed mean and standard deviation. The differential-entropy inequality is described (and generalized) in our section 7. This inequality is maybe not regarded as a classical inequality in the usual textbooks on inequalities but it is very important for applications e.g. in Information Theory (see [4] and [15]).

2. TWO "SETVALUED" GENERALIZATIONS OF SOME CLASSICAL INEQUALITIES

First we state a somewhat generalized form of an inequality of Peetre-Persson[25].

THEOREM 1 *Let* $F: R^n_+ \to R_+$ *be non-negative and convex,* $G: D \to P(R_+)$ and $f: D \to R^n_+$. *Then the function*

$$f_1(x) = \inf_{a \in G(x)} \left(aF\left(\frac{f(x)}{a}\right)\right)$$

is subadditive if <u>*one*</u> *of the following conditions holds:*

(1) $f(x)$ *is affine and, for all* $a \in G(x)$ *and* $b \in G(y)$, $a+b \in G(x+y)$

(2) $F(u)$ *is isotone,* $f(x)$ *is subadditive and if* $a \in G(x)$ *and* $b \in G(y)$, *then* $a+b \in G(x+y)$.

(3) $F(0) = 0$, $F(u)$ *is isotone,* $f(x)$ *is subadditive and if* $a \in G(x)$ *and* $b \in G(y)$, *then there exists* $c \geq a+b$ *such that* $c \in G(x+y)$.

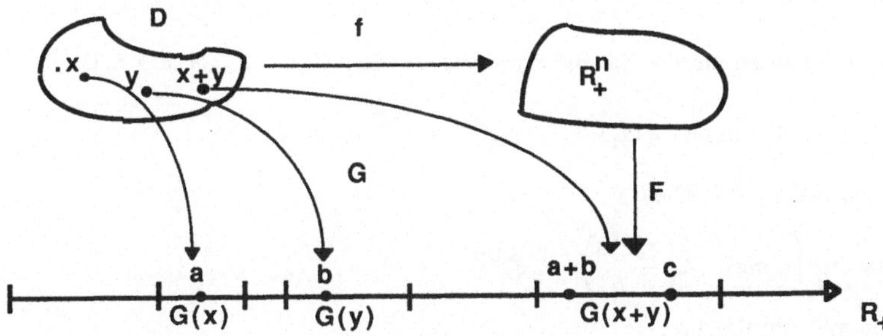

The proof of Theorem 1 can be carried out by only making obvious modifications of the proof in [25]. Compare also with our proof of the following dual version of Theorem 1:

THEOREM 2 *Let* $F: R^n_+ \to R_+$ *be non-negative and concave,* $G: D \to P(R_+)$ *and* $f: D \to P(R_+)$. *Then the function*

$$f_2(x) = \sup_{a \in G(x)} \left(aF\left(\frac{f(x)}{a}\right)\right)$$

is superadditive if <u>*one*</u> *of the conditions* (1)-(3) *in Theorem 1 holds with "subadditive" replaced by "superadditive".*

<u>Proof</u>: First we let the assumptions in the (modified) condition (3) be satisfied. Assume that $x,y \in D$, $a \in G(x)$, $b \in G(y)$ and choose $c \in G(x+y)$ such that $c \geq a+b$. Then

$$\frac{f(x+y)}{c} \geq \frac{f(x)+f(y)}{c} = \frac{a}{c}\frac{f(x)}{a} + \frac{b}{c}\frac{f(y)}{b}$$

and

$$F\left(\frac{f(x+y)}{c}\right) \geq \frac{a}{c}F\left(\frac{f(x)}{a}\right) + \frac{b}{c}F\left(\frac{f(y)}{b}\right) + \frac{c-b-a}{c}F(0) = \frac{a}{c}F\left(\frac{f(x)}{a}\right) + \frac{b}{c}F\left(\frac{f(y)}{b}\right).$$

Thus, for any fixed ε, $0 < \varepsilon < 1/2$, and some $c \in G(x+y)$, we have

$$cF\left(\frac{f(x+y)}{c}\right) \geq (1-\varepsilon)f_2(x) + (1-\varepsilon)f_2(y).$$

By taking supremum once more and letting $\varepsilon \to 0+$ we find that

$$f_2(x+y) \geq f_2(x) + f_2(y).$$

The proofs of the remaining cases are quite similar.

Remark: The proof shows that Theorem 2 holds also if the condition "$F(0) = 0$" is weakened to "$F(0) \geq 0$".

First we present an application for the extreme case when $G(x) = R_+$ for all $x \in D$.

Example Let $D = R_+^n$, $f(x) = x = (x_1, x_2, \ldots, x_n)$, $G(x) = R_+$ for all $x \in D$ and $F(u) = \max(1, \sum_1^n u_k^p)$, $p \geq 1$. Then

$$f_1(x) = \inf_{a \in R_+} \left(a \max\left(1, \sum_1^n \left(\frac{x_k}{a}\right)^p\right)\right) = \left(\sum_1^n x_k^p\right)^{1/p}$$

and Theorem 1 implies that

$$(2.1) \quad \left(\sum_1^n (x_k+y_k)^p\right)^{1/p} \leq \left(\sum_1^n (x_k)^p\right)^{1/p} + \left(\sum_1^n (y_k)^p\right)^{1/p}.$$

In a similar way Theorem 2 implies that (2.1) holds in the reversed direction if $0 < p < 1$. Moreover, by instead considering $F(u) = 1 + \sum_1^n \Phi(u_k)$, where $\Phi: R_+ \to R_+$, is convex, we obtain

$$f_1(x) = \inf_{a \in R_+} a\left(1 + \sum_1^n \Phi\left(\frac{x_k}{a}\right)\right)$$

and this is the usual Amemiya description of the Orlicz-norm of x. Thus Theorem 1 also contains a generalization of (2.1) to the corresponding Orlicz-space case.

In the next section we present some applications for the other extreme case $G(x) = \{g(x)\}$ (the singleton case).

3. THE SINGLETON CASE WITH APPLICATIONS

We restate Theorems 1 and 2 for the "single-valued" case.

PROPOSITION 3 *Let* $F: R_+^n \to R_+$ be convex, $f: D \to R_+^n$ *and* $g: D \to R_+$. *Then*

$$(3.1)\quad g(x+y)F\left(\frac{f(x+y)}{g(x+y)}\right) \le g(x)F\left(\frac{f(x)}{g(x)}\right) + g(y)F\left(\frac{f(y)}{g(y)}\right)$$

if one of the following conditions holds:
(a) $f(x)$ *and* $g(x)$ *are affine.*
(b) $F(u)$ *is isotone,* $f(x)$ *is subadditive and* $g(x)$ *is affine.*
(c) $F(0) = 0$, $F(u)$ *is isotone,* $f(x)$ *is subadditive and* $g(x)$ *is superadditive.*
Moreover (3.1) *holds in the reversed direction if* F *is concave and if* (a), (b) *or* (c) *holds with "subadditive" replaced by "superadditive".*

Proof: Apply Theorems 1 and 2 with $G(x) = \{g(x)\}$.

By applying Proposition 3 with D = L, $F(u) = u^p$ and using induction and the well-known subadditivity (superadditive) properties of the functionals $f(x) = (B(x^\alpha))^{1/\alpha}$ and $g(x) = C(x^\beta))^{1/\beta}$ (see e.g. [25, Lemma 1.2] and compare also with our Theorem 6) we obtain the following generalization of Beckenbach-Dresher's inequality (see [2] and [13]):

COROLLARY 1 (Peetre-Persson) *Let* $B, C: \Gamma \to R_+$ *be isotone linear functionals and let the functions* $z_i: E \to R_+, i = 1,2,\ldots ,N$, *satisfy that* $z_i^\alpha, z_i^\beta, (\sum_1^N z_i)^\alpha, (\sum_1^N z_i)^\beta \in \Gamma$.

(a) *If* $\beta \le 1 \le \alpha$, $\beta \ne 0$, $\alpha \ne \beta$ *and* $p \ge 1$, *then*

$$(3.2)\quad \frac{(B((\sum_1^N z_i)^\alpha))^{p/\alpha}}{(C((\sum_1^N z_i)^\beta))^{(p-1)/\beta}} \le \sum_1^N \frac{(B(z_i^\alpha))^{p/\alpha}}{(C(z_i^\beta))^{(p-1)/\beta}} .$$

(b) *If* $\alpha,\beta \le 1$ $\alpha,\beta \ne 0$ *and* $0 < p \le 1$, *then* (3.2) *holds in the reversed direction.*

See [25]. Another proof of a special case of a) can also be found in [26].

Another obvious consequence of Proposition 3 (a) is the following statement:

COROLLARY 2 (Matkowski). *A function* $F: R_+^n \to R_+$ *is convex if and only if*

$$(3.3)\quad (x+y)F\left(\frac{\mathbf{x}+\mathbf{y}}{x+y}\right) \le xF\left(\frac{\mathbf{x}}{x}\right) + yF\left(\frac{\mathbf{y}}{y}\right)$$

holds for all $\mathbf{x},\mathbf{y} \in R_+^n$ *and all* $x,y \in R_+$. *The function* F *is concave if and only if* (3.3) *holds in the reversed direction.*

See [22]. Now by using (3.3), induction and a standard density argument for $L^1(\Omega)$-spaces we also have the following corollary:

COROLLARY 3 *If* $F: R_+^n \to R_+$ *is convex and if* $f_1, f_2, \ldots, f_n, g \in L^1(\Omega)$, *then*

$$(3.4)\quad \int_\Omega g d\mu F\left(\frac{\int_\Omega f_1 d\mu}{\int_\Omega g d\mu}, \ldots, \frac{\int_\Omega f_n d\mu}{\int_\Omega g d\mu}\right) \le \int_\Omega F\left(\frac{f_1}{g}, \ldots, \frac{f_n}{g}\right) g d\mu.$$

This inequality holds in the reversed direction if F *is concave* ($f_i \ge 0$, $g > 0$).

Compare with [22]. We finish this section by pointing out some examples of applications of the results stated in this section.

Example 1 By applying Corollary 1 with B = C and $p = \alpha/(\alpha-\beta)$ and using symmetry we find that

$$G_B(\alpha,\beta;x+y) \le G_B(\alpha,\beta;x) + G_B(\alpha,\beta;y)$$

if $0 \le \beta \le 1 \le \alpha$ or if $0 \le \alpha \le 1 \le \beta$. This inequality holds in the reversed direction if $\beta \le 0 \le \alpha \le 1$ or if $\alpha \le 0 \le \beta \le 1$.

Remark For the case $B(x) = \int_\Omega x \, d\mu$ the statement in Example 1 coincides with the version of *Beckenbach–Dresher's inequality* we presented in the introduction.

It is obviously possible to iterate the procedure in Proposition 3. In particular by using the special case considered in Example 1 we obtain

Example 2 If B and C are isotone linear functionals, if $0 \le \beta \le 1 \le \alpha$, $\gamma \le 0 \le \delta \le 1$ and if $F: R_+ \to R_+$, is convex, isotone and $F(0) = 0$, then

$$G_C(\gamma,\delta;x+y)F\left(\frac{G_B(\alpha,\beta;x+y)}{G_C(\gamma,\delta;x+y)}\right) \le G_C(\gamma,\delta;x)F\left(\frac{G_B(\alpha,\beta;x)}{G_C(\gamma,\delta;x)}\right) + G_C(\gamma,\delta;y)F\left(\frac{G_B(\alpha,\beta;y)}{G_C(\gamma,\delta;y)}\right).$$

In particular if $F(u) = u^p$, $p = (\alpha-\beta)/(\alpha-\beta-\gamma+\delta)$, and $B(x) = C(x) = \int_\Omega x(t)dt$ we obtain the following somewhat curious extention of Beckenbach–Dresher's inequality (1.7): If $0 \le \beta \le 1 \le \alpha$, $\gamma \le 0 \le \delta \le 1$, $\alpha \ne \beta$, $\gamma \ne \delta$, $\alpha-\beta-\gamma+\delta \ge 0$, $x(t), y(t) \ge 0$, then

$$\left(\frac{\int_\Omega (x+y)^\alpha d\mu \int_\Omega (x+y)^\delta d\mu}{\int_\Omega (x+y)^\beta d\mu \int_\Omega (x+y)^\gamma d\mu}\right)^{\frac{1}{\alpha-\beta-\gamma+\delta}} \le \left(\frac{\int_\Omega x^\alpha d\mu \int_\Omega x^\delta d\mu}{\int_\Omega x^\beta d\mu \int_\Omega x^\gamma d\mu}\right)^{\frac{1}{\alpha-\beta-\gamma+\delta}} + \left(\frac{\int_\Omega y^\alpha d\mu \int_\Omega y^\delta d\mu}{\int_\Omega y^\beta d\mu \int_\Omega y^\gamma d\mu}\right)^{\frac{1}{\alpha-\beta-\gamma+\delta}}$$

Example 3 Let $\mu(\Omega) = 1$ and $g = 1$. Then, according to Corollary 3, we obtain the following n–dimensional version of *Jensen's inequality* : If F: $R_+^n \to R_+$ is convex and if $\mu(\Omega) = 1$, then

$$F(\int_\Omega f_1 d\mu, \ldots , \int_\Omega f_n d\mu) \le \int_\Omega F(f_1, \ldots , f_n) d\mu.$$

The inequality holds in the reversed direction if F is concave.

Example 4 Let $F(u_1, u_2, \ldots , u_n) = u_1^{1/p_1} u_2^{1/p_2} \ldots u_n^{1/p_n}$, where $\sum_1^n \frac{1}{p_i} = 1$ $(p_i > 1)$. We note that F is concave and, thus, by Corollary 3 used with f_i replaced by $|f_i|^{p_i}$, we obtain that

$$(\int_\Omega |f_1|^{p_1} d\mu)^{1/p_1} (\int_\Omega |f_1|^{p_2} d\mu)^{1/p_2} \ldots (\int_\Omega |f_1|^{p_n} d\mu)^{1/p_n} \ge \int_\Omega |f_1 f_2 \ldots f_n| d\mu.$$

We conclude that also this n-dimensional version of *Hölder's inequality* is a special case of Corollary 3.

Example 5 Let $F(u_1, u_2, \ldots , u_n) = (u_1^{1/p_1} + 1)^{p_1} (u_2^{1/p_2} + 1)^{p_2} \ldots (u_n^{1/p_n} + 1)^{p_n}$, where $p_i > 1$. Since F is concave we can use Corollary 3 to obtain the following generalized version of *Minkowski's inequality* :

$$\int_\Omega \prod_1^n (f_i + g^{1/p_i})^{p_i} \left(\frac{g}{\int_\Omega g d\mu} \right)^{1-n} d\mu \le \prod_1^n ((\int_\Omega f_i^{p_i} d\mu)^{1/p_i} + (\int_\Omega g d\mu)^{1/p_i})^{p_i}.$$

By replacing g by $|f_0|^p$ and putting $n = 1$ and $p = p_1$ in this inequality we obtain the version of Minkowski's inequality we presented in our introduction.

Remark More information concerning the close connection between convexity and inequalities can be found in the forthcoming paper [20].

4. TWO GENERALIZATIONS OF HÖLDER'S INEQUALITY

Let p(t) and W(t) denote positive functions on (0,b), b > 0, and let p be defined by

$$\frac{1}{p} = \frac{1}{I_W} \int_0^b \frac{1}{p(t)} W(t) dt, \quad \text{where } I_W = \int_0^b W(t) dt < \infty.$$

Assume that $\{\omega_t(s)\}$, $t \in [0,b)$, $s \in \Omega$, is a family of weight functions such that $\omega_t(s)$ is measurable on $[0,b) X \Omega$ and define (the generalized geometric mean)

$$\omega = \omega(s) = \exp\left(\frac{1}{I_W} \int_0^b \log \omega_t(s) W(t) dt \right).$$

First we present a generalization of Hölder's inequality to a case where infinite many spaces is involved.

THEOREM 4 (Nikolova-Persson). *Let X be a BFS such that* $X'' \equiv X$ and let $Z_t = X^{p(t)}(A,\omega_t)$. *Assume that* $x_t(s) \in Z_t$ *and* $W(t)\log \|x_t(s)\|_{Z_t}$ *and* $W(t)\log \|x_t(s)\|_A$, $s \in \Omega$, *are integrable over* $[0,b)$. *Then*

$$y = y(s) = \exp\left(\frac{1}{I_W}\int_0^b \log \|x_t(s)\|_A W(t)dt\right)$$

belongs to $X^p(\omega)$ *and*

$$\|y\|_{X^{p}(\omega)} \le \exp\left(\frac{1}{I_W}\int_0^b \log \|x_t(s)\|_{Z_t} W(t)dt\right).$$

Theorem 1 was discovered by us in connection to our work with interpolation between infinite many spaces (see [23]). A surprisingly simple proof has also been presented in [24] .

Remark. We don't know if Theorem 1 holds also without the assumption $X'' \equiv X$. However the proof presented in [24] shows that the theorem holds also without this assumption if we instead assume that $(\alpha_t(s)\omega_t(s))^{p(t)}, t \in [0,b)$, are strongly measurable functions (taking values in X).

COROLLARY . *Let* $z_i \in X^{p_i}(A)$, $0 < p_i < \infty$, $0 < \alpha_i < 1$, $i = 1,2,\ldots ,N$, $\sum_1^N \alpha_i = 1$ *and* $\frac{1}{p} = \sum_1^N \frac{\alpha_i}{p_i}$. *Then*

$$\|\prod_1^N(\|z_i(s)\|_A)^{\alpha_i}\|_{X^p} \le \prod_1^N(\|z_i(s)\|_{X^{p_i}(A)})^{\alpha_i}.$$

Proof: Let $0 = a_1 < a_2 < \ldots < a_{N+1} = b$ where $\alpha_i = (a_{i+1} - a_i)/b$, $i = 1,2,\ldots ,N$.
Apply Theorem 4 and the last remark with $\omega(s) \equiv 1$, $W(t) \equiv 1$, $p(t) = p_i$ and $x_t(s) = z_i(s)$ on $[a_i, a_{i+1})$, $i = 1,2,\ldots ,N$, and the proof follows.

Example. *In particular, if* $A = R$, $p_i/\alpha_i = q_i$, $|z_i|^{\alpha_i} = |x_i|$, $i = 1,2,\ldots ,N$, *then the estimate in Corollary 1 can be written as*

$$\|\prod_1^N x_i\|_{X^p} \le \prod_1^N \|x_i\|_{X^{q_i}},$$

where $\frac{1}{p} = \sum_1^N \frac{1}{q_i}$, $0 < q_i < \infty$.

Remark. Concerning this well-known version of Hölder's inequality for finite many X^p-spaces see e.g. [18] and the references given there.

Remark. It is also possible to obtain other classical inequalities as special cases of Theorem 4 e.g. the generalizations of Minkowski's and Beckenbach-Dresher's inequalities presented in [29] (see [24]).

In this section we also include the following generalized and completely symmetric version of Hölder's inequality:

THEOREM 5 *Let* p,q,r *be real numbers* $\neq 0$ *such that* $1/p + 1/q = 1/r$ *and let* x *and* y *denote positive functions. Let* $B_p(x) = (B(x^p))^{1/p}$, *where* B *denotes an isotone linear functional defined on* x *and* y *and suitable potenses of these functions. Then*

(a $B_r(xy) \leq B_p(x)B_q(y)$

if $p > 0, q > 0, r > 0$ *or if* $p < 0, q > 0, r < 0$ *or if* $p > 0, q < 0, r < 0$.

(b) $B_r(xy) \geq B_p(x)B_q(y)$

if $p > 0, q < 0, r > 0$ *or if* $p < 0, q > 0, r > 0$ *or if* $p < 0, q < 0, r < 0$.

Theorem 5 can be proved by using similar arguments as those in the proof of the special case considered in [29] (compare also with [1]). For the sake of completness we include a proof (which essentially only depends on the convexity of the exponential function).

Proof: First we consider the case $p > 0, q > 0, r > 0$. Let $B_p(x) = B_q(y) = 1$. Since $f(t) = \exp(t)$ is convex we have

$$|x(t)y(t)|^r = \exp(r(\ln|x(t)| + \ln|y(t)|)) = \exp\left(\frac{r}{p}\ln|x(t)|^p + \frac{r}{q}\ln|y(t)|^q\right) =$$

$$= \frac{r}{p}|x(t)|^p + \frac{r}{q}|y(t)|^q.$$

Thus, according to the fact that B is a linear and isotone functional, we find that

$$(B_r(xy))^r \leq \frac{r}{p}B_p^p(x) + \frac{r}{q}B_q^q(y) = \frac{r}{p} + \frac{r}{q} = 1, \text{ i.e. } B_r(xy) \leq 1.$$

For the general case we use this estimate with x and y replaced by $x/B_p(x)$ and $y/B_q(y)$, respectively. Next we let $p > 0, q < 0, r > 0$. By using the estimate we just have proved we find that

$$B_p(x) = B_p\left(xy\frac{1}{y}\right) \leq B_r(xy)B_{-q}\left(\frac{1}{y}\right) = B_r(xy)(B_q(y))^{-1}, \text{ i.e. (b) holds.}$$

By symmetry we find that (b) holds also for the case $p < 0, q > 0, r > 0$. For the cases $p < 0, q > 0, r < 0$, and $p > 0, q < 0, r < 0$, we use the obtained results with r,p,q replaced by -r,-p,-q, respectively, and obtain

$$B_r(xy) = (B_{-r}((xy)^{-1}))^{-1} \leq (B_{-p}(x^{-1}))^{-1}\,(B_{-q}(y^{-1}))^{-1} = B_p(x)B_q(y), \text{ i.e. (b) holds.}$$

The proof of the remaining case $p< 0, q < 0, r < 0$, is quite similar.

Remark By using the relation $B_r(x) = (B_{-r}(x^{-1}))^{-1}$ and induction we see that Theorem 5 implies that if

$$\prod_1^n z_i = 1, \sum_1^n \frac{1}{p_i} = 0, p_i \neq 0, i = 1,2,\ldots ,n,$$

and if exactly one of p_i is negative (respectively positive), then

$$\prod_1^n B_{p_i}(z_i) \geq 1 \text{ (respectively } \prod_1^n B_{p_i}(z_i) \leq 1).$$

A SHARP GENERALIZED VERSION OF MINKOWSKI'S INEQUALITY

Let B denote an isotone and linear functional defined on the positive functions x, y, x^p and y^p. We introduce the following measure of the "deviation" between x and y:

$$d(x,y) = B_p\left(\frac{x}{B_p(x)} - \frac{y}{B_p(y)}\right), B_p(x) = (B(|x|^p))^{1/p}, \ 0 < B_p(x), B_p(y) < \infty.$$

We note that $0 \leq d(x,y) \leq 2$, $a(x,y) = 0$ iff $y = cx$, $c > 0$, and $d(x,y) = 2$ iff $y = cx$, $c < 0$. In particular this means that the coefficients b_i in the formulation of our next theorem satisfies $-1 \leq b_i \leq 1$ and thus that the theorem may be regarded as a sharp generalized form of Minkowski's inequality.

THEOREM 6 *Let* $0 < p < \infty$ *and let* y_i *be functions such that* $0 < B_p(y_i) < \infty$. If $y = \sum_1^n y_i$, $a_i = d(y_i, y)$, $b_i = 2\left(1-\left(\frac{a_i}{2}\right)^q\right)^{1/q} - 1$ *if* $1 < p \leq 2$ *and* $b_i = 2\left(1-\left(\frac{a_i}{2}\right)^p\right)^{1/p} - 1$ *if* $p \geq 2$, $i = 1,2\ldots ,n$, *then*

$$B_p(\sum_1^n y_i) \leq \sum_1^n b_i B_p(y_i). \tag{5.1}$$

Remark We can define $\theta_i = \arccos b_i$ and compare (5.1) with the well-known formula

$$\|\sum_1^n x_i\|_{R^3} = \sum_1^n b_i \|x_i\|_{R^3}, b_i = \cos\theta_i,$$

where x_i are vectors in R^3 and θ_i are the angles between x_i and $\sum_1^n x_i$, $i = 1,2,\ldots ,n$.

Example Let $0 < \varepsilon < 1$ and $y = -(1-\varepsilon)x$. Then an elementary calculation shows that $a_1 = d(x,x+y) = 0$, i.e. $b_1 = 1$, and $a_2 = d(y,x+y) = 2$, i.e. $b_2 = -1$. Thus, by Theorem 6, $B_p(x+y) \leq B_p(x) - B_p(y)$ if $y = -(1-\varepsilon)x$ (for the L^p-case we have even equality).

For the proof of Theorem 6 we need the following crucial lemma:

LEMMA 1 *Let* Y *be a uniformly convex space,* $y_i \in Y$, $y = \sum_1^n y_i$, $a_i = d(y_i, y)$, $i = 1,2,\dots,n$, *where* $d(x,y) = \left\| \frac{x}{\|x\|_Y} - \frac{y}{\|y\|_Y} \right\|_Y$. *Then*

$$\left\| \sum_1^n y_i \right\|_Y \leq \sum_1^n (1-2\delta(a_i)) \|y_i\|_Y,$$

where $\delta(\varepsilon)$ *denotes the modulus of convexity in the definition of uniformly convex spaces.*

A proof of Lemma 1 can be found in Clarcson´s original paper [10].

We also need the following generalization of Clarcson´s inequality, which can be of independent interest

LEMMA 2 *If* $1 < p \leq 2$ *and if* B *is an isotone and linear functional, then*

$$B_p^q\left(\frac{x+y}{2}\right) + B_p^q\left(\frac{x-y}{2}\right) \leq \left(\frac{1}{2}B_p^p(x) + \frac{1}{2}B_p^p(y)\right)^{q-1}$$

If $p \geq 2$, *then this inequality holds in the reversed direction.*

The proof of this lemma can be carried out by only making suitable modifications of the proof for the special case where $B_p(x) = \|x\|_X^p$ (see [29]) so we omit the details.

More information concerning Clarcson´s inequality can also be found in [11], [12] and [19] (and the references given in these papers).

Proof of Theorem 6. Let B^p, $p > 1$, denote the normed space equipped with the norm $\|x\|_{B^p} = B_p(x)$. Let $0 < \varepsilon < 1/2$, $\|x\|_{B^p} = \|y\|_{B^p} = 1$ and $\|x-y\|_{B^p} \geq \varepsilon > 0$. If $1 < p \leq 2$, then, by Lemma 2, we find that

$$\left(\left\|\frac{x+y}{2}\right\|_{B^p}\right)^q + \left(\frac{\varepsilon}{2}\right)^q \leq 1, \text{ i.e. } \left\|\frac{x+y}{2}\right\|_{B^p} \leq 1-\delta,\ \delta = 1-\left(1-\left(\frac{\varepsilon}{2}\right)^q\right)^{1/q}.$$

In the same way we find that if $p \geq 2$, then

$$\left\|\frac{x+y}{2}\right\|_{B^p} \leq 1-\delta,\ \delta = 1-\left(1-\left(\frac{\varepsilon}{2}\right)^p\right)^{1/p}.$$

Thus we have proved that B^p is a uniformly convex space where the corresponding modulus of convexity $\delta = \delta(\varepsilon)$ can be chosen as above. Finally, we apply Lemma 1 and the proof follows.

We finish this section by pointing out the following special case of Theorem 6:

Example If $1 < p \le 2$, $y_i \in L^s(\mu)$, $p \le s \le q$, $d_i = d(y_i, \sum_1^n y_i)$ and $b_i = 2\left(1-\left(\frac{a_i}{2}\right)^q\right)^{1/q} - 1$, $i = 1,2,\dots,n$, then

$$\|\sum_1^n y_i\|_{L^s(\mu)} \le \sum_1^n b_i \|y_i\|_{L^s(\mu)}.$$

For the cases s = p and s = q this statement can also be found in [10, p.405]. See also [29].

6. RELATIONS BETWEEN GENERALIZED VERSIONS OF SOME CLASSICAL INEQUALITIES

Let $B,C : \Gamma \to R_+$ denote arbitrary isotone linear functionals. In this section we present some relations between the following generalized forms of classical inequalities, which hold simultanously for all isotonic linear functionals.

Hölder's inequality (H): If $p > 1$, then $B(xy) \le B_p(x)B_q(y)$.
Lyaponov's inequality (L) : If $0 < \gamma < \beta < \alpha$, then

$$B(x^\beta) \le (B(x^\gamma))^{(\alpha-\beta/(\alpha-\gamma)}(B(x^\alpha)^{(\beta-\gamma)/(\alpha-\gamma)}.$$

Generalized AG-inequality (P): $G_B(\alpha,\beta;x)$ is nondecreasing in α (β fixed) and in β (α fixed).
See [25] and our Proposition 10. In particular the special case $G_B(0,0;x) \le G_B(1,0;x)$ is a generalization of the usual AG-inequality.
Minkowski's inequality (M): If $p \ge 1$, then $B_p(x+y) \le B_p(x) + B_p(y)$.
Beckenbach-Dresher's inequality (BD): If $p \ge 1$ and $\beta \le 1 \le \alpha$, $\beta \ne 0$, then

$$\frac{(B((x+y)^\alpha))^{p/\alpha}}{(C((x+y)^\beta))^{(p-1)/\beta}} \le \frac{(B(x^\alpha))^{p/\alpha}}{(C(x^\beta))^{(p-1)/\beta}} + \frac{(B(y^\alpha))^{p/\alpha}}{(C(y^\beta))^{(p-1)/\beta}}.$$

Remark By choosing $p = \alpha/(\alpha-\beta)$, $B = C$ and using the symmetry and continuity properties of $G_B(\alpha,\beta;x)$ we find that (BD) implies that the inequality

(p) $\quad G_B(\alpha,\beta;x+y) \le G_B(\alpha,\beta;x) + G_B(\alpha,\beta;y)$

holds if $0 < \beta \le 1 \le \alpha$ or if $0 < \alpha \le 1 \le \beta$.

THEOREM 7 *The following holds:*

$$\begin{array}{c} (M) \\ \Uparrow \\ (P) \Leftrightarrow (H) \Leftrightarrow (L) \\ \Downarrow \\ (BD) \end{array}$$

Remark Theorem 7 may be seen as a generalization of some results in [21, p. 457].

Remark According to Theorem 7 and our proof of Theorem 5 we see that all the inequalities (H), (L), (P), (M), (BD), and (p) easily follows from the convexity of the function exp(t).

Proof: (P) ⇒ (H). Assume that (P) holds. Let $0 < \theta < 1$ and consider the functional $B_0(x) = B(x\omega)$, where ω denotes a positive weight function. Then

$$\frac{1}{B(\omega)}B\left(\left(\frac{z}{\omega}\right)^{\theta}\omega\right) = \frac{B_0\left(\left(\frac{z}{\omega}\right)^{\theta}\right)}{B_0(1)} = \left(G_{B_0}\left(\theta,0;\frac{z}{\omega}\right)\right)^{\theta} \leq \left(G_{B_0}\left(1,0;\frac{z}{\omega}\right)\right)^{\theta} = \left(\frac{B_0\left(\frac{z}{\omega}\right)}{B_0(1)}\right)^{\theta} =$$

$$(B(z))^{\theta}(B(\omega))^{-\theta} \text{ i.e. } B(z^{\theta}\omega^{1-\theta}) \leq (B(z))^{\theta}(B(\omega))^{1-\theta}.$$

We put $x = z^{\theta}w^{1-\theta}$, $y = w^{1-\theta}$, $\theta = 1/p$ and find that (H) holds.

(H) ⇒ (L). This implication follows by using (H) with $x = z^{\gamma/p}$ and $y = z^{\alpha/q}$, where $\frac{1}{p} = \frac{\alpha-\beta}{\alpha-\gamma}$ and $\frac{1}{q} = \frac{\beta-\gamma}{\alpha-\gamma}$.

(L) ⇒ (P). Let (L) hold. We define $F(a) = B(x^a)$, $a \in R$, and note that

$$F(\beta) \leq (F(\gamma))^{\theta}(F(\alpha))^{1-\theta}, \ \theta = \frac{\alpha-\beta}{\alpha-\gamma}, \ \gamma < \beta < \alpha \text{ or } \alpha < \beta < \gamma.$$

This means that

$$G_B(\alpha,\beta;x) = \left(\frac{F(\alpha)}{F(\beta)}\right) \geq \left(\frac{F(\alpha)}{F(\gamma)}\right)^{\frac{1}{\alpha-\gamma}} = G_B(\alpha,\gamma;x) \text{ if } \gamma < \beta < \alpha$$

and

$$G_B(\alpha,\beta;x) = \left(\frac{F(\alpha)}{F(\beta)}\right) \leq \left(\frac{F(\alpha)}{F(\gamma)}\right)^{\frac{1}{\alpha-\gamma}} = G_B(\alpha,\gamma;x) \text{ if } \alpha < \beta < \gamma$$

Hence by using the symmetry and continuity properties of $G_B(\alpha,\beta;x)$ (see [25]) we find that (P) holds.

(H) ⇒ (M). Write $(x+y)^p = (x+y)^{p-1}x + (x+y)^{p-1}y$, apply B and use (H).

(H) ⇒ (BD) Let $\alpha_1, \alpha_2, \beta_1, \beta_2$ denote positive numbers. According to (H) (applied with $B(\alpha) = \alpha_1 + \alpha_2$) we have

$$\text{(6.1)} \quad \frac{(\alpha_1+\alpha_2)^p}{(\beta_1+\beta_2)^{p-1}} \leq \frac{\alpha_1^p}{\beta_1^{p-1}} + \frac{\alpha_2^p}{\beta_2^{p-1}}.$$

Since (H) ⇒ (M) we also have

$$(6.2)\quad (B((x+y)^{\alpha}))^{1/\alpha} \le (B(x^{\alpha}))^{1/\alpha} + (B(y^{\alpha}))^{1/\alpha}$$

and

$$(6.3)\quad (C((x+y)^{\beta}))^{1/\beta} \ge (C(x^{\beta}))^{1/\beta} + (C(y^{\beta}))^{1/\beta}.$$

By using (6.1)-(6.3) we obtain that

$$\frac{(B((x+y)^{\alpha}))^{p/\alpha}}{(C((x+y)^{\beta}))^{(p-1)/\beta}} \le \frac{((B(x^{\alpha}))^{1/\alpha} + (B(y^{\alpha}))^{1/\alpha})^{p}}{((C(x^{\beta}))^{1/\beta} + (C(y^{\beta}))^{1/\beta})^{p-1}} \le \frac{(B(x^{\alpha}))^{p/\alpha}}{(C(x^{\beta}))^{(p-1)/\beta}} + \frac{(B(y^{\alpha}))^{p/\alpha}}{(C(y^{\beta}))^{(p-1)/\beta}}.$$

The proof is complete

7. A GENERALIZATION OF THE DIFFERENTIALENTROPY INEQUALITY

Let X be a stochastic variable with the density function $f_X(x) = f(x)$, $-\infty < x < \infty$. Then the differentialentropy H(X) is defined by

$$H(X) = -\int_{-\infty}^{\infty} f(x)\ln f(x)dx.$$

<u>Example</u> If $f(x) = \frac{1}{\sigma\sqrt{2\pi}}\exp\left(\frac{x^2}{2\sigma^2}\right), -\infty < x < \infty$ (the normal distribution), then

$$H(X) = \ln(\sigma\sqrt{2\pi e}).$$

According to the **differential–entropy inequality** it holds that

$$H(X) \le \ln(\sigma\sqrt{2\pi e})$$

for every stochastic variable with the fixed variance σ^2.

We introduce the following **scale** of entropies:

$$H(\alpha;X) = \frac{1}{1-\alpha}\ln(\int_{-\infty}^{\infty} (f(x))^{\alpha}dx, \alpha > 0, \alpha \ne 1,$$

and note that the continuity property of generalized Gini means (see [25]) implies that

$$\lim_{\alpha\to 1} H(\alpha;X) = H(X).$$

We also define the **exponential entropies** $E(\alpha;X)$ by $E(\alpha;X) = \exp H(\alpha;X)$. We remark that these entropies are of special interest for some important problems in Information Theory e.g. in the theory of quantization of data (see [4], [15] and [17]).

Let $\Gamma(x)$ and $B(p,r)$ denote the usual Gamma and Beta functions, respectively (see e.g. [31]). For convenience of writing we also introduce the function $T(a)$ defined by

$$T(a) = \int_{-1}^{1} (1+x^2)^a \, dx \, , \; a \in \mathbb{R}.$$

The following precize estimates for the exponential entropies can in particular be seen as a generalization of the differential entropy inequality:

THEOREM 8 (Koski-Persson). *Consider any continuous probability distribution on the real line with the density* $f(x)$ *with finite standard deviation* σ_0.

a) *If* $\alpha > 1$, *then*

$$E(\alpha;f) \leq \left(\frac{2\alpha}{3\alpha-1}\right)^{\frac{1}{1-\alpha}} \sqrt{\frac{3\alpha-1}{\alpha-1}} \, B\left(\frac{\alpha}{\alpha-1}, \frac{1}{2}\right) \sigma_0 .$$

Equality is obtained if and only if $f(x)$ *coincides (almost everywhere) with*

$$f^*(x) = \begin{cases} \dfrac{c}{a}\left(1-\dfrac{x^2}{a^2}\right)^{\frac{1}{\alpha-1}} & , |x| < a, \\ 0 & , |x| \geq a, \end{cases}$$

where $c = \dfrac{1}{B\left(\frac{\alpha}{\alpha-1}, \frac{1}{2}\right)}$ *and* $a = \sqrt{\dfrac{3\alpha-1}{\alpha-1}}\,\sigma_0$.

b) *If* $1 > \alpha > 1/3$, *then*

$$E(\alpha;f) \leq \left(\frac{2\alpha}{3\alpha-1}\right)^{\frac{1}{1-\alpha}} \sqrt{\frac{3\alpha-1}{1-\alpha}} \, B\left(\frac{1+\alpha}{2(1-\alpha)}, \frac{1}{2}\right) \sigma_0 .$$

Equality is obtained if and only if $f(x)$ *coincides (almost everywhere) with*

$$f^*(x) = \frac{c}{a}\left(1+\frac{x^2}{a^2 r}\right)^{-\frac{r+1}{2}}, \quad -\infty < x < \infty,$$

where $r = \dfrac{1+\alpha}{1-\alpha}, c = \dfrac{\Gamma\left(\frac{r+1}{2}\right)}{\sqrt{\pi r}\,\Gamma\left(\frac{r}{2}\right)}$ *and* $a = \sqrt{\dfrac{r-2}{r}}\,\sigma_0$.

c) *If* $0 \le a \le 1/3$ *and if* f *satisfies the "tail condition"*

$$f(x) \le \alpha^{\frac{1}{\alpha-1}} \frac{c}{a}\left(1+\frac{x^2}{a^2}\right)^{\frac{1}{\alpha-1}} \quad \text{for } |x| \ge a,$$

where

$$c = \frac{1}{T\left(\frac{1}{\alpha-1}\right)} \text{ and } a = \sqrt{\frac{T\left(\frac{1}{\alpha-1}\right)}{T\left(\frac{\alpha}{\alpha-1}\right)-T\left(\frac{1}{\alpha-1}\right)}}\,\sigma_0,$$

then

$$E(\alpha; f) \le \left(T\left(\frac{1}{\alpha-1}\right)\right)^{\frac{\alpha}{\alpha-1}}\left(T\left(\frac{\alpha}{\alpha-1}\right)\right)^{\frac{1}{1-\alpha}}\sqrt{\frac{T\left(\frac{1}{\alpha-1}\right)}{T\left(\frac{\alpha}{\alpha-1}\right)-T\left(\frac{1}{\alpha-1}\right)}}\,\sigma_0.$$

Equality is obtained if and only if f(x) coincides (almost everywhere) with

$$f^*(x) = \begin{cases} \frac{c}{a}\left(1+\frac{x^2}{a^2}\right) & , |x| < a, \\ 0 & , |x| \ge a. \end{cases}$$

A detailed proof of Theorem 7 together with some applications can be found in [15]

Remark. By making some straightforward calculations we find that the constants on the right hand sides of a) and b) both converge to $(2\pi e)^{1/2}\,\sigma_0$ if $\alpha \to 1+$ and $a \to 1-$, respectively. Mreover, some other elementary calculations show that the corresponding distributions converge to the distribution function of the normal distribution (see [15]).

Remark. The optimal distributions discussed in this section is mostly of the types classified in the usual Pearson's system of frequency curves. More information about generalized entropies, its applications and relations to generalized Gini means has recently been obtained in [15], [17] and [25].

8. ON THE BEST CONSTANT IN A VARIANT OF HARDY'S INEQUALITY AND CONCLUDING REMARKS

We consider a nonincreasing and nonnegative function $x = x(t)$ on $(0,\infty)$. First we recall the following variant of Hardy's inequality:

If $0 < p,r,s < \infty$ *and* $0 < b < 1/p$, *then there exists a finite constant* $K = K(p,r,s,b)$, *such that*

$$(8.1)\quad \int_0^\infty t^{-sb+s/p}\left(\int_t^\infty (x(u)u^b)^r \frac{du}{u}\right)^{s/r}\frac{dt}{t} \le K\int_0^\infty (x(t)t^{1/p})^s\frac{dt}{t}.$$

A proof of (a weighted version of) this estimate can be found in [28] (see also [27]). The best possible constant is not known in general. However, the following result was recently proved by J. Bergh [5]:
If $0 < p < 1$, *then*

$$(8.2)\quad \int_0^\infty\left(\frac{1}{t}\int_t^\infty x(u)du\right)^p dt \le \frac{\pi p}{\sin\pi p}\int_0^\infty (x(t))^p dt.$$

The constant $\pi p/\sin\pi p$ *cannot be improved.*

Here we present the following generalization of this statement:

PROPOSITION 9 *If* $r = 1$, *then the best possible constant in* (8.1) *is equal to*

$\frac{1}{b^s}\frac{s}{p}B\left(\frac{s}{p} - bs, 1+bs\right)$. *Equality occur if* $x(t) \equiv \chi_{[0,c]}$ *for any* $c > 0$.

Proof The proof of Proposition 9 can be carried out by only making obvious modifications of the proof by J. Bergh so we omit the details.

Remark For the case $b = 1$ also V.I. Burenkov [6] has independently (and with a quite different proof) obtained the result in Proposition 9 (and, thus, Bergh's result). For the case $s = p$, $b = 1$, Proposition 9 coincides with Bergh's result.

Some new information concerning Generalized Gini means have been obtained in [24], [25] and [30]. Here we mention only the following result which we partly already have used previously in this paper:

PROPOSITION 10 (Peetre-Persson) *Let* $-\infty < \alpha,\beta < \infty$ *and consider the generalized Gini means* $G(\alpha,\beta;x) = G_A(\alpha,\beta;x)$. *Then*

(a) $G(\alpha,\beta;x) = \exp\frac{1}{\alpha-\beta}\int_\beta^\alpha \ln G(a,a;x)da.$

(b) $G(\alpha,\beta;x)$ *is non-decreasing in* α *(*β *fixed) and in* β *(*α *fixed).*

(c) $G(\alpha,\beta;x)$ *is continuous in* α *and* β.

A proof of the representation formula in (a) can be found in [25]. One simple proof of (b) (and, thus, of a generalized AG-inequality) only consists of using the representation formula and a generalized form of Schwarz´ inequality (see [25]).

Here we only remark that it is in many cases easy to obtain appropriate limiting cases of known inequalities by using these continuity and monotonity properties of Generalized Gini means. For example by replacing $|x(t)|$ by $|x(t)|^p$ in Hardy´s inequality (1.8) by considering $G_A\left(\frac{1}{p},0;x\right)$ with $A(x) = \int_0^t x(u)du$ we obtain in this way the usual *Carleman´s inequality*

$$\int_0^\infty \exp\left(\frac{1}{t}\int_0^t \ln|x(u)|du\right)dt \le e\int_0^\infty |x(t)|dt$$

as a limiting case (of (1.8)) as $p \to \infty$. In a similar way we find that if $x(t)$ has its support on $[0,1]$, then the inequality

$$\int_0^1 \ln\frac{x(t)}{1-t}dt \le \int_0^1 \ln x'(t)\, dt$$

may be regarded as a limiting case of (8.2) as $p \to 0$ (for a simple direct proof of this inequality see also [5]). Moreover, the limiting case of (1.7) for $\alpha = \beta = 1$ (and positive $x(t)$ and $y(t)$) reads:

$$\exp\left(\frac{\int_\Omega (x+y)\ln(x+y)d\mu}{\int_\Omega (x+y)d\mu}\right) \le \exp\left(\frac{\int_\Omega x\ln x d\mu}{\int_\Omega x d\mu}\right) + \exp\left(\frac{\int_\Omega y\ln y d\mu}{\int_\Omega y d\mu}\right)$$

and the corresponding inequality for the other extremal case $\alpha = \beta = 0$ reads:

$$\exp\left(\frac{1}{\mu(\Omega)}\int_\Omega \ln(x+y)d\mu\right) \ge \exp\left(\frac{1}{\mu(\Omega)}\int_\Omega \ln x d\mu\right) + \exp\left(\frac{1}{\mu(\Omega)}\int_\Omega \ln y d\mu\right).$$

As special cases of these inequalities we obtain the following inequalities for positive sequences:

$$\left(\prod_1^n (x_k+y_k)^{x_k+y_k}\right)^{1/\sum_1^n (x_k+y_k)} \le \left(\prod_1^n x_k^{x_k}\right)^{1/\sum_1^n x_k} + \left(\prod_1^n y_k^{y_k}\right)^{1/\sum_1^n y_k}$$

and

$$\left(\prod_1^n (x_k+y_k)\right)^{1/n} \geq \left(\prod_1^n x_k\right)^{1/n} + \left(\prod_1^n y_k\right)^{1/n}$$

respectively. For another proof of the last inequality see also [3,p.26].

It is well-known that several inequalities can be proved by using interpolation. Some new results in this connection concerning classical inequalities and its applications can be found in [16] and [19]. Finally we remark that we in our sections 2 and 3 have given examples of applications of our general inequalities in Theorems 1 and 2 only for the extremal cases. Sigrid Sjöstrand, Lund (personal communication) has recently pointed out to me some new applications also in "intermediate" cases. However, this new interesting possibility seems not to be fully investigated yet.

REFERENCES

[1] J. Aczél and E.F. Beckenbach, *On Hölder's inequality*, General inequalities 2, Proceedings second international conf. Oberwolfach (1978), Birkhauser, Basel, 1980, 145-150.

[2] E.F. Beckenbach, *A class of mean-value functions*, Am. Math. Monthly 57, 1950, 1-6.

[3] E.F. Beckenbach and R. Bellman, *Inequalities*, Springer Verlag, 1983.

[4] R.E. Blahut, *Principles and Practice of Information Theory*, Addison-Wesley, 1987.

[5] J. Bergh, *Hardy's inequality-A complement*. Math. Z. 202, 1989, 147-149.

[6] V.I. Burenkov, *Function Spaces : Main integral inequalities connected with the spaces* L^p, Moscow publishing House of the University of friendship of Nations, 1989.

[7] J.M. Borwein and P.B. Borwein, *Pi and* the AGM, Wiley, 1987.

[8] P.S. Bullen, D.S. Mitrinovic and P.M. Vasic, *Means and Their Inequalities*, Reidel Publ., 1988.

[9] P.S. Bullen, D.S. Mitrinovic and P.M. Vasic, *Recent Advances in Geometric Inequalities*, Kluwer Acad. Publ., 1989.

[10] J.A. Clarkson, *Uniformly convex spaces*, Trans. Amer. Math. Soc. 40, 1936, 396-414.

[11] F. Cobos, *Clarcson's inequalities for Sobolev spaces*, Math. Japonica 31, No 1, 1986, 17-22.

[12] F. Cobos and D.E. Edmonds , *Clarcson's Inequalities, Besov Spaces and Triebel-Sobolev spaces*, Zeitschrift für Analysis und ihre Anwendungen 7, 1988, 229-232.

[13] M. Dresher, *Moment Spaces and inequalities*, Duke Math. J. 20, 1953, 261-271.

[14] G.H. Hardy, J.E. Littlewood and G. Pólya, *Inequalities*, Cambridge University Press, 1978.

[15] T.Koski and L.E. Persson, *On quantizer distortion and the upper bound for exponential entropies* , IEEE Transactions on Information Theory (to appear).

[16] T.Koski and L.E. Persson, *On quantization of data and interpolation*(to appear)
[17] T.Koski and L.E. Persson, *Some properties of generalized entropies with applications to compression of data,* Journal of Information Sciences (to appear).
[18] L. Maligranda and L.E. Persson, *Generalized duality of some Banach Function spaces,* Indagationes Math., Vol. 92, No. 3, September 1989, 323-338.
[19] L. Maligranda and L.E. Persson, *On some* L^p*-inequalities and interpolation*(in preparation).
[20] L. Maligranda and L.E. Persson, *Convexity and Inequalities* (in preparation).
[21] A.W. Marshall and I. Olkin, *Inequalities: Theory of Majorization and Its Applications,* Academic Press, 1979.
[22] J. Matkowski, *A convexity condition generalizing Hölder's and Minkowski's inequalities and symmetric norms in* R^n, To appear in Acq. Math.
[23] L.I. Nikolova and L.E. Persson, *On interpolation between* X^p*-Spaces,*In: Function Spaces, Differential operators and Nonlinear Analysis, Pitman's Research Notes in Math., Ser. 211, 1989, 89-107.
[24] L.I. Nikolova and L.E. Persson, *Some properties of* X^p*-Spaces.* Research report 5, Dept of Math., Luleå University, 1989. Teubner Texte zur Mathematik series (to appear during 1990).
[25] J. Peetre and L.E. Persson, *A general Beckenbach's Inequality with Applications,* In: Function Spaces, Differential Operators and Nonlinear Analysis, Pitman's Research Notes in Math., Ser. 211, 1989, 125-139.
[26] J.E. Pecaric and P.R. Beesack, *On Jessen's Inequality for Convex Functions* 2, J. Math. Anal. Appl. 118, 1986, 125-144.
[27] L.E. Persson, *Interpolation with a parameter function,* Math. Scand. 59, 1986, 199-222.
[28] L.E. Persson, *On a weak-type theorem with applications,* Proc. London Math. Soc. (3) 38, 1979, 295-308.
[29] L.E. Persson, *Some elementary inequalities in connection with* X^p*-Spaces,* Publishing House of the Bulgarian Academy of Sciences, 1988, 367-376.
[30] L.E. Persson and S. Sjöstrand, *On generalized Gini means and scales of means ,* Research report 1, Dept. of Math., Luleå University, 1990. Submitted.
[31] L. Råde and B. Westergren, *BETA β, Mathematics Handbook,* Studentlitteratur, 1988.

ASPECTS OF NONLINEAR SCATTERING THEORY

Gary F. Roach, Glasgow

1. Introduction

Scattering theory means different things to different people. Broadly speaking it can be thought of as the study of the interaction of an evolutionary process with a nonhomogeneous medium. Certainly, it has played a central role in mathematical physics over the years with perhaps the earliest investigation of such phenomena being attributed to Leonard da Vinci who studied the scattering of light into the geometrical shadow of an opaque body. Subsequently other scattering processes have been discovered and often utilised in such varied fields as acoustics, quantum mechanics and medical diagnosis.

If u, the total field describing the evolutionary process, is thought of as the sum of u_i, an incident field, and u_s, a scattered field, then two classes of problem present themselves. First there is the so called direct scattering problem in which the aim is to determine u_s from a knowledge of u_i. the equations governing the process and the various auxiliary conditions, such as boundary and initial conditions, which might be imposed on the process. The second and perhaps more challenging class of problems is the inverse scattering problem in which the intention is to determine the nature of the inhomogeneous medium from a, possibly incomplete, knowledge of the asymptotic form of u_s, that is to reconstruct the governing equation and its domain of definition from such data.

The above description is rather simplified and covers a vast range of physical and mathematical details. Clearly it would be impossible to do justice to them all in these lectures. The intention therefore will be to give a survey of the field, adopting a unified approach and concentrating mainly on direct scattering problems. Consequently, much of the material could be well known, particularly to those to whom it is well known. Nevertheless, it is hoped that these lectures will offer a sufficiently persuasive introduction for those whose primary research interests lie elsewhere to begin working in this very active area.

Mathematically, scattering processes can be realised as a consequence of perturbing either the symbolic form of an associated operator (so called potential scattering) or its domain (so called target scattering) or indeed both. A first step in developing a framework for investigating nonlinear problems is to examine and extend as far as possible the theory developed for corresponding linear problems. For this reason to begin with two typical problems in linear scattering theory are considered. First an initial value problem arising in potential scattering is examined; this will introduce many of the techniques used later. Next the rather more intuitively acceptable problem of the scattering of an acoustic wave by a rigid target is discussed; this initial boundary value problem offers different challenges and gives a first indication of the profound effect which the geometry of the medium can have on a propagating wave profile. With this introduction nonlinear scattering processes are then examined.

The results and ideas presented in the following sections are based on the research of a large number of authors. In this connection we would particularly mention the contributions of Kato, [4], Kuroda [6], Ikebe [2] and Wilcox [16] to linear scattering and of Reed and Simon [12], Segal [13] and Strauss [15] for nonlinear problems together with the references cited in these various works.

2. Potential scattering

As motivation for much of the detailed discussions which will follow consider the evolutionary process governed by the following model.

$$u_t(x,t) - u_x(s,t) = 0, \quad x \in \mathbb{R} = (-\infty.\infty), \quad t>0 \tag{2.1}$$

$$u(x,0) = f(x), \quad x \in \mathbb{R} \tag{2.2}$$

It will be convenient to analyse this problem in a Hilbert setting. The evolutionary process governed by (2.1) - (2.2) is then described by a curve $t \to u(.,t) \equiv u(t) \in H$ and has an associated operator realisation: determine $u \equiv u(.,t) \equiv H$ satisfying

$$u_t + iA_0u = 0, \quad u \in D(A_0) \tag{2.3}$$

$$u(.,0) \equiv u(0) = f \in H \tag{2.4}$$

$$A_0: H \to H \equiv L_2(\mathbb{R})$$

$$A_0u := L_0u := -i\frac{\partial u}{\partial x}, \quad u \in D(A_0)$$

$$D(A_0) = \{u \in H: L_0u \in H\}.$$

Proceeding formally, (2.3) has a solution in the form

$$u(x,t) = \exp(-itA_0)f(x) =: (U_tf)(x). \tag{2.5}$$

For such a solution to be of any use an interpretation of U_t is required. To this end the properties of the Fourier Transform in $\mathbb{R}$

$$(Fu)(k) := \hat{u}(k) = \frac{1}{\sqrt{2\pi}}\int_{\mathbb{R}} e^{-ikx}\, u(x)dx \tag{2.6}$$

indicate that

$$(F(A_0u))(k) = k\hat{u}(k) \tag{2.7}$$

$$(F(U_tu))(k) = \exp(-itx)\hat{u}(k) \tag{2.8}$$

$$(U_tf)(x) = \frac{1}{\sqrt{2\pi}}\int_{\mathbb{R}} \exp(ik(x-t))\hat{f}(k)dk = f(x-t). \tag{2.9}$$

Consequently we see that

$$u(x,t) = (U_tf)(x) = f(x-t)$$

a solution which could have determined directly by inspection of (2.1), (2.2). However, with a view to generalisations we notice

(i) F diagonalises A_0 in the sense that F enables A_0 on u(x) to be replaced by multiplication of k on $\hat{u}(k)$.

(ii) A_0 is symmetric on H and consequently possesses self adjoint extensions.

(iii) A_0 is unbounded on H.

(iv) The spectrum of $A_0 := \sigma(A_0) = \sigma_c(A_0)$ = the continuous spectrum of A_0 $= (-\infty,\infty)$.

Now consider, instead of (2.3) - (2.4), the problem

$$v_t + iAv = 0, \quad v \in D(A), \quad v \equiv v(.,t) = v(t) \tag{2.10}$$

$$v(.,0) \equiv v(0) = g \tag{2.11}$$

where

$$A : H \to H \equiv L_2(\mathbb{R}) \tag{2.12}$$

$$Av = Lv$$

$$D(A) := \{v \in H : Lv \in H\}.$$

Here L may be regarded as perturbation of L_0 involving additional terms. Preceding formally (2.11) - (2.12) has a solution in the form

$$v(x,t) = \exp(-itA)g(x) := (V_t g)(x). \tag{2.13}$$

However, in general, since V_t depends crucially on the nature of L it is not so easy to determine as U_t. Nevertheless, the solution u of (2.3) - (2.4) and v of (2.11) - (2.12) can be compared provided the following assumptions hold.

<u>A1</u>: A is symmetric on it and has self adjoint extensions.

<u>A2</u>: A and A_0 are unitarily equivalent.

In this case a possible comparison is afforded by examining

$$\|v(.,t)-u(.,t)\|_H = \|v-u\|_H = \|V_t g - U_t f\|_H = \|g - V_t^* U_t g\| =: \|g - \Omega(t) f\|_H. \tag{2.14}$$

To remove the time dependence we take the limit as $t \to \pm\infty$ to obtain

$$\lim_{t\to\pm\infty} \|v-u\|_H = \lim_{t\to\pm\infty} \|g-\Omega(t)f\|_H = \|g-\Omega_\pm f\|_H \tag{2.15}$$

where in arriving at (2.15) we have assumed

$\underline{A3}$: $\Omega_\pm f \equiv \Omega_\pm(A,A_0)f := \lim_{t\to\pm\infty} \Omega(t)f$

exists as an element of H for all $f \in H$.

Thus in terms of strong limits we can write

$$\Omega_\pm := \text{s-}\lim_{t\to\pm\infty} \Omega(t), \quad \Omega(t) := V_t^* U_t. \tag{2.16}$$

To fix ideas even further consider the specific case when

$$Lv(x,t) := -i\frac{\partial v}{\partial x}(x,t) + q(x)v(x,t) = (L_0+q(x)))v(x,t) \tag{2.17}$$

where q is real valued and bounded.

Introduce

$$W : H \to H \tag{2.18}$$

$$Wf(x) := \exp(-i\rho(x))f(x), \quad f \in H$$

where ρ is a sufficiently smooth function yet to be determined.

Clearly W is a unitary operator and direct calculation yields

$$A = WA_0W^{-1}$$

that is, A and A_0 are unitarily equivalent. Further calculation indicates

$$(FW^{-1}Af)(k) = k(FW^{-1}f)(k). \tag{2.19}$$

Consequently, introducing the "generalised" Fourier transform

$$T := FW^{-1} \tag{2.20}$$

we see that T replaces A on f by multiplication of k on $\tilde{f} =: Tf$, that is T diagonalises A whereas F alone was required to diagonalise A_0.

With

$$U_t := \exp(itA_0), \quad V_t := \exp(itA) \tag{2.21}$$

the quantities $\Omega_\pm$ can be computed to yield

$$(\Omega_\pm f)(x) = \lim_{t\to\pm\infty} (V_tU_tf)(x) = \lim_{t\to\pm\infty} (\Omega(t)f(x)$$
$$= \exp\left\{i\int_x^{\pm\infty} q(\xi)d\xi\right\}f(x) \tag{2.22}$$

and we see that in this particular case $\Omega_\pm$ are operators of multiplication by the quantity $\exp\left\{i\int_x^{\pm\infty} q(\xi)d\xi\right\}$. Furthermore they are unitary operators and provide the result

$$A = \Omega_{\pm} A_0 \Omega_{\pm}^{-1}. \qquad (2.23)$$

Moreover, direct calculation shows that

$$F\Omega_{\pm}^{-1} Af = kF\Omega_{\pm}^{-1} f. \qquad (2.24)$$

Setting

$$G := F\Omega_{\pm}^{-1}$$

we see from (2.24) that

$$GAf = kGf \qquad (2.25)$$

which implies that G is a "generalised" Fourier transform which diagonalises A.

Interesting though this example may be a more systematic approach is required in order to tackle more general problems especially those involving boundary as well as initial conditions. However, before moving in this direction we notice from the above example that in any such approach we shall be required to perform at least the following steps.

(a) Provide an operator realisation of the scattering process in which A_0 and A are self adjoint.
 If A were not self adjoint then V_t in (2.13) would not be unitary. Consequently the quantities $\Omega_{\pm}$ would not provide a unitary equivalence between A_0 and A and the above method would fail to provide a means of obtaining the spectral properties of A from those of A_0.

(b) Define the spectrum of a self adjoint operator and investigate the existence of its various subdivisions.

(c) Establish the spectral properties of A_0.

(d) Prove the existence of the quantities $\Omega_{\pm}$ and investigate their properties.

(e) Obtain the spectral properties of A.

If this programme can be carried through then expansion theorems can be established for problems involving A_0 and A. In addition information regarding the asymptotic behaviour of solutions can be obtained and compared.

With the above example as a guide it would appear that evolutionary processes can conveniently be studied in terms of

A "free" system modelled by

$$u_t = iA_0 u, \quad u \in D(A_0) \subset H$$
$$u(0) = f \in H \tag{2.26}$$

and

A "perturbed" system modelled by

$$v_t = -iAv, \quad v \in D(A) \subset H$$
$$v(0) - g \in H \tag{2.27}$$

where

u,v = state (vectors) of the system

f,g = initial state (vectors) of the system

A_0,A = linear, self-adjoint operators on a Hilbert space H.

For definiteness we shall think of the operators as having the symbolic forms

$$A_0 = -\Delta \qquad A = -\Delta + V$$

where V, the perturbation of A_0, will be referred to as a potential. A reason for this terminology and also for certain other concepts introduced later can be given by considering a particle moving in one direction, interacting with an obstacle and being deflected to move off in another direction. If the obstacle were not present the motion (evolution) of the particle would be governed by the free system (2.26) whilst in the presence of an obstacle its evolution is governed by the perturbed system (2.27). A characteristic feature of scattering in this sense is that if the history of the particle is traced back in time then it moves further from the obstacle (the scattering centre). Therefore it seems plausible that it ought to behave more like a free particle the further back in time we go. Mathematically this can be expressed by requiring that there should exist $f_- \in H$ such that the free process, as governed by (2.26),

$$u(t) = e^{-itA_0} f_-$$

and the perturbed scattering process, as governed by (2.27),

$$v(t) = e^{-itA} g$$

should satisfy

$$\lim_{t \to -\infty} \| e^{-itA} g - e^{-itA_0} f_- \|_H = 0 \qquad (2.28)$$

We shall regard (2.28) as the mathematical expression for the intuitive idea that originally the particle moves freely.

Similarly we can argue that eventually the particle should also move freely, a fact that we can express mathematically by requiring that there should exist $f_+ \in H$ such that

$$\lim_{t \to +\infty} \| e^{-itA} g - e^{itA_0} f_+ \|_H = 0. \qquad (2.29)$$

In an actual experiment involving the scattering of a particle it is very difficult, if indeed possible, to work with the initial condition g, Instead a particle is prepared a long way from the obstacle in what is considered to be a good approximation to free motion. It is then allowed to evolve in the given system and, after a long time, it is again detected in a state which is regarded to be a good approximation to free motion. Therefore the quantities in the experiment with which there are direct contact are

the prepared state: $e^{-itA_0} f_-$

the detected state: $e^{-itA_0} f_+$

Thus the initial condition $f_- \in H$ in (2.26) determines the initial free motion of the particle and $f_+ \in H$ in (2.26) determines the detected or eventual free motion of the particle.

Scattering theory is concerned with the problem of being able to predict what will be measured knowing what had been prepared. Mathematically this means scattering theory must indicate the conditions under which each $f_- \in H$ uniquely determines an $f_+ \in H$ and whenever possible provide a method for computing f_+ given f_-. In many cases it turns out that for a suitable class of potentials this problem is solvable in the form

$$f_+ = Sf_- \tag{2.30}$$

where S, the scattering operator, is unitary. In this case it is convenient to discuss the problem in two parts.

Part 1:

Given $f_- \in H$ determine $g \in H$ such that

$$\lim_{t \to -\infty} \|e^{-itA}g - e^{-itA_0}f_-\| = \lim_{t \to -\infty} \|V_t g - U_t f_-\| = 0.$$

This means that for any $f_- \in H$ we expect to be able to prepare $e^{-itA_0}f_-$, an initial state of the system, which initiates an evolutionary process in the system, of the form $e^{-itA}g$, which converges strongly to the initial state $e^{-itA_0}f_-$ as $t \to -\infty$. Of course the question remains as to whether or not there exists a state $e^{-itA}g$ which agrees asymptotically with $e^{-itA_0}f_-$ as $t \to -\infty$. An argument suggesting that this should be the case can be given by considering again the motion of a particle; for large enough time (in the past) the particle is in a region of space where the effects of the obstacle are negligible and the motion of the particle can be satisfactorily approximated by the state function $e^{-itA_0}f_-$.

Part 2:

Given $g \in H$, determined in Part 1, find $f_+ \in H$ such that

$$\lim_{t \to +\infty} \|e^{-itA}g - e^{-itA_0}f_+\| = \lim_{t \to +\infty} \|V_t g - U_t f_+\| = 0.$$

This implies that the evolutionary process, $e^{-itA}g$, defined by the quantity g determined in Part 1 as a consequence of an initial or free process $e^{-itA_0}f_-$ once again becomes a free process described by $e^{-itA}f_+$.

Whether or not the problems presented in Part 1 and Part 2 are solvable depends on the nature of the potential V. Results in this direction have been obtained by Cook [1] and Kuroda [6,7].

The arguments used above for describing particle motions will be used for more general evolutionary processes. Consequently we introduce

<u>Definition 2.1</u> (Asymptotic Condition)

As $t \to \pm\infty$ perturbed states behave, in the sense described in Parts 1 and 2, as free states.

<u>Definition 2.2</u> (Scattering states)

Scattering states are elements $g \in H$ such that the Asymptotic Condition holds.

We notice that if $g \in H$ is an eigenvector of A then there exists $\mu \in \mathbb{R}$ such that $Ag = \mu g$. Consequently

$$v = V_t g = \exp(-itA)g = \exp(it\mu)g$$

and v is simply a multiple of the eigenvector g and as such describes a bound (non-scattering) state. Therefore it is to be expected that scattering states are associated with $\sigma_c(A)$.

<u>Definition 2.3</u>

A family of operators $\{U_t\}$ defined on a Hilbert space H by

$$U_t : \mathbb{R} \to B(H), \qquad t \in \mathbb{R}$$

and satisfying

(i) Strong Continuity:

$$s - \lim_{\tau \to 0}(U_{t+\tau} - U_t) = 0, \qquad t \in \mathbb{R}$$

(ii) Unitarity:

$$U_t^* = U_t^{-1}, \qquad t \in \mathbb{R}.$$

(iii) Group property:

$$U_t U_s = U_s U_t = U_{t+s}, \qquad s,t \in \mathbb{R}$$

$$U_0 = I$$

is called a strongly continuous, one parameter, unitary group of operators. We shall also need the result

Theorem 2.4 (Stone's Theorem) [14]

Let

(i) $\{U_t\}$, $-\infty < t < \infty$, a strongly continuous, one parameter unitary group of operators on a Hilbert space H.

(ii) A be a linear operator (an infinitesimal generator of $\{U_t\}$) satisfying

(a) $D(A) := \{f \in H: \text{s-lim}[i\tau^{-1}(U_\tau I)f]$ exists as $\tau \to 0$

(b) $Af = \text{s-lim}[i\tau^{-1}(U_\tau - I)f], \quad f \in D(A).$

Then

(iii) D(A) is dense in H

(iv) A is self adjoint.

Corollary 2.5

Let

(i) $\Psi_t(\lambda) := \exp(-it\lambda), \quad \lambda, t \in \mathbb{R}.$

(ii) $A: H \to H$, densely defined, self adjoint.

then $\{\Psi_t\}$, $-\infty < t < \infty$ is a strongly continuous, one-parameter unitary group with infinitesimal generator A.

Furthermore we shall assume that with each operator $A : H \to H$ there can be associated a set of scattering states, M(A), satisfying

(i) $M(A) \subset H$ is a subspace.

(ii) M(A) is invariant with respect to the group $\{-itA)\}$.

We can now say that scattering is associated with

an "initial" state $f_- \in M(A_0)$ at $t = 0$

a "final" state $f_+ \in M(A_0)$ at $t = 0$

The asymptotic condition together with the property of unitarity implies

$$\lim_{t\to -\infty} \|V_t g - U_t g_-\| = \lim_{t\to -\infty} \|g - V_t^* U_t f_-\| = 0 \tag{2.31}$$

and this requires that

$$\text{s} - \lim_{t\to -\infty} V_t^* U_t$$

should exist on $M(A_0)$.

A similar result involving f_+ and the limit $t \to +\infty$ also obtains. This leads to

<u>Definition 2.6</u> (Wave operators)

$$\Omega_{\pm} \equiv \Omega_{\pm}(A,A_0) := s - \lim_{t\to\pm\infty} V_t^* U_t P(A_0)$$

where

$$P(A_0) : H \to M(A_0) \qquad P(A_0) = P(A_0)^* = P(A_0)^2.$$

The concept of a scattering process introduced above can be conveniently thought of as trajectories in H as illustrated in the diagram.

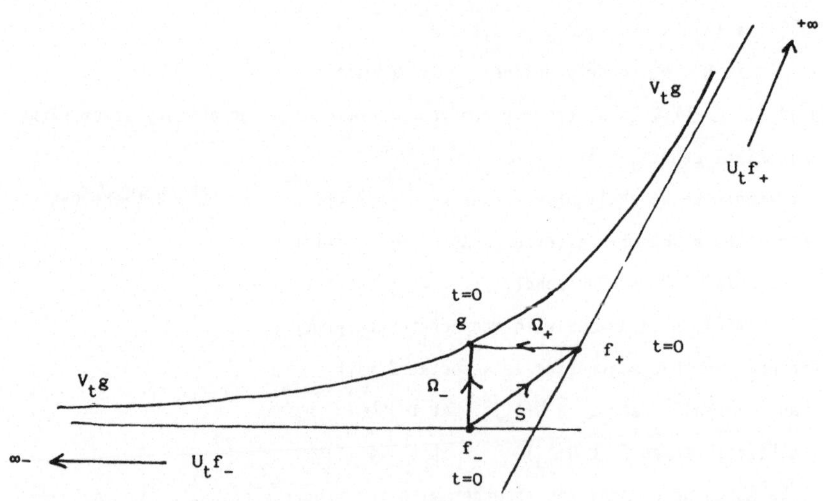

Straight lines = free evolutions.

Curved line = perturbed evolution.

The asymptotic condition, of which (2.31) is a typical part indicates that the wave operator $\Omega_-(\Omega_+)$ applied to $f_- \in M(A_0)$ at $t = 0$ yields an initial state $g \in M(A)$ at $t = 0$ for the perturbed system which will evolve from a prepared state $[\exp(-itA_0)]f_-$ in the remote past and which will converge, in the distant future, to the state $[\exp(-itA_0)]f_+$.

Conditions ensuring the existence of the wave operators $\Omega_\pm$ can be found in the works of Cook [1], Jauch [3], Kato [4] and Kuroda [7]. Furthermore, it is sometimes possible to prove that the wave operators have identical ranges, that is

$$R(\Omega_+) = R(\Omega_-).$$

In this case the wave operators are said to be asymptotically complete. Results in this connection can be found in Kato [4], Kato and Kuroda [5].

3. Target scattering

In this section we consider the scattering of an acoustic wave by a rigid obstacle. This process can be modelled by the d'Alembert equation

$$\left[\frac{\partial^2}{\partial t^2} - \Delta\right]u(\underset{\sim}{x},t) = 0, \quad \underset{\sim}{x} \in \mathbb{R}^n \tag{3.1}$$

together with initial and boundary conditions as appropriate. This particular problem has been comprehensively analysed by Wilcox [16] and much of the presentation here is based on this work.

We consider first the case when the target is absent; that is, the free problem. This will be analysed in $L_2(\mathbb{R}^n)$ with the usual structure denoted by $(.,.)$. It will be convenient to introduce

$$L_2^m(\mathbb{R}^n) := L_2(\mathbb{R}^n) \cap \{u : D^\alpha u \in L_2(\mathbb{R}^n), |\alpha| \le m\} \tag{3.2}$$

$$L_2(\Delta,\mathbb{R}^n) := L_2(\mathbb{R}^n) \cap \{u : \Delta u \in L_2(\mathbb{R}^n)\} \tag{3.3}$$

the latter being endowed with the inner product

$$(u,v)_\Delta := (u,v) + (\Delta u,\Delta v) \tag{3.4}$$

where $(.,.)$ denotes the usual $L_2(\mathbb{R}^n)$ inner product.

If we define

$$A_0 : L_2(\mathbb{R}^n) \to L_2(\mathbb{R}^n) \tag{3.5}$$

$$Au = -\Delta u, \quad u \in D(A_0) = L_2(\Delta,\mathbb{R}^n)$$

then the free problem can be cast in the form: determine $u_0(.,t) \equiv u_0 \in L_2(\mathbb{R}^n)$ satisfying

$$(D_0^2+A_0)u_0 = 0, \quad t \in \mathbb{R}, \quad D_0 = \partial/\partial t \tag{3.6}$$

$$u_0(0) = f \in L_2(\mathbb{R}^n)$$

$$D_0u_0(0) = g \in L_2(\mathbb{R}^n).$$

Formally, this has a solution

$$u_0(t) = (\cos A_0^{1/2})f + (A_0^{-1/2}\sin t\ A_0^{1/2})g. \tag{3.7}$$

This solution is interpreted by means of the spectral theorem since

Theorem 3.1 [16]

The operator A_0 is self adjoint and non-negative on $L_2(\mathbb{R}^n)$.

When the scattering obstacle is immersed in the incident field boundary conditions have to be allowed for.

Let $B \subset \mathbb{R}^n$ denote the open region occupied by the target and let $\Omega = \mathbb{R}^n\backslash\bar{B}$ denote the open region exterior to B and $\partial\Omega$ its boundary. We shall assume that $\bar{B} = B \cup \partial\Omega$ is bounded and that $\partial\Omega$ is smooth.

The target scattering (perturbed) problem will be analysed in $L_2(\Omega)$ endowed with the usual inner product. Furthermore we introduce

$$L_2^m(\Omega) := L_2(\Omega) \cap \{u:D^\alpha u \in L_2(\Omega), \quad |\alpha|\le m\} \tag{3.8}$$

$$L_2(\Delta,\Omega) := L_2(\Omega) \cap \{u:\Delta u \in L_2(\Omega)\} \tag{3.9}$$

$$L_2^1(\Delta,\Omega) := L_2^1(\Omega) \cap L_2(\Delta,\Omega) \tag{3.10}$$

with the following inner products

$L_2^m(\Omega)$: $\qquad (u,v)_m := \sum_{|\alpha|\le m} (D^\alpha u,D^\alpha v)$

$L_2(\Delta,\Omega)$: $\qquad (u,v)_\Delta := (u,v) + (\Delta u,\Delta v)$

$L_2^1(\Delta,\Omega)$: $\qquad (u,v)_{1,\Delta} := (u,v)_1 + (\Delta u,\Delta v)$

where

$$(u,v) = \int_\Omega u(x)\bar{v}(x)dx$$

is the usual inner product in $L_2(\Omega)$.

An element $u \in L_2^1(\Delta,\Omega)$ is said to satisfy a generalised Neumann condition if

$$\int_\Omega \{v(x)\Delta u(x) + \nabla v(x).\nabla v(x)\}d\tau = 0, \qquad v \in L_2^1(\Omega). \tag{3.11}$$

Consequently we shall define

$$L_2^N(\Delta,\Omega) := L_2^1(\Delta,\Omega) \cap \{u\colon u \text{ satisfies } (3.11)\} \tag{3.12}$$

Furthermore if we define

$$A\colon L_2(\Omega) \to L_2(\Omega) \tag{3.13}$$

$$Au = -\Delta u, \qquad u \in D(A) := L_2^N(\Delta,\Omega)$$

then the perturbed problem can be stated in the form: determine $u(.,t) \equiv u(t) \in L_2(\Omega)$ satisfying

$$(D_0^2+A)u = 0 \quad \text{in } \Omega, \quad t \in \mathbb{R} \tag{3.14}$$

$$u \in D(A) = L_2^N(\Delta,\Omega)$$

$$u(0) = f \in L_2(\Omega), \quad D_0u(0) = g \in L_2(\Omega).$$

Corresponding to Theorem 3.1 there can be established

<u>Theorem 3.2</u> [16]

A is a non negative, self-adjoint operator on $L_2(\Omega)$. Consequently a solution of (3.14) written in the form

$$u(t) = \{\cos tA^{1/2})f + (A^{-1/2}\sin t\, A^{1/2})g \tag{3.15}$$

can be made meaningful by using the spectral theorem to interpret the solution operators $(\cos t\, A^{1/2})$ and $(A^{-1/2}\sin t\, A^{1/2})$. However, by virtue of the generality of this theorem it is often the case that very little specific information can be obtained about the wave function $u(x,t)$. This can be overcome by developing an eigenfunction expansion for A. Then, if required the relation between the spectral family and eigenfunctions of A can be developed.

With a view to developing eigenfunction expansions we notice that

$$w_0(x,p) := \frac{1}{(2\pi)^{n/2}} e^{ix.p} \qquad ,x \in \mathbb{R}^n \tag{3.16}$$

satisfies

$$(\Delta+|p|^2)w_0(x,p) = 0, \quad x,p \in \mathbb{R}^n.$$

Hence

$$A_0 w_0(x,p) = -\Delta w_0(x,p) = |p|^2 w_0(x,p) \tag{3.17}$$

and we conclude that $w_0(x,p)$ in (3.16) is an eigenfunction of A_0. However it is a generalised eigenfunction since

$$w_0(.,p) \notin L_2(\mathbb{R}^n) \tag{3.18}$$

Nevertheless, the Plancherel theory of Fourier Transforms yields the following results:

$$\hat{f}(p) = \lim_{M\to\infty} \int_{|x|\leq M} \overline{w_0(x,p)}\, f(x)dx =: (Ff)(p) \tag{3.19}$$

$$f(x) = \lim_{M\to\infty} \int w_0(x,p)\hat{f}(p)dp =: (F^*\hat{f})(x) \tag{3.20}$$

$$\Phi(A_0)f(x) = \lim_{M\to\infty} \int_{|p|\leq M} w_0(x,p)\Phi(|p|^2\hat{f}((p)dp \tag{3.21}$$

where the limit must be interpreted in the $L_2(\mathbb{R}^n)$ sense. Thus we have obtained a spectral representation for A_0 as required.

The quantity $w_0(x,p)$ can also be interpreted as a steady state solution of the d'Alembert equation

$$\left[\frac{\partial^2}{\partial t^2} - \Delta\right]u(x,t:|p|) = 0 \tag{3.22}$$

if we assume a time dependence in the form

$$u(x,t:|p|) = e^{i|p|t}w_0(x,p). \tag{3.23}$$

Thus $w_0(x,p)$ can be interpreted as a plane wave propagating in the direction p.

For the perturbed problem a corresponding eigenfunction for the operator A would then be a steady wave $w(x,p)$ which is produced when the obstacle B is immersed in the plane wave $w_0(x,p)$. Mathematically this means that $w(x,p)$ should satisfy

$$(\Delta+|p|^2)w(x,p) = 0, \qquad x \in \Omega \tag{3.24}$$
$$w \in L_2^N(\Delta,\Omega).$$

In this Neumann problem w represents the total wave and is assumed to have the form

$$w(x,p) = w_0(x,p) + w'(x,p) \tag{3.25}$$

where

w_0 = the incident wave

w' = the scattered wave.

The scattered wave w' will be required to satisfy either an outgoing or an incoming radiation condition. Depending on which radiation condition is employed the total wave w, also called a distorted plane wave, will be called appropriately either outgoing or incoming and will be denoted by w_+ and w_- respectively.

Employing the Plancherel theory once again it is possible to introduce generalised Fourier transforms

$$(F_\pm f)(p) = \hat{f}_\pm(p) = \lim_{R\to\infty} \int_{B(R)} \overline{w_\pm(x,p)} f(x)dx \tag{3.26}$$
$$(F_\pm^* \hat{f})(x) = f(x) = \lim_{k\to\infty} \int_{|p|\leq R} w_\pm(x,p)\hat{f}(p)dp \tag{3.27}$$

where

$$B(R) = \Omega \cap \{x:|x|<R\}$$

and the limits are to be interpreted in the $L_2(\Omega)$ sense. Furthermore, it can be shown that if ϕ is Lebesgue measurable then

$$\Phi(A)f(x) = \lim_{R\to\infty} \int_{|p|\leq R} w_\pm(x,p)\Phi(|p|^2)\hat{f}_\pm(p)dp. \tag{3.28}$$

Consequently (3.26) - (3.27) provide the required spectral representation for A.

In order to investigate the asymptotic properties of the free and perturbed waves as $t \to \infty$ it will be convenient to rewrite the solutions (3.7) and (3.15). To this end set

$$h_0 = f_0 + iA_0^{-1/2}g_0, \qquad h = f + iA^{-1/2}g$$
$$v_0(.,t) = \{\exp(-itA_0^{1/2})\}h_0(.), \quad v(.,t) = \{\exp(-itA^{1/2})\}h(.) \tag{3.29}$$

then a straightforward calculation allows the solutions (3.7) and (3.15) to be written in the form

$$u_0(x,t) = R_e(v_0(x,t)) \tag{3.30}$$

$$u(x,t) = R_e(v(x,t)). \tag{3.31}$$

Arguments based on the local decay of energy [16] suggest that all the distorted waves should tend asymptotically to a free wave, that is

$$\lim_{t\to\infty} \| e^{-itA^{1/2}}h - e^{-itA_0^{1/2}}h_0 \|_{L_2(\Omega)} = 0 \tag{3.32}$$

or equivalently, since $h_0 \in L_2(\mathbb{R}^n)$

$$\lim_{t\to\infty} \| J_\Omega e^{-itA^{1/2}}h - e^{-itA_0^{1/2}}h_0 \|_{L_2(\mathbb{R}^n)} \tag{3.33}$$

where

$$J_\Omega : L_2(\Omega) \to L_2(\mathbb{R}^n)$$

$$J_\Omega u(x) = \begin{cases} u(x), & x \in \mathbb{R}^n \\ 0, & x \in \mathbb{R}^n \backslash \Omega. \end{cases}$$

Recognising the unitarity of the operators involved (3.33) can be written

$$\lim_{t\to\infty} \| e^{itA_0^{1/2}} J_\Omega e^{-itA^{1/2}}h - h_0 \|_{L_2(\mathbb{R}^n)} = 0. \tag{3.34}$$

Furthermore, (3.34) is equivalent to requiring that

$$W_+ \equiv W_+(A_0^{1/2}, A^{1/2}, J_\Omega) := s - \lim_{t\to\infty} e^{-itA_0^{1/2}} J_\Omega e^{-itA^{1/2}} \tag{3.35}$$

should exist. W_+ is called the wave operator associated with the triple $(A_0{}^{1/2}, A^{1/2}J_\Omega)$ and its existence is entirely equivalent to the asymptotic condition (3.33) with

$$h_0 = W_+h. \tag{3.36}$$

Similarly, we can define the wave operator

$$W_- \equiv W_-(A_0^{1/2}, A^{1/2}, J_\Omega) := s - \lim_{t\to-\infty} e^{itA_0^{1/2}} J_\Omega e^{-itA^{1/2}} \tag{3.37}$$

The wave operators

$$W_{\pm}: L_2(\Omega) \to L_2(\mathbb{R}^n) \tag{3.38}$$

are unitary and provide the unitary equivalence between A and A_0. Furthermore it can be shown [16] that

$$W_+ = F^* F_-, \qquad W_- = F^* F_+. \tag{3.39}$$

This construction can be carried out for more general equations and more general initial and boundary conditions [8,9].

This we see that in scattering theory there would appear to be four basic problems; one free problem and three perturbed problems with the following typical structure

operator	conditions
A_0	(ic)
$A_1=A_0+V$	(ic)
$B_0=A_0$	(ic)+(bc)
$B_1=A_1$	(ic)+(bc)

and related by means of the various wave operators according to the diagram

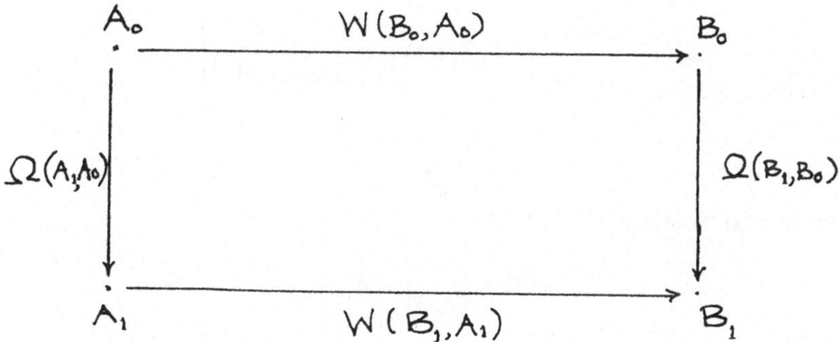

4. Nonlinear scattering theory

As we have seen scattering theories investigate the behaviour in the distant past and in the distant future of a system evolving in time. In practice this is done by comparing the behaviour of the given system with an associated free system. The process is called nonlinear if a system evolves in a nonlinear manner. It turns out that the nonlinear effects can often be quite well accounted for by taking as the free problem an associated linear problem which has been fully analysed by the methods outlined in the previous sections.

As a typical nonlinear system consider the following initial value problem for the nonlinear Klein-Gordon equation.

$$u_{tt}(x,t) - \Delta u(x,t) + m^2 u(x,t) = -q|u(x,t)|^{p-1}u(x,t) \qquad (4.1)$$

$$u(x,0) = f(x)$$

$$u_t(x,0) = g(x)$$

where $x \in \mathbb{R}^n$ and $t \in \mathbb{R}$.

An abstract setting for this problem can be obtained by first writing it as a first order system. Let

$$v(x,t) = u_t(x,t)$$

then (4.1) reduces to

$$\phi'(t) - L\phi(t) = N(\phi(t)) \qquad (4.2)$$

where

$$\phi(t) = \begin{bmatrix} u(x,t) \\ v(x,t) \end{bmatrix}, \qquad N(\phi(t) = \begin{bmatrix} 0 \\ -q|u(x,t)|^{p-1} \end{bmatrix}$$

$$L = \begin{bmatrix} 0 & I \\ \Delta - m^2 & 0 \end{bmatrix}$$

and the initial condition

$$\phi(0) = \begin{bmatrix} f(x) \\ g(x) \end{bmatrix} \equiv \phi_0$$

The operator on $L_2(\mathbb{R}^n)$ defined by

$$B^2 = (-\Delta + m^2)$$

is positive and self-adjoint. Furthermore the operator L is self-adjoint on the Hilbert space

$$H_B = D(B) \times L_2(\mathbb{R}^n) \qquad (4.3)$$

endowed with the scalar product

$$((f_1,f_2),(g_1,g_2))_B := (Bf_1,Bg_1) + (f_2,g_2). \tag{4.4}$$

Now let

$$A := iL \tag{4.5}$$

then it can be verified that A is symmetric on H_B with domain $D(A) = D(B^2) \times D(B)$. Furthermore A is closed since B and B^2 are closed. Therefore the original problem has been reformulated as follows: Find $\phi(t) \in H_B$, $T \in \mathbb{R}$ satisfying

$$\frac{d\phi(t)}{dt} = -iA\phi(t) + N(\phi(t)) \tag{4.6}$$

$$\phi(0) = \phi_0. \tag{4.7}$$

This is the form of the abstract problem with which we shall be mainly concerned in the following sections. A scattering theory for a nonlinear system of the form (4.6) - (4.7) can be developed by comparing its evolution with that of the free system

$$\frac{d\phi(t)}{dt} = -iA\phi(t) \tag{4.8}$$

$$\phi(0) = \tilde{\phi}_0 \tag{4.9}$$

for large positive and negative times. However, the abstract theory is not so satisfactorily developed as in the linear case. Furthermore the applications are not so well understood; like all nonlinear problems the equations must, to a large extent, be dealt with individually in order to make full use of their various special properties. Nevertheless the abstract theory, which originated in the work of Segal [13], does form the basis for much of the recent work in nonlinear scattering theory. As might be expected many problems remain both in the abstract and particular settings.

To develop a scattering theory for a system of the form

$$\phi'(t) = -iA\phi(t) + N(\phi(t)) \tag{4.10}$$

$$\phi(0) = \phi_0 \tag{4.11}$$

the approach will be to reformulate (4.10), (4.11) as an equivalent integral equation. Direct integration yields

$$\phi(t) = e^{-itA}\phi_0 + \int_0^t U(t-s)N(\phi(s))ds. \tag{4.12}$$

We notice

(i) $\phi(t)$ given by (4.12) satisfies (4.10) and (4.11)

(ii) the first term in (4.12) is a solution of the free problem (4.8), (4.9) with data ϕ_0 at $t = 0$.

Similarly (4.10) can be integrated to yield

$$\phi(t) = U(t)\phi_\tau + \int_\tau^t U(t-s)N(\phi(s))ds. \tag{4.13}$$

We notice that

(i) $\phi(t)$ given by (4.13) satisfies (4.10) with data $U(\tau)\phi_\tau$ at $t = \tau$.

(ii) the first term in (4.13) satisfies the free problem (4.8) with data ϕ_τ at $t = 0$.

(iii) from (4.13)

$$\phi(0) := \phi_0 = \phi_\tau + \int_\tau^0 U(-s)N(\phi(s))ds. \tag{4.14}$$

In developing linear scattering theory we looked for conditions under which the solutions $\phi(t)$ of the "perturbed" system were related to the solutions $\psi_\pm(t) = U(t)f_\pm$ under the asymptotic condition

$$\|\phi(t)-\psi_\pm(t)\| = \|\phi(t)-U(t)f_\pm\| \to 0 \text{ as } t \to \pm\infty. \tag{4.15}$$

The same approach will be adopted for nonlinear problems. Motivation for this can be seen by formally taking the limits $\tau = \pm\infty$ in (4.13). This yields the two integral equations

$$\phi(t) = U(t)\phi_\pm + \int_{\pm\infty}^t U(t-s)N(\phi(s))ds. \tag{4.16}$$

We notice

(i) the first term on the right hand side of (4.16) is a "free" solution with data $\phi_\pm$ at $t = 0$.

(ii) for t very large negative we have the possibility of approaching the "prepared" state $U(t)\phi_-$.

(iii) for t very large positive we have the possibility of approaching the "eventual" state $U(t)\phi_+$.

The problems centred on (4.16) are as follows:

Let $M(A) \subset H$ denote a set of initial data for (4.16). Determine conditions which ensure

(i) (4.16) has a unique global solution for $\phi_\pm \in M(A)$.

(ii) $\phi(t) \in M(A)$ for each value of t.

(iii) if $\phi(t)$ satisfies

$$\phi(t) = U(t)\phi_- + \int_{-\infty}^{t} U(t-s)N(\phi(s))ds$$

then

$$\|\phi(t) - U(t)\phi_-\| \to 0 \text{ as } t \to -\infty.$$

(iv) there exists $\phi_+ \in M(A)$ such that for $\phi(t)$ as in (iii)

$$\|\phi(t) - U(t)\phi_+\| \to 0 \text{ as } t \to +\infty.$$

(v) the scattering operator

$$S: \phi_- \to \phi_+ \tag{4.17}$$

exists and is (1-1) and continuous.

(vi) the wave operators

$$\Omega_\pm: \phi_\pm \to \phi_0 \tag{4.18}$$

exist and are (1-1).

(vii) range of $\Omega_+ := R(\Omega_+) = R(\Omega_-)$.

This is the so-called asymptotic completeness of the wave operators. Complete results in this connection are known in only a few cases. However, if the initial condition ϕ_- is small enough global existence of the (4.16) can be settled. A scattering theory for the abstract problem (4.10), (4.11) can then be developed provided the solutions $U(t)\phi_-$ of the free problem decay sufficiently rapidly and N has a high enough degree.

Typical result which can be obtained for the problems (i) - (vii) listed above are outlined in the next three sections.

5. Nonlinear scattering with small data

Let H be a Hilbert space with norm $\|.\|$ and A a linear, self adjoint operator on H. Further, let $\|.\|_1$ and $\|.\|_2$ denote two other "norms" defined on H which satisfy all the usual norm axioms except that $\|\phi\|_1 = 0$ need not imply $\phi = 0$ and $\|.\|_2$ might possibly be unbounded. We shall assume

A1: There exist constants $c > 0$, $d > 0$ so that for $\phi \in H$

$$\|\phi\|_1 \leq c\|\phi\| \quad , \quad \phi \in H.$$

A2: There exist constants $c_1 > 0$, $d > 0$ so that for $\phi \in H$

$$\|U(t)\phi\|_1 \leq c_1 t^{-d}\|\phi\|_2 \quad , \ |t| \geq 1.$$

A3: There exist $\beta > 0$, $\delta > 0$ and $q \geq 1$ with $dq > 1$ so that

$$\|N(\phi_1)-N(\phi_2)\| \leq \beta\{\|\phi_1\|_1+\|\phi_2\|_1\}^q\|\phi_1-\phi_2\|$$

$$\|N(\phi_1)-N(\phi_2)\|_2 \leq \beta\{[\|\phi_1\|_1+\|\phi_2\|_1]^{q-1}\|\phi_1-\phi_2\|_1+[\|\phi_1\|_1+\|\phi_2\|_1]^q\|\phi_1-\phi_2\|\}$$

for all $\phi_i \in H$ satisfying $\|\phi_i\| \leq \delta$, $i = 1,2$.

In order to define scattering states and an associated norm we introduce

$$\|\phi\|_3 = \sup_{-\infty \leq t \leq \infty} \{\|\phi(t)\| + (1+|t|)^d\|\phi(t)\|_1\}.$$

We now define (with a slight abuse of notation)

$$M(A) = \{\phi \in H: \|U(t)\phi\|_3 < \infty\}$$

$$\|\phi\|_M = \|U(t)\phi\|_3.$$

Thus we see that the scattering states are those which have a certain acceptable decay under free propagation.

The following results can now be established [12], [13], [15].

Theorem 5.1

Let

(i) A: H → H be a self-adjoint, linear operator.

(ii) N: H → H a nonlinear operator.

(iii) $\|.\|_1$, $\|.\|_2$ be two auxiliary norms on H so that

$$A_1, A_2, A_3 \text{ hold.}$$

Then there exists $\eta_0 > 0$ such that for all $\phi_- \in M(A)$ with $\|\phi_-\|_M \leq \eta_0$, the integral equation

$$\phi(t) = U(t)\phi_- + \int_{-\infty}^{t} U(t-s)N(\phi(s))ds \tag{5.1}$$

has a unique global solution ϕ which is continuous and satisfies $\|\phi\|_3 \leq 2\eta_0$. Furthermore

(a) $\phi(t) \in M(A)$ for each t.

(b) $\|\phi(t) - U(t)\phi_-\| \to 0$ as $t \to -\infty$.

Theorem 5.1 settles global existence for small data. Corresponding results for the scattering operator are given by [15].

<u>Theorem 5.2</u>

Assume all the hypotheses of Theorem 5.1 and let $\phi(t)$ be a solution of (5.1) corresponding to $\phi_- \in M(A)$ with $\|\phi_-\|_M \leq \eta_0$. Then for sufficiently small η_0

(a) there exists $\phi_+ \in M(A)$ with $\|\phi_+\|_M \leq 2\eta_0$ such that

$$\|\phi(t) - U(t)\phi_+\| \to 0 \text{ as } t \to +\infty.$$

(b) $S: \phi_- \to \phi_+$ is (1-1) and continuous, with respect to $\|.\|$, mapping of $\{\psi \in M: \|\psi\|_M \leq \eta_0\}$ into $\{\psi \in M: \|\psi\|_M \leq 2\eta_0\}$.

(c) for $q \neq 1$

$$\|U(-t)\phi(t) - \phi_+\|_M \to 0 \text{ as } t \to +\infty.$$

(d) S is continuous with respect to $\|.\|_M$.

Two aspects of these Theorems should be noted: the hypotheses of the theorems required neither a priori estimates on the solutions of the non linear equations nor the use of "energy" inequalities. All that was required was that the solution of the associated free equation should decay sufficiently rapidly, and that the degree of non linearity in the given problem be sufficiently high.

This development of a scattering theory for small data was initiated by the pioneering work of Segal [13] in his investigations of the Klein-Gordon equation. Subsequently Strauss [15] simplified much of Segal's work and presented an abstract formulation of the problem.

6. Existence of wave operators

Let Y be a Banach space and I an interval of real numbers. We introduce the notation

$L_p(I,Y)$ = Lebesgue space of functions: $I \to Y$

$C(I,Y)$ = space of strongly continuous functions: $I \to Y$

$B(I,Y)$ = space of bounded functions: $I \to Y$

H = Hilbert space with norm $\|.\| = (H,\|.\|)$

H_1 = Hilbert space H with norm $\|.\|_1 = (H,\|.\|_1)$

H_2 = Hilbert space H with norm $\|.\|_2 = (H_1,\|.\|_2)$

Z = $L_{p+1}(\mathbb{R},H_1) \cap B(\mathbb{R},H_1)$.

We also make the following assumptions.

A4: Let G denote a functional which maps a neighbourhood of zero in H_1 into $\mathbb{R}$. The functional G is lower semicontinuous and continuous at the zero element.

A5: Whenever I is a time interval, $\tau \in I$, $f \in H$ and $\phi \in Z$, with $\|\phi\|_Z$ sufficiently small, a solution of

$$\phi(t) = U(t)\phi_\tau + \int_\tau^t U(t-s)N(\phi(s))ds \tag{6.1}$$

then $\phi \in C(I,H)$ with

$$\frac{1}{2}\|\phi(t)\|^2 + G(\phi(t))$$

independent of t.

With this preparation the following result can be stated, [15].

Theorem 6.1

Assume A1 to A5. If $\phi_- \in \mathbb{M} \subset H$ then there exists a time $T > -\infty$ and a unique solution of ϕ of

$$\phi(t) = U(t)\phi_- + \int_{-\infty}^t U(t-s)N(\phi(s))ds \tag{6.2}$$

in the time interval $I := (-\infty,T)$ such that

$$\phi \in C(I,H) \cap L_{p+1}(I,H_1)$$

and

$$\|\phi(t) - U(t)\phi_-\| \to 0 \quad \text{as } t \to -\infty \tag{6.3}$$

$$\frac{1}{2}\|\phi(t)\|^2 + G(\phi(t)) = \frac{1}{2}\|\phi_-\|^2. \tag{6.4}$$

Furthermore the mapping

$$\Omega_- : \phi_- \mapsto \phi(0)$$

is a (1-1) mapping of $M(A)$ into $M(A)$ which is uniformly continuous on compacta in $M(A)$.

Similarly, for $\phi_+ \in M \subset H$

$$\|\phi(t) - U(t)\phi_+\| \to 0 \quad \text{as } t \to +\infty \tag{6.5}$$

$$\frac{1}{2}\|\phi(t)\|^2 + G(\phi(t)) = \frac{1}{2}\|\phi_+\|^2 \tag{6.6}$$

and the mapping

$$\Omega_+ : \phi_+ \to \phi(0)$$

is a (1-1) mapping of $M(A)$ into $M(A)$ which is uniformly continuous on compacta in $M(A)$.

7. Asymptotic completeness

We assume that the hypotheses of the previous two sections hold. Consequently, we can assert that the wave operators

$$\Omega_\pm : \phi_\pm \to \phi(0) \tag{7.1}$$

exist as (1-1) continuous maps of $M(A)$ into itself.

If we can now prove that

$$R(\Omega_+) = R(\Omega_-) \tag{7.2}$$

then it will be possible to define the scattering operator

$$S = (\Omega_+)^{-1}\Omega_- : \phi_- \to \phi_+. \tag{7.3}$$

An indication of how the asymptotic completeness problem posed by (7.2) may be approached can be given as follows.

Let $\phi_0 \in R(\Omega_-)$ then there exists a $\phi_- \in M(A)$ such that $\Omega_-\phi_- = \phi_0$ and a $\phi(t)$ satisfying

$$\phi(t) = U(t)\phi_- + \int_{-\infty}^{t} U(t-s)N(\phi(s))ds \qquad (7.4)$$

so that

$$\phi(0) = \phi_0$$

and

$$\|\phi(t) - U(t)\phi_-\| \to 0 \quad \text{as } t \to -\infty.$$

It is worth remarking that (7.4) can be rewritten in the form

$$\phi(t) = U(t)\left\{\phi_- + \int_{-\infty}^{t} U(-s)N(\phi(s))ds\right\} + \int_0^t U(t-s)N(\phi(s)ds \qquad (7.5)$$

from which we can conclude

$$\phi(0) = \phi_0 = \left\{\phi_- + \int_{-\infty}^{0} U(-s)N(\phi(s))ds. \qquad (7.6)\right.$$

What now has to be done is prove that there exists $\phi_+ \in M(A)$ such that for $\phi(t)$ defined by (7.4)

$$\|\phi(t) - U(t)\phi_+\| = \|U(-t)\phi(t) - \phi_+\| \to 0 \quad \text{as } t \to +\infty \qquad (7.7)$$

the equality following by virtue of the unitarity of $U(t)$.

We now notice that (7.7) holds if and only if

$$\phi_+ = \lim_{t\to+\infty} U(-t)\phi(t) \qquad (7.8)$$

and hence the existence of ϕ_+ is settled.

With a view to establishing (7.8) consider the Cauchy sequence, with respect to t, generated by the right hand side of (7.8). Using A3 we obtain

$$\begin{aligned} \|U(t_1)\phi(t_1) - U(-t_2)\phi(t_2)\| &= \left\|\int_{t_1}^{t_2} e^{isA}N(\phi(s))ds\right\| \\ &\le \int_{t_1}^{t_2} \|N(\phi(s))\|ds \\ &\le \beta\int_{t_1}^{t_2} \|\phi\|_1^q\|\phi\|ds. \end{aligned} \qquad (7.9)$$

In general an upper bound for $\|\phi\|$ appearing in (7.9) can be obtained by means of an energy argument. Therefore, to ensure that the right hand side of (7.9) tends to zero, as required, as $t \to \infty$ we need to be able to show that

$$\int_{-\infty}^{\infty} \|\phi\|_1^q ds < \infty. \tag{7.10}$$

That is we need an a priori estimate on the solutions of the nonlinear equations which indicate that the solutions decay sufficiently rapidly in the $\|.\|_1$ norm as $t \to \pm\infty$. If, in addition, we want to establish the continuity of the scattering operator then we will need to be able to estimate the decay constants. In general no such a priori results are available. However, for particular equations some progress has been made and one of the most complete results is the following due to Morawetz and Strauss [10].

Theorem 7.1

Let

(i) $u(x,t)$ denote the solution of the free problem

$$u_{tt}(x,t) - \Delta u(x,t) + m^2 u(x,t) = 0, \quad x \in \mathbb{R}^3 \tag{7.11}$$

$$u(x,0) = f(x), \quad u_t(x,0) = g(x). \tag{7.12}$$

(ii) F denote the closure of $C_0^\infty(\mathbb{R}^3) \times C_0^\infty(\mathbb{R}^3)$ with respect to the norm $\|.\|_4$ defined by

$$\|(f,g)\|_4 := \sup_t\left\{\int_{\mathbb{R}^3}\left[u_t^2+|\nabla u|^2+m^2u^2\right]dx\right\} + \sup_t \sup_x\left\{|u(x,t)|^2\right\} + \int_{-\infty}^{\infty}\sup_x\left\{|u(x,t)|^2 dt\right\}. \tag{7.13}$$

Then the scattering operator for

$$u_{tt}(x,t) - \Delta u(x,t) + m^2 u(x,t) = -u^3(x,t), \quad x \in \mathbb{R}^3 \tag{7.14}$$

with data (7.12) exists and is a (1-1) continuous mapping of F into itself.

If $(f,g) \in F$ and ∇f has finite energy then the solution $\tilde{u}(x,t)$ of (7.14) with data (f, g) satisfies

$$\|\tilde{u}(x,t)\|_\infty \le c(1+|t|)^{-3/2} \tag{7.15}$$

8. Concluding remarks

This has been a very rapid tour through a technically demanding area. Nevertheless, it is hoped that just the simple statements of some of the results which can be obtained will be sufficiently attractive to persuade people to work in the field. There are considerable opportunities for making significant contributions to the area. For instance in the development of an abstract scattering theory without the restriction of small data and high degree of nonlinearity. This could then be followed by application to specific equations and systems of equations, of particular interest would be the behaviour of the wave and scattering operators as functions of the various parameters. The question of asymptotic completeness will always be a difficult one and any progress in this direction would always be welcome. There are two areas which are perhaps more promising than most for making early contributions. The first concerns the influence of boundary conditions. A considerable amount of work has been done in this direction for linear scattering theory and in this connection we would cite the work of Wilcox [16] and his co-workers. There would seem to be good prospects for performing a similarly detailed investigation for nonlinear scattering problems. The second centres on inverse scattering problems. If the "free" or linear problem is described in terms of a group $V_0(t)$ and the "perturbed" or nonlinear problem is described in terms of a group $V(t)$ then the inverse problem is concerned with the determination of $V(t)$ from a knowledge of $V_0(t)$ and S. There are indications that for nonlinear scattering problems certain results can be obtained sometimes more readily than for linear scattering problems. Results in this direction have been obtained by Morawetz and Strauss [11]. However, there still remains the detailed considerations of boundary effect in this class of problem.

REFERENCES

[1] J. M. Cook: *Convergence to the Moller wave matrix.* J. Math. & Phys. **3** (1957), 82-87.

[2] T. Ikebe: *Eigenfunction expansions associated with the Schrödinger operator and their applications to scattering theory.* Arch. Rational Mech. Anal. **5** (1960), 1-35.

[3] J. M. Jauch: *Theory of the scattering operator.* Helv. Phys. Acta **31** (1958), 127-158.

[4] T. Kato: *Perturbation theory for linear operators.* Springer, Berlin, 1966.

[5] T. Kato and S. T. Kuroda: *A remark on the unitarity property of the scattering operator.* Nuovo Cimento **14** (1959), 1102-1107.

[6] S. T. Kuroda: *On the existence and unitarity property of the scattering operator.* Nuovo Cimento **12** (1959), 431-454.

[7] S. T. Kuroda: *Perturbation of continuous spectra by unbounded operators. I.* J. Math. Soc. Japan **11** (1959), 246-262.

[8] R. Leis and G. F. Roach: *A transmission problem for the plate equation.* Proc. Roy. Soc. Edin. **99A** (1985), 285-312.

[9] R. Leis and G. F. Roach: *An initial boundary value problem for the Schrödinger equation with long range potential.* Proc. Roy. Cos. Lond. **A417** (1988), 353-362.

[10] C. Morawetz and W. Strauss: *Decay and scattering of solutions of a nonlinear relativistic wave equation.* Comm. Pure Appl. Math. **25** (1972), 1-31.

[11] C. Morawetz and W. Strauss: *On a nonlinear scattering operator.* Comm. Pure Appl. Math. **26** (1973), 47-54.

[12] M. Reed and B. Simon: *Methods of modern mathematical physics. Vol. I Functional analysis. Vol. II Fourier analysis and self-adjointness. Vol. III Operator algebras.* Academic Press, New York, 1972.

[13] I. Segal: *Non-linear semigroups.* Ann. Math. **78** (1963), 339-364.

[14] M. H. Stone: *Linear·transformations in Hilbert space and their applications to analysis.* Amer. Math. Soc. Colloq. Publ. **15** (1932).

[15] W. A. Strauss: *Nonlinear scattering theory at low energy.* J. Funct. Anal. **41** (1981), 110-133.

[16] C. H. Wilcox: *Scattering theory for the d'Alembert equation in exterior domains.* Lecture Notes in Mathematics No. **442**, Springer, Berlin, 1975.

The weak Dirichlet and Neumann problem for the Laplacian in L^q for bounded and exterior domains. Applications.

Christian G. Simader

The purpose of these lectures is to present a rather elementary and selfcontained approach to the weak first and second boundary value problem for the Laplacian in L^q where $1 < q < \infty$. These problems are basic for a lot of applications in mathematical physics, like as e.g. Stokes' problem. From the viewpoint of applications it is necessary to consider as well bounded as exterior domains. Our approach rests on two variational inequalities in L^q and a type of regularity argument. The results presented here are part of a joint work with H. Sohr (Paderborn/FRG) [9].

1. Notations. Throughout this paper $G \subseteq \mathbb{R}^n$ $(n \geq 2)$ denotes a domain, i.e. G is open and connected. G is called an exterior domain if G is a domain and if there exists a bounded open set $\emptyset \neq K \subset \mathbb{R}^n$ such that $G = \mathbb{R}^n \setminus \bar{K}$. Without loss of generality we may assume $0 \in K$. If $G \subseteq \mathbb{R}^n$ is a domain we write $\partial G \in C^1$ if the boundary is of class C^1. If A,B are subsets of $\mathbb{R}^n$ we write $A \subset\subset B$ if A and B are open, $\bar{A}$ is compact and $\bar{A} \subset B$. For $x \in \mathbb{R}^n$ and $r > 0$ by $B_r(x)$ we denote the open ball with radius r centered a x. If $x = 0$ we use the abbreviation $B_r := B_r(0)$. If $M \subset \mathbb{R}^n$ is a Lebesgue measurable subset of $\mathbb{R}^n$ by $|M|$ we denote its Lebesgue measure. Let $1 < q < \infty$ and let q' be defined by $\frac{1}{q} + \frac{1}{q'} = 1$, that is $q' = \frac{q}{q-1}$. Observe $(q')' = q$. For a domain $G \subset \mathbb{R}^n$ by $L^q(G)$ we denote the usual (real) Lebesgue space equipped with norm $\|u\|_{L^q(G)} := \|u\|_{q,G} := (\int_G |u(x)|^q dx)^{1/q}$. For $f \in L^q(G)$ and $g \in L^{q'}(G)$ we write $\langle f,g\rangle := \int_G f(x)g(x)dx$. If $f \in (L^q(G))^n$, $g \in (L^{q'}(G))^n$ are vector fields we use the same notation $\langle f,g\rangle := \sum_{i=1}^{n} \langle f_i,g_i\rangle$. Beside the usual space

$$L^q_{loc}(G) := \{f : G \longrightarrow \mathbb{R} : f \text{ measurable in } G \text{ and } f|_K \in L^q(K) \text{ for each } K \subset\subset G\}$$

we use the convenient abbreviation

$$L^q_{loc}(\bar{G}) := \{f: G \rightarrow \mathbb{R} : f \text{ measurable in } G \text{ and } f|_{G\cap B_R} \in L^q(G\cap B_R) \text{ for each } R > 0\}.$$

We write in the sequel $G_R := G \cap B_R$. Observe that for G bounded $L^q_{loc}(\bar{G}) = L^q(G)$.

So the notation $L^q_{loc}(\bar G)$ is interesting only in connection with unbounded domains. In the same sense we use the notation $C^\infty_o(\bar G) := \{\phi|_{\bar G} : \phi \in C^\infty_o(\mathbb{R}^n)\}$. Observe again for G bounded that $C^\infty_o(\bar G) = C^\infty(\bar G)$. For $i = 1,..,n$ by $\partial_i := \frac{\partial}{\partial x_i}$ we denote the partial derivates, $\nabla := \nabla_n := (\partial_1,..,\partial_n)$ denotes the gradient and $\Delta := \Delta_n := \partial_1^2 + .. + \partial_n^2$ the Laplacian. If X is a Banach space by X^* we denote the dual space equipped with norm

$$\|x^*\|_{X^*} := \sup_{0 \neq x \in X} \frac{|x^*(x)|}{\|x\|_X} \quad \text{for } x^* \in X^*.$$

2. <u>Sobolev spaces.</u> For $1 \leq q < \infty$ and a domain $G \subset \mathbb{R}^n$ by $H^{1,q}(G) := \{p \in L^q(G): \partial_i p \in L^q(G),\ i = 1,..,n\}$ we denote the usual Sobolev space equipped with norm $\|p\|_{1,q} := (\|p\|_q^q + \|\nabla p\|_q^q)^{1/q}$ where $\|\nabla p\|_q := (\sum_{i=1}^n \|\partial_i p\|_q^q)^{1/q}$. Observe that this norm is equivalent to the norm $(\int_G |\nabla p(x)|^q dx)^{1/q}$ where $|\nabla p| = (\sum_i (\partial_i p)^2)^{1/2}$. Here $\partial_i p$ denotes the weak (= distributional) derivative of p. For the well known properties of these spaces we refer e.g. to Ne as [6] or Kufner, John, Fu ik [5]. As usual $H^{1,q}_o(G) := \overline{C^\infty_o(G)}^{\|\cdot\|_{1,q}}$. Considering for G bounded the space $H^{1,q}_o(G)$, because of the elementary Poincaré-inequality

$$\text{(2.1)} \qquad \|p\|_q \leq C(G)\ \|\nabla p\|_q \quad \text{for } p \in C^\infty_o(G)$$

a norm being equivalent to $\|p\|_{1,q}$ is defined by $\|\nabla p\|_q$. If G is a bounded domain say with boundary $\partial G \in C^1$ (or if G is convex, see e.g. [4]) then the general Poincaré-inequality

$$\text{(2.2)} \qquad \|p\|_q \leq C(G)\|\nabla p\|_q \quad \text{for } p \in H^{1,q}(G) \text{ with } \int_G p\,dy = 0$$

holds true. Considering the quotient space $H^{1,q}(G)/\mathbb{R}$ (identifying elements whose difference is constant) then again by $\|\nabla p\|_q$ an equivalent norm on $H^{1,q}(G)/\mathbb{R}$ is defined. This procedure is no longer possible for G unbounded. But from the viewpoint of applications we have to use Sobolev spaces equipped with the (order homogeneous) norm $\|\nabla p\|_q$. For this purpose we define for a domain $G \subset \mathbb{R}^n$ with boundary $\partial G \in C^1$ and for $1 \leq q < \infty$

(2.3) $E^q(G) := \{\nabla p : p \in L^q_{loc}(\bar G),\ \nabla p \in L^q(G)^n\}$.

This space is equipped with norm $\|\nabla p\|_q$. Observe that for G bounded we have $E^q(G) = \{\nabla p : p \in H^{1,q}(G)\}$. For technical reasons we need the following lemma (compare [8], Lemma 2.2) admitting an elementary proof using solely the Poincaré-inequality (2.2) for balls:

Lemma 2.1. Let $n \geq 2$ and let $G \subset \mathbb{R}^n$ be a domain. Let $1 \leq q < \infty$ and let a sequence $(p_i) \subset H^{1,q}_{loc}(G)$ be given such that (∇p_i) is a Cauchy sequence in $L^q_{loc}(G)^n$. Then there exists a sequence $(c_i) \subset \mathbb{R}$ and some $p \in H^{1,q}_{loc}(G)$ such that $(p_i - c_i)$ converges in $H^{1,q}_{loc}(G)$ to p. The sequence (c_i) may be chosen independently of q.

Using this lemma and the fact that for $\partial G \in C^1$ we may conclude from $p \in L^1_{loc}(G)$ with $\nabla p \in L^q(G)$ that $p \in L^q_{loc}(\bar{G})$ (see e.g. Necas [6], p. 114), it is not too hard to see that $E^q(G)$ is complete. Since $E^q(G)$ for $1 < q < \infty$ may be regarded as a closed subspace of the reflexive space $L^q(G)^n$ it is reflexive too. So we end with

Theorem 2.2. Let $G \subset \mathbb{R}^n$ be a domain with boundary $\partial G \in C^1$ and let $1 < q < \infty$. Then $E^q(G)$ is a reflexive Banach space. For $q = 2$ $E^2(G)$ is a Hilbert space with inner product $\langle \nabla p, \nabla \phi \rangle$ for $\nabla p, \nabla \phi \in E^2(G)$.

Next we study approximation properties. If G is bounded with $\partial G \in C^1$ from the fact $E^q(G) = \{\nabla p : p \in H^{1,q}(G)\}$ and the classical density result for Sobolev spaces $H^{1,q}(G) = \overline{C^\infty(\bar{G})}^{\|\cdot\|_{1,q}}$ (compare e.g. Ne as [6], p. 67) we immediately derive

Theorem 2.3. Let $G \subset \mathbb{R}^n$ be a bounded domain with boundary $\partial G \in C^1$ and let $1 < q < \infty$. Then

$$(2.4)\quad E^q(G) = \overline{\{\nabla p : p \in C^\infty(\bar{G})\}}^{\|\cdot\|_q}$$

For exterior domains we get

Theorem 2.4. Let $G \subset \mathbb{R}^n$ be an exterior domain with boundary $\partial G \in C^1$ and let $1 < q < \infty$. Then

$$(2.5)\quad E^q(G) = \overline{\{\nabla p : p \in C^\infty_0(\bar{G})\}}^{\|\cdot\|_q}$$

Proof. i) For $k \in \mathbb{N}$ let $R_k := \{x \in \mathbb{R}^n : k < |x| < 2k\}$. Then there is $k_0 \in \mathbb{N}$ such that $R_k \subset\subset G$ for $k \geq k_0$. For $k = 1$, R_1 is a C^1-domain and the Poincaré-inequality (2.2) holds for q with $1 < q < \infty$ and for R_1 with a certain constant $C_1 = C_1(q) > 0$. If $k \in \mathbb{N}$ and $u \in H^{1,q}(R_k)$ with $\int_{R_1} u dy = 0$ then with $p(x) := u(kx)$ for $x \in R_1$ we have $p \in H^{1,q}(R_1)$. Further

$$\int_{R_1} p(x)dx = k^{-n} \int_{R_k} u(y)dy = 0, \qquad \|p\|_{q,R_1} = k^{-n/q}\|u\|_{q,R_k},$$

$\|\nabla p\|_{q,R_1} = k^{1-n/q}\|\nabla u\|_{q,R_k}$. From (2.2) valid for p and R_1 we derive

$$\text{(2.6)} \qquad \|u\|_{q,R_k} \le k\cdot C_1 \|\nabla u\|_{q,R_k} \qquad \text{for } u \in H^{1,q}(R_k) \text{ with } \int_{R_k} u\, dy = 0.$$

ii) Choose $\rho \in C_0^\infty(\mathbb{R}^n)$, $0 \le \rho \le 1$ with $\rho(x) = 1$ for $|x| \le 1$ and $\rho(x) = 0$ for $|x| \ge 2$. For $k \in \mathbb{N}$ put $\rho_k(x) := \rho(k^{-1}x)$. Then supp $\nabla\rho_k \subset R_k$ and $|\nabla\rho_k(x)| \le M\cdot k^{-1}$ where $M = \max_{z\in\mathbb{R}^n} |\nabla\rho(z)|$. Take now some $p : G \to \mathbb{R}$ such that $\nabla p \in E^q(G)$. For $k \ge k_0$ put $h_k(x) := \rho_k(x)(p(x)-c_k)$ where $c_k := |R_k|^{-1} \int_{R_k} pdy$. By the properties of p and ρ_k we see $h_k|_{G_{2R}} \in H^{1,q}(G_{2R})$ where $G_{2R} := G \cap B_{2R}$.
Further $\nabla h_k = \rho_k\nabla p + \nabla\rho_k(p-c_k)$. Clearly $\|\nabla p-\rho_k\nabla p\|_q \to 0$. Since supp $\nabla\rho_k \subset R_k$ we derive from (2.6)

$$\|\nabla\rho_k(p-c_k)\|_{q,G} \le M\cdot k^{-1}\|p-c_k\|_{q,R_k} \le M\cdot C_1\|\nabla p\|_{q,R_k} \to 0$$

for $k \to \infty$. Therefore $\|\nabla p-\nabla h_k\|_{q,G} \to 0$. Since

$p|_{G_{2k}} \in H^{1,q}(G_{2k}) = \overline{C^\infty(\bar G_{2k})}^{\|\cdot\|_{1,q}}$ there exists $\phi_k \in C^\infty(\bar G_{2k})$ such that

$$\|\nabla p-\nabla\phi_k\|_{q,G_{2k}} \le \|p - \phi_k\|_{1,q,G_{2k}} \le k^{-1}.$$

Put $p_k := \rho_k(\phi_k-d_k)$ where $d_k := |R_k|^{-1} \int_{R_k} \phi_k dy$.

Then $p_k \in C_0^\infty(\bar G)$ is vanishing outside G_{2k} and again by (2.6)

$$\|\nabla h_k-\nabla p_k\|_q \le \|\rho_k(\nabla p-\nabla\phi_k)\|_{q,G_{2k}} + M\cdot k^{-1}\|(p-c_k)-(\phi_k-d_k)\|_{q,R_k}$$

$$\le \|\nabla p-\nabla\phi_k\|_{q,G_{2k}} + MC_1\|\nabla p-\nabla\phi_k\|_{q,R_k} \le (1+MC_1)\cdot\frac{1}{k}$$

Altogether we get $\|\nabla p-\nabla p_k\|_{q,G} \to 0$. ■

As an interesting corollary we derive

<u>Corollary 2.5.</u> Let $G \subset \mathbb{R}^n$ be an exterior domain with boundary $\partial G \in C^1$ and let $1 < q < \infty$.

Then

(2.7) $\quad E^q(G) = \overline{\{\nabla p : p \in H^{1,q}(G)\}}^{\|\cdot\|_q}$

Proof. Clearly $\{\nabla p : p \in C_o^\infty(\bar{G})\} \subset \{\nabla p : p \in H^{1,q}(G)\} \subset E^q(G)$ and (2.7) follows from (2.5). ■

As we will see later for exterior domains we have $\{\nabla p : p \in H^{1,q}(G)\} \subsetneqq E^q(G)$. In connection with the Dirichlet problem with homogeneous boundary data we consider for $1 < q < \infty$ and $G \subset \mathbb{R}^n$ an open set the spaces

(2.8) $\quad E_o^q(G) := \overline{\{\nabla\phi : \phi \in C_o^\infty(G)\}}^{\|\cdot\|_q}$

Clearly $E_o^q(G) \subset E^q(G)$. If G is bounded we immediately see by means of (2.1) that $E_o^q(G) = \{\nabla p : p \in H_o^{1,q}(G)\}$. This representation no longer holds for exterior domains. A useful partial substitute for (2.1) is given by

Theorem 2.6. Let $G \subset \mathbb{R}^n$ be an exterior domain, $G = \mathbb{R}^n \backslash \bar{K}$ where $\phi \neq K \subset\subset \mathbb{R}^n$. Suppose $0 \in K$. Let $1 < q < \infty$. Then there is a constant $C = C(q,G,n) > 0$ such that for each $R > 0$ with $K \subset\subset B_R$

(2.9) $\quad \|p\|_{p,G_R} \leq CR^{n/q + 1/q'}\|\nabla p\|_{q,G}$

holds for $p \in C_o^\infty(G)$, where $q' = \frac{q}{q-1}$.

Proof. Since $0 \in K$ and K is open there is $\delta > 0$ such that $B_\delta \subset\subset K$. Let $p \in C_o^\infty(G)$. Then p vanishes in a neighborhood of B_δ. Let $S := \{\zeta \in \mathbb{R}^n : |\zeta| = 1\}$ denote the unit sphere. For $0 \neq x \in \mathbb{R}^n$ write $x = r\zeta$, $r = |x|$, $\zeta = \frac{x}{|x|} \in S$. Then

$$p(x) = p(r\zeta) = p(r\zeta) - p(0) = \int_0^r \sum_{i=1}^n (\partial_i p)(t\zeta)\zeta_i dt$$

and by Hölder's inequality

$$|p(r\zeta)|^q \leq r^{q-1} \int_0^r |\nabla p(t\zeta)|^q dt$$

Since p vanishes in B_δ we get after integrating with respect to $\zeta \in S$ for $0 < \delta < r \leq R$

$$\int_S |p(r\zeta)|^q d\omega_\zeta \le R^{q-1} \int_\delta^R \frac{t^{n-1}}{t^{n-1}} \int_S |\nabla p(t\zeta)|^q d\omega_\zeta dt \le \delta^{1-n} R^{q-1} \|\nabla p\|^q_{q,G_R}$$

Multiplying by r^{n-1} and integrating with respect to $r \in [0,R]$ yields (2.9). ∎

As an immediate consequence we get

Theorem 2.7. Let $G \subset \mathbb{R}^n$ be an exterior domain and let $1 < q < \infty$. Then

(2.10) $\quad E_o^q(G) = \{\nabla p \in E^q(G)$: there exists a sequence $(p_i) \subset C_o^\infty(G)$ such that $\|\nabla p - \nabla p_i\|_q \to 0$ and $\|p-p_i\|_{q,G_R} \to 0$ for each $R > 0\}$

Proof. i) If $\nabla p \in E_o^q(G)$ then by (2.8) there is a sequence $(p_i) \subset C_o^\infty(G)$ such that $\|\nabla p_i - \nabla p_j\|_q \to 0$. By (2.9) (p_i) is a Cauchy sequence in G_k for each fixed $k \in \mathbb{N}$. Denote the $L^q(G_k)$-limit of (p_i) by $p^{(k)}$. Then after eventually changing $p^{(k+1)}$ on a subset $N_k \subset G_k$ of measure zero we may assume $p^{(k+1)}|_{G_k} = p^{(k)}$. So we get a measurable $p : G \to \mathbb{R}$ such that $p|_{G_R} \in L^q(G_R)$ for $R > 0$ and for $\phi \in C_o^\infty(G)$ we conclude

$$\langle p,\partial_j\phi\rangle = \lim_{i\to\infty}\langle p_i,\partial_j\phi\rangle = -\lim_{i\to\infty}\langle\partial_j p_i,\phi\rangle$$

telling us that the distributional gradient of p is given by the L^q-limit of the sequence (∇p_i). That is $\nabla p \in E^q(G)$ and the above approximation property holds.

ii) If conversely ∇p belongs to the set at the right hand side of (2.10), by (2.8) we see $\nabla p \in E_o^q(G)$. ∎

By the Sobolev embedding theorem ([6], p.69, [5], p.282) we see that for any domain $G \subset \mathbb{R}^n$ (bounded or unbounded) and $1 < q < n$ holds

(2.11) $\nabla p \in E_o^q(G) \Rightarrow p \in L^{q^*}(G)$ where $q^* = \frac{nq}{n-q}$.

To study conversely the case $q > n$ we first consider the Morrey-estimate (compare e.g. [1], p.242): Let $G \subset \mathbb{R}^n$ be an open set and $0 < \alpha \le 1$ and let $p \in C_o^\infty(G)$ with the property that there is a constant $M \ge 0$ such that

(2.12) $\quad \int_{G\cap B_r(x_o)} |\nabla p| dx \le M r^{n-1+\alpha}$

holds for all $x_o \in G$ and $r > 0$. Then there is a constant $C = C(n,\alpha) > 0$

independent of p such that for $x_1,x_2 \in G$

(2.13) $|p(x_1)-p(x_2)| \le C \cdot M\, |x_1-x_2|^{\alpha}.$

For last estimate compare in addition e.g. [6], p. 73 or [5], p. 289.
If now $\nabla p \in E_o^q(G)$ with $q > n$ then

$$\int_{G\cap B_r(x_o)} |\nabla p|dx \le \|\nabla p\|_{q,B_r(x_o)} |B_r(x_o)|^{1/q'} \le C_1 \|\nabla p\|_q\, r^{n/q'}$$

where $C_1 = C_1(n,q) > 0$. Since $\frac{n}{q'} = n - \frac{n}{q} = n-1 + (1-\frac{n}{q})$

(2.12) holds with $0 < \alpha := 1 - \frac{n}{q} < 1$. By definition there is a sequence $(p_i) \subset C_o^\infty(G)$ such that $\|\nabla p - \nabla p_i\|_q \to 0$. By (2.9) $\|p-p_i\|_{q,G_k} \to 0$ for each $k \in \mathbb{N}$. So we may select a subsequence again denoted by (p_i) such that $p_i \to p$ a.e. in G. With $M := C_1 \sup_{i\in\mathbb{N}} \|\nabla p_i\|_q < \infty$ (2.13) holds for the p_i and at the end for p and almost all $x_1,x_2 \in G$. After changing p on a set of measure zero (2.13) holds for all $x \in G$. So p is Hölder-continuous with Hölder exponent $\alpha = 1 - \frac{n}{q}$. Suppose now that $G \subset \mathbb{R}^n$ is an exterior domain, $G = \mathbb{R}^n \backslash K$ where $0 \in K \subset\subset \mathbb{R}^n$. Given $\nabla p \in E_o^q(G)$ we may extend p by zero to the whole $\mathbb{R}^n$ leading to $\nabla p \in E^q(\mathbb{R}^n)$. (2.13) holds for this extension too and because of $p(0) = 0$ we get

(2.14) $|p(x)| \le CM\, |x|^{1-n/q}.$

By no means p need to vanish near $x = \infty$, neither pointwise nor in any L^s-mean. Conversely let $\varphi \in C^\infty(\mathbb{R}^n)$, $0 \le \varphi \le 1$, $\varphi(x) = 0$ for $|x| \le R$ and $\varphi(x) = 1$ for $|x| \ge 2R$ where $R > 0$ is such that $K \subset B_R$. Let $q > n$ and $0 < \lambda < 1 - \frac{n}{q}$ and put $p(x) := \varphi(x)|x|^\lambda$. Then p vanishes in a neighborhood of ∂G. Let $\rho \in C_o^\infty(\mathbb{R}^n)$, $0 \le \rho \le 1$, $\rho(x) = 1$ for $|x| \le 1$, $\rho(x) = 0$ for $|x| \ge 2$ and for $k \in \mathbb{N}$ let $\rho_k(x) := \rho(k^{-1}x)$ and put $p_k := \rho_k\, p$. Then $p_k \in C_o^\infty(\mathbb{R}^n)$.
Since $|\nabla p(x)| \le c|x|^{\lambda-1}$ for $|x| \ge 2R$ we see $|\nabla p| \in L^q(G)$ for $q > n$ and $\|\nabla p(1-\rho_k)\|_q \to 0 (k \to \infty)$. Since $|\nabla \rho_k(x)| \le \frac{c}{|x|}$ for $k \le |x| \le 2k$ we see $\|p \cdot \nabla \rho_k\|_q \to 0$ and therefore $\|\nabla p - \nabla p_k\|_q \to 0$, that is $p \in E_o^q(G)$.

3. Some auxiliary tools. First we need some facts on harmonic functions. If $G \subset \mathbb{R}^n$ is an open set and $u \in C^\infty(G)$, $\Delta u = 0$, then we have the two mean value properties: If $x \in G$, $R > 0$ such that $B_R(x) \subset\subset G$, then

(3.1) $u(x) = \frac{1}{\omega_n} \int_S u(x+r\zeta)d\omega_\zeta$ for $0 < r \le R$ where $S = \{\zeta \in \mathbb{R}^n : |\zeta| = 1\}$ and

(3.2) $u(x) = |B_r(x)|^{-1} \int_{B_r(x)} u(y)dy$ for $0 < r \le R$

We consider Friedrichs' mollifier with a r a d i a l depending kernel: $j \in C_0^\infty(\mathbb{R}^n)$, $j(z) = j(|z|)$, $0 \le j(z)$, $j(z) = 0$ for $|z| \ge 1$ and $\int_{\mathbb{R}^n} j(z)dz = 1$ (with suitable $c \ge 0$ choose e.g. $j(z) := c \exp[(1-|z|^2)^{-1}$~ for $|z| < 1$ and $j(z) = 0$ for $|z| \ge 1$). For $\epsilon > 0$ put $j_\epsilon(z) := \epsilon^{-n} j(\frac{z}{\epsilon})$ and for $f \in L^1(G)$ put

(3.3) $f_\epsilon(x) := \int_G j_\epsilon(x-y)\, f(y)dy = (j_\epsilon * f)(x)$.

As is well known (see e.g. [6], p.58 or [5],p.72)

$$f_\epsilon \in C^\infty(\mathbb{R}^n) \text{ and } \|f-f_\epsilon\|_{L^1(G)} \to 0$$

Suppose now that u is harmonic in G and let $x \in G, \epsilon_o > 0$ such that $B_{\epsilon_o}(x) \subset\subset G$. Then introducing polar coordinates $y = x+r\zeta$, $\zeta \in S$, we get for $0 < \epsilon \le \epsilon_o$

$$u_\epsilon(x) = \int j_\epsilon(x-y)u(y)dy = \int_0^\epsilon r^{n-1}\, j_\epsilon(r) \int_S u(x+r\zeta)d\omega_\zeta dr$$

By (3.1) $\int u(x+r\zeta)d\omega_\zeta = \omega_n u(x)$ and

$$\int_0^\epsilon r^{n-1} j_\epsilon(r)dr\omega_n = \int j_\epsilon(z)dz = 1. \text{ Therefore}$$

(3.4) $u_\epsilon(x) = u(x)$ for harmonic u, $B_\epsilon(x) \subset\subset G$.

Last observation admits a rather simple proof of

<u>Theorem 3.1</u> (Weyl's lemma). Let $G \subset \mathbb{R}^n$ be open and $u \in L^1_{loc}(G)$ such that

(3.5) $\int_G u(x)\Delta\phi(x)dx = 0$ for $\phi \in C_0^\infty(G)$

Then u coincides a. e. in G with a harmonic C^∞-function.

<u>Proof.</u> Because there exists a sequence (G_k) with $G_k \subset\subset G$, $G_k \subset\subset G_{k+1}$, $G = \bigcup_{k=1}^\infty G_k$ it suffices to proof the theorem for any $G' \subset\subset G$. Choose a set G" such that $G' \subset\subset G'' \subset\subset G$. Let $\epsilon_o := \frac{1}{2}\min(\text{dist}(G',\partial G''), \text{dist}(G'',\partial G)) > 0$. Then for $y \in G''$ and $0 < \epsilon \le \epsilon_o$ with $\phi(x) := j_\epsilon(y-x)$ we see $\phi \in C_0^\infty(G)$ and therefore by (3.5)

$$0 = \int u(x)\Delta_x j_\epsilon(x-y)dx = \int u(x)\Delta_y j_\epsilon(x-y)dx = \Delta u_\epsilon(y).$$

Therefore u_ϵ is harmonic in G''. Let $0 < \delta \le \epsilon_0$ and $x \in G'$. Then by (3.4)

$$u_\epsilon(x) = u_{\epsilon\delta}(x) = u_{\delta\epsilon}(x) = u_\delta(x)$$

since the convolutions commute. But then for $x \in G'$ $u_\epsilon(x)$ does not depend on ϵ. Since $u \in L^1(G')$ and $\|u-u_\epsilon\|_{L^1(G')} \to 0$ we conclude $u = u_\epsilon$ a.e. in G', proving the theorem since $u_\epsilon \in C^\infty(G')$ and $\Delta u_\epsilon = 0$. ■

An easy consequence is now

Theorem 3.2. Let $1 < q < \infty$. Then

$$(3.6) \qquad L^q(\mathbb{R}^n) = \overline{\{\Delta\phi : \phi \in C_0^\infty(\mathbb{R}^n)\}}^{\|\cdot\|_q}$$

Proof. Denote by M_q the right hand side of (3.6) and suppose $M_q \subsetneq L^q(\mathbb{R}^n)$. By the Hahn-Banach theorem there exists $F^* \in L^q(\mathbb{R}^n)^*$ with $\|F^*\|_* > 0$ and $F^*|_{M_q} = 0$. Since $L^q(\mathbb{R}^n)^* \cong L^{q'}(\mathbb{R}^n)$ isometrically isomorphic $(q' = \frac{q}{q-1})$ there is $f \in L^{q'}(\mathbb{R}^n)$, $\|f\|_{q'} = \|F\|_* > 0$ such that $F(g) = \langle f,g\rangle$ for $g \in L^q(\mathbb{R}^n)$. Since $F|_{M_q} = 0$ we conclude $\langle f,\Delta\phi\rangle = 0$ for $\phi \in C_0^\infty(\mathbb{R}^n)$. By Theorem 3.1 f is harmonic (eventually after change on a set of measure zero) and for $x \in \mathbb{R}^n$ and arbitrary $r > 0$ by (3.2)

$$|f(x)| \le |B_r(x)|^{-1} \int_{B_r(x)} |f(y)|dy \le |B_r(x)|^{-1/q'} \|f\|_{q',\mathbb{R}^n} \to 0$$

for $r \to \infty$. Therefore $f \equiv 0$ contradicting $\|f\|_{q'} > 0$. ■

With the fundamental solution

$$S(z) := \begin{cases} \frac{2}{(n-2)\omega_n} |z|^{2-n} & \text{for } n \ge 3 \\ -\frac{1}{2\pi} \ln|z| & \text{for } n = 2 \end{cases}$$

we have for $u \in C_0^\infty(\mathbb{R}^n)$ the representation

$$(3.7) \qquad u(x) = - \int S(x-y)\Delta u(y)dy$$

This formula is basic to derive L^q-estimates for second derivatives of u via

<u>Theorem 3.3</u> (Calderon-Zygmund estimate). Let $n \geq 2$, $S_n := \{z \in \mathbb{R}^n : |z| = 1\}$ and let $K : S_n \to \mathbb{R}$ be a continuous function with the property $\int_{S_n} K(z)d\omega_z = 0$. Let $1 < q < \infty$, $f \in L^q(\mathbb{R}^n)$ and define for $\epsilon > 0$

$$(T_\epsilon f)(x) := \int_{\{y \in \mathbb{R}^n \,:\, |y-x| \geq \epsilon\}} \frac{K(\frac{x-y}{|x-y|})}{|x-y|^n} f(y)dy$$

Then $Tf := \lim\limits_{\epsilon \to 0} T_\epsilon f$ exists in $L^q(\mathbb{R}^n)$ and there is a constant $C = C(n,q,K) > 0$ such that

(3.8) $\quad \|Tf\|_q \leq C\|f\|_q.$

For a proof see e.g. [1], p.277, [10], p.39

<u>Theorem 3.4.</u> Let $1 < q < \infty$. Then there exists a constant $C = C(n,q) > 0$ such that for $u \in C_0^\infty(\mathbb{R}^n)$

(3.9) $\quad \left(\sum\limits_{j,k=1}^{n} \|\partial_j\partial_k u\|_q^q \right)^{1/q} \leq C\|\Delta u\|_q$

<u>Proof.</u> By partial integration we derive from (3.7)

$$\partial_j u(x) = - \int S(x-y)\, \Delta\partial_j u(y)dy \text{ and therefore}$$

$$\begin{aligned} \partial_k\partial_j u(x) &= - \int \partial_{x_k} S(x-y)\, \partial_j\Delta u(y)dy = \\ &= - \lim_{\substack{\epsilon \to 0 \\ \epsilon > 0}} \int_{\{y : |y-x| > \epsilon\}} \partial_{x_k} S(x-y)\, \partial_j\Delta u(y)dy \end{aligned}$$

Let $\epsilon > 0$ be fixed. Integrating by parts leads to

$$\begin{aligned} I_\epsilon(x) := -\int_{\{y : |y-x| > \epsilon\}} \partial_{x_k} S(x-y)\, \partial_j\Delta u(y)dy &= \int_{\{y : |y-x| = \epsilon\}} \partial_{x_k} S(x-y) \frac{y_j - x_j}{|y-x|} \Delta u(y)dw_y \\ &+ \int_{\{y : |y-x| > \epsilon\}} \partial_{x_k}\partial_{y_j} S(x-y)\, \Delta u(y)dy =: D_\epsilon(x) + T_\epsilon(x) \end{aligned}$$

For $n \geq 2$ and $x \neq y$ we have $\partial_{x_k} S(x-y) = \frac{1}{\omega_n} \frac{y_k - x_k}{|y-x|^n}$

and therefore writing $y = x + \epsilon\zeta$, $\zeta \in S_n$,

$$D_\epsilon(x) = \frac{1}{\omega_n} \int_{S_n} \zeta_k\zeta_j(\Delta u)(x+\epsilon\zeta)d\omega_\zeta$$

Since $u \in C_0^\infty(\mathbb{R}^n)$, D_ϵ has compact support too and is bounded. Since $\lim_{\epsilon\to 0} D_\epsilon(x) = \frac{\Delta u(x)}{n}\delta_{jk}$ we conclude by Lebesgue's theorem

$$(3.10) \qquad \|\Delta u \cdot \frac{\delta_{jk}}{n} - D_\epsilon\|_q \to 0.$$

Further $\partial_{x_k}\partial_{y_j} S(x-y) = \frac{1}{\omega_n}\left[\frac{\delta_{jk}}{|y-x|^n} - n\frac{(y_k-x_k)(y_j-x_j)}{|y-x|^{n+2}}\right]$

Writing $K(z) := \frac{1}{\omega_n}(\delta_{jk} - nz_jz_k)$ for $z \in S_n$ we get

$$T_\epsilon(x) = \int_{\{y\in\mathbb{R}^n:|y-x|>\epsilon\}} \frac{K(\frac{x-y}{|x-y|})}{|x-y|^n}\Delta u(y)dy$$

If $k = j$, then $\omega_n \int_{S_n} K(z)d\omega_z = \int_{S_n} d\omega_z - n\int_{S_n} z_j^2 d\omega_z = 0$

If $k \neq j$, then $\omega_n \int_{S_n} K(z)d\omega_z = -n\int_{S_n} z_jz_k d\omega_z = 0$

By Theorem 3.3 we then derive the existence of the L^q-limit T of T_ϵ and by (3.8) $\|T\|_q \le C\|\Delta u\|_q$. Combining this with (3.10) we are finished. ■

4. Main Theorems. A consequence of the following theorems is the weak solvability of the Dirichlet and of the Neumann problem in L^q for the Laplacian under the assumptions given there.

Theorem 4.1 ("Neumann problem"). Let $n \ge 2$ and let $G \subset \mathbb{R}^n$ be either a bounded or an exterior domain with boundary $\partial G \in C^1$ and let $1 < q < \infty$, $q' := \frac{q}{q-1}$. Then:

a) There exists a constant $C = C(G,q) > 0$ such that

$$(4.1) \qquad \|\nabla p\|_q \le C \sup_{0\neq\nabla\phi\in E^{q'}(G)} \frac{|\langle\nabla p,\nabla\phi\rangle|}{\|\nabla\phi\|_{q'}} \quad \text{for all } \nabla p \in E^q(G)$$

b) For $F^* \in (E^{q'}(G))^*$, $\|F^*\|_{(E^{q'})^*} := \sup_{0\neq\nabla\phi\in E^{q'}(G)} \frac{|F^*(\nabla\phi)|}{\|\nabla\phi\|_{q'}}$

there exists a unique $\nabla p \in E^q(G)$ such that

(4.2) $F^*(\nabla\phi) = \langle\nabla p, \nabla\phi\rangle$ for all $\nabla\phi \in E^{q'}(G)$

and

(4.3) $\|\nabla p\|_q \le C\, \|F^*\|_{(E^{q'})^*} \le C\, \|\nabla p\|_q$

with the same constant $C > 0$ as in (4.1).

Theorem 4.2 ("Dirichlet problem"). Let $n \ge 2$ and $G \subset \mathbb{R}^n$ be either bounded or an exterior domain and assume $\partial G \in C^1$.

a) If G is bounded, let $1 < q < \infty$.
   If G is an exterior domain and if $n \ge 3$ let $1 < q < n$ and if $n = 2$ let $1 < q \le 2$. Then there exists a constant $C = C(G,q) > 0$ such that

(4.4) $$\|\nabla p\|_q \le C \sup_{0 \ne \nabla\phi \in E_o^{q'}(G)} \frac{|\langle\nabla p, \nabla\phi\rangle|}{\|\nabla\phi\|_{q'}}$$

holds for all $\nabla p \in E_o^q(G)$.

b) If G is bounded, let $1 < q < \infty$.
   If G is an exterior domain and if $n \ge 3$, let $\frac{n}{n-1} < q < n$ and if $n = 2$ let $q = 2$. Then for $F^* \in (E_o^{q'}(G))^*$, $q' = \frac{q}{q-1}$, $\|F^*\|_{(E^{q'})^*} := \sup_{0 \ne \nabla\phi \in E_o^{q'}(G)} \frac{|F^*(\nabla\phi)|}{\|\nabla\phi\|_{q'}}$

there exists a unique $\nabla p \in E_o^q(G)$ such that

(4.5) $F^*(\nabla\phi) = \langle\nabla p, \nabla\phi\rangle$ for all $\nabla\phi \in E_o^{q'}(G)$

and

(4.6) $\|\nabla p\|_q \le C\|F^*\|_{(E_o^{q'})^*} \le C\|\nabla p\|_q$

with the constant $C > 0$ from (4.4).

In case $n = q = 2$ for the exterior domain too a) and b) are trivially satisfied by the Frechet-Riesz theorem. If e.g. $n \ge 3$ and $q \ge n$ in case of an exterior domain there is a one-dimensional exceptional space such that (4.4) don't hold. This case has to be treated separately and demands a more detailed analysis.

5. A priori estimates. Roughly spoken the proof of (4.1) resp. (4.4) is based on local estimates of the same type and at the end is performed by a partition of unity. The local estimates are derived from estimates in the whole space (interior estimates) and in the half-space (estimates up to the boundary). The case of the half-space is reduced in both cases to that of the whole space by means of reflection arguments. It turns out that the uniqueness results of Theorem 5.18 are decisive. These in turn are based on certain "regularity" properties, that is, e.g. under the assumptions of Theorem 4.1 we may conclude if $\nabla p \in E^q(G)$ for a q with $1 < q < \infty$ and $<\nabla p, \nabla\phi> = 0$ for all $\nabla\phi \in E^{q'}(G)$, then $\nabla p \in L^2(G)$, analogously for the Dirichlet problem. For this reason we proof in Lemma 5.2 the estimates as well as the regularity property. The idea how part b) in Theorems 4.1 and 4.2 is derived from part a) by purely functional analytic considerations may be read off from Lemma 5.1.

In the following let $G \subset \mathbb{R}^n$ be a domain and $1 < s < \infty$. For i = 0 we write $E^s_0(G)$ (compare 2.8) and for i = 1 let $E^s_1(G) := E^s(G)$.

We say that G has property $P^i_a(s)$ for i = 0 or 1

if there exists a constant $C_s = C(s,G) > 0$ such that

(5.1.s.i) $$\|\nabla p\|_s \le C_s \sup_{0 \ne \nabla\phi \in E^{s'}_i(G)} \frac{<\nabla p, \nabla\phi>}{\|\nabla\phi\|_{s'}},$$

holds for all $\nabla p \in E^s_i(G)$, w here $s' = \frac{s}{s-1}$.

We say that G has the property $P^i_b(s)$ for i = 0 or 1 if

the map $\sigma^i_s : E^s_i(G) \to (E^{s'}_i(G))^*$ defined by $\sigma^i_s : \nabla p \to <\nabla p, .>$ (that is $(\sigma^i_s(\nabla p))(\nabla\phi) = <\nabla p, \nabla\phi>$ for $\nabla p \in E^s_i(G)$ and $\nabla\phi \in E^{s'}_i(G)$) is a bijection and there is a constant $\tilde{C}_s = \tilde{C}(S,G) > 0$ such that for $\nabla p \in E^s_i(G)$

(5.2.s.i) $$\tilde{C}_s \|\nabla p\|_s \le \|\sigma^i_s(\nabla p)\|_{(E^{s'}_i(G))^*} \le \|\nabla p\|_s$$

Lemma 5.1 Let $G \subset \mathbb{R}^n$ be a domain, $1 < q < \infty$ and $q' := \frac{q}{q-1}$. For i = 0 or 1 holds: G has the property $P^i_a(s)$ for s = q and s = q' if and only if G has the property $P^i_b(s)$ for s = q and s = q'.

<u>Proof.</u> We abbreviate $E^s_i := E^s_i(G)$. Observe $(s')' = s$.

i) Suppose G has the property $P^i_a(s)$ for $s = q$ and $s = q'$.

By (5.2.s.i) we conclude for $\nabla p \in E^s_i$

$$(5.3.s.i)\quad C_s^{-1}\,\|\nabla p\|_s \le \sup_{0\ne\nabla\phi\in E^{s'}_i} \frac{|\langle\nabla p,\nabla\phi\rangle|}{\|\nabla\phi\|_{s'}} = \|\sigma^i_s(\nabla p)\|_{(E^{s'})^*} \le \|\nabla p\|_s$$

Therefore $\sigma^i_s(E^s_i)$ is a closed linear subspace of $(E^{s'})^*$. Suppose $\sigma^i_s(E^s_i) \subsetneqq (E^{s'}_i)^*$. By the Hahn-Banach theorem there exists $F^{**} \in (E^{s'}_i)^{**}$ such that $F^{**} \ne 0$ but $F^{**}|_{\sigma^i_s(E^s_i)} = 0$. Since $E^{s'}_i$ may be regarded as a closed subspace of the reflexive space $L^{s'}(G)^n$, it is reflexive too and we may identify $(E^{s'}_i)^{**}$ with $E^{s'}_i$. Then there exists a unique $\nabla\phi \in E^{s'}_i$ such that $F^{**}(F^*) = F^*(\nabla\phi)$ for all $F^* \in (E^{s'}_i)^*$ and $\|\nabla\phi\|_{s'} = \|F^{**}\|_{(E^{s'}_i)^{**}} > 0$.

But for each $\nabla p \in E^s_i$ we then have $0 = (\sigma^i_s(\nabla p))(\nabla\phi) = \langle\nabla p,\nabla\phi\rangle$ and therefore by (5.1.s'.i) we conclude $\|\nabla\phi\|_{s'} = 0$ what is a contradiction.

ii) Suppose conversely that $P^i_b(s)$ holds for $s = q$ and $s = q'$. Then because of $\sigma^i(E^{s'}) = (E^s)^*$ and (5.2.s'.i)

$$\|\nabla p\|_s = \sup_{0\ne F^*\in(E^s_i)^*} \frac{F^*(\nabla p)}{\|F^*\|_{(E^s_i)^*}} \le \sup_{0\ne\nabla\phi\in E^{s'}_i} \frac{|\sigma^i_{s'}(\nabla\phi))(\nabla p)|}{\tilde{C}_{s'}\|\nabla\phi\|_{s'}} = \tilde{C}^{-1}_{s'} \sup_{0\ne\nabla\phi\in E^{s'}_i} \frac{\langle\nabla p,\nabla\phi\rangle}{\|\nabla\phi\|_{s'}}$$

Therefore (5.1.s.i) holds with $C_s = \tilde{C}^{-1}_{s'}$. ■

In the terminology used above e.g. Theorem 4.1 tells that if G is bounded or an exterior domain with $\partial G \in C^1$ then G has property $P^2_a(q)$ and $P^2_b(q)$ for all $1 < q < \infty$. Analogously we may understand Theorem 4.2. The proof of Theorems 4.1 and 4.2 is given via a number of steps. In fact we will prove more than section 4 says. First we show that the whole space and the half-space have property $P^i_a(q)$ for $1 < q < \infty$ and $i = 0$ and $i = 1$ (and by Lemma 5.1 they have property $P^i_b(q)$ too). Then we will prove by a perturbation argument that a sufficiently small "bended" half-space (the "smalness" depends on q) has still property $P^i_a(q)$ for $i = 0, 1$. The following lemma constitutes the basis for all subsequent estimates. Solely in the proof of Lemma 5.2 we need estimate (3.9), a consequence of the Calderon-Zygmund-Theorem. Conversely, Remark 5.3 tells us that (3.9) is equivalent to the assertion of Lemma 5.2. According to (2.3) we have for $1 < s < \infty$ that $E^s(\mathbb{R}^n) = \{\nabla p : p \in L^s_{loc}(\mathbb{R}^n),\ \nabla p \in L^s(\mathbb{R}^n)\}$.

<u>Lemma 5.2.</u> Let $1 < q < \infty$, $1 < r < \infty$ and suppose $\nabla p \in E^r(\mathbb{R}^n)$ and

$$(5.4) \qquad \sup_{0 \neq v \in C_o^\infty(\mathbb{R}^n)} \frac{|\langle \nabla p, \nabla v \rangle|}{\|\nabla v\|_{q'}} < \infty.$$

Then $\nabla p \in E^q(\mathbb{R}^n)$ and there is a constant $C_1 = C_1(n,q) > 0$ such that

$$(5.5) \qquad \|\nabla p\|_q \leq C_1 \sup_{0 \neq v \in C_o^\infty(\mathbb{R}^n)} \frac{|\langle \nabla p, \nabla v \rangle|}{\|\nabla v\|_{q'}}.$$

<u>Proof.</u> For $i = 1,\ldots,n$ we conclude with $C_1 := C(n,q')^{-1}$ and $C(n,q')$ via Theorem 3.4 by means of (3.9)

$$(5.6) \qquad \infty > \sup_{0 \neq v \in C_o^\infty(\mathbb{R}^n)} \frac{|\langle \nabla p, \nabla v \rangle|}{\|\nabla v\|_{q'}} \geq \sup_{0 \neq u \in C_o^\infty(\mathbb{R}^n)} \frac{|\langle \nabla p, \nabla \partial_i u \rangle|}{\|\nabla \partial_i u\|_{q'}} =$$

$$= \sup_{0 \neq u \in C_o^\infty(\mathbb{R}^n)} \frac{|\langle \partial_i p, \Delta u \rangle|}{\|\nabla \partial_i u\|_{q'}} \geq C_1 \sup_{0 \neq u \in C_o^\infty(\mathbb{R}^n)} \frac{|\langle \partial_i p, \Delta u \rangle|}{\|\Delta u\|_{q'}}.$$

From this we conclude that the linear functional $F^*(f) := \langle \partial_i p, f \rangle$ for $f \in M := \Delta u : u \in C_o^\infty(\mathbb{R}^n) \subset L^{q'}(\mathbb{R}^n)$ is well defined and continuous. By Theorem 3.2 M is dense in $L^{q'}(\mathbb{R}^n)$ with respect to $L^{q'}$-norm. Therefore this functional may be uniquely and norm-preserving extended to a continuous linear functional on the whole space $L^{q'}(\mathbb{R}^n)$. Therefore there is a unique $g \in L^q(\mathbb{R}^n)$ such that $\langle \partial_i p, \Delta u \rangle = \langle g, \Delta u \rangle$ for all $u \in C_o^\infty(\mathbb{R}^n)$. From Weyl's lemma (Theorem 3.1) follows that $W := \partial_i p - g$ is harmonic on $\mathbb{R}^n$. For fixed $x \in \mathbb{R}^n$ and $R > 0$ by (3.2)

$$W(x) = |B_R(x)|^{-1}\Big(\int_{B_R(x)} \partial_i p(y)dy - \int_{B_R(x)} g(y)dy\Big)$$

and by Hölder's inequality

$$|W(x)| \leq |B_R(x)|^{-\frac{1}{r}} \|\partial_i p\|_r + |B_R(x)|^{-\frac{1}{q}} \|g\|_q \to 0 \ (R \to \infty).$$

Therefore $\partial_i p = g \in L^q(\mathbb{R}^n)$ and again by Theorem 3.4

(5.7) $$\sup_{0\neq u\in C_o^\infty(\mathbb{R}^n)} \frac{|\langle\partial_i p,\Delta u\rangle|}{\|\Delta u\|_{q'}} = \sup_{0\neq f\in L^{q'}(\mathbb{R}^n)} \frac{|\langle\partial_i p,f\rangle|}{\|f\|_{q'}} = \|\partial_i p\|_{q'}.$$

Combining (5.6) and (5.7) yields (5.5).
It remains to show $p \in L^q_{loc}(\mathbb{R}^n)$. Since $p \in L^r_{loc}(\mathbb{R}^n) \subset L^1_{loc}(\mathbb{R}^n)$ and given any ball $B_R \subset \mathbb{R}^n$ we see for the mollified functions $c_\epsilon := |B_R|^{-1} \int_{B_R} p_\epsilon(y)dy \to c := |B_R|^{-1} \int_{B_R} p(y)dy$. Put $\tilde{p}_\epsilon := p_\epsilon - c_\epsilon$. By (2.2) with a constant $\gamma = \gamma(R,q) > 0$

$$\|\tilde{p}_{\epsilon'}-\tilde{p}_{\epsilon''}\|_{q,B_R} \leq \gamma\|\nabla(\tilde{p}_{\epsilon'}-\tilde{p}_{\epsilon''})\|_{q,B_R} \leq \gamma\|\nabla p_{\epsilon'}-\nabla p_{\epsilon''}\|_q \to 0.$$

Since (c_ϵ) converges in $\mathbb{R}$ we conclude that $(p_{\epsilon'})$ forms a Cauchy-sequence in $L^q(B_R)$ and has the limit $p_1 \in L^q(B_R) \subset L^1(B_R)$ and therefore $\|p_1-p\|_{L^1(B_R)} = 0$. So $p = p_1 \in L^q(B_R)$. ■

<u>Remark 5.3:</u> Suppose $1 < q < \infty$ and (5.5) holds for all $p \in C_o^\infty(\mathbb{R}^n)$. Then (3.9) holds for all $p \in C_o^\infty(\mathbb{R}^n)$ too: Let $1 \leq i \leq n$. Then by (5.5)

$$\|\nabla\partial_i p\|_q \leq C \sup_{0\neq v\in C_o^\infty(\mathbb{R}^n)} \frac{|\langle\nabla\partial_i p,\nabla v\rangle|}{\|\nabla v\|_{q'}} = C \sup_{0\neq v\in C_o^\infty(\mathbb{R}^n)} \frac{|\langle\Delta p,\partial_i v\rangle|}{\|\nabla v\|_{q'}} \leq C\|\Delta p\|_q$$

immediately leading to (3.9). ■

An immediate consequence is the following density property.

<u>Corollary 5.4.</u> Let $1 < q < \infty$. Then $E^\infty(\mathbb{R}^n) := \{\nabla v : v \in C_o^\infty(\mathbb{R}^n)\}$ is dense in $E^q(\mathbb{R}^n)$ with respect to $\|\nabla.\|_q$-norm.

<u>Proof.</u> Suppose $E^\infty(\mathbb{R}^n)$ is not dense in $E^q(\mathbb{R}^n)$. Then there exists $F^* \in (E^q(\mathbb{R}^n))^*$ with $F^*|_{E^\infty} = 0$, $\|F^*\|_{(E^q)^*} > 0$. By Theorem 5.2 we conclude that $\mathbb{R}^n$ has property $P_a^1(s)$ for $s = q$ and q'. Therefore by Lemma 5.1 there exists a unique $\nabla u \in E^{q'}(\mathbb{R}^n)$ with $\|\nabla u\|_{q'} > 0$ such that $F^*(\nabla p) = \langle\nabla u,\nabla p\rangle$ for all $\nabla p \in E^q(\mathbb{R}^n)$. But since $F^*(\nabla v) = 0$ for $\nabla v \in E^\infty(\mathbb{R}^n)$ by Lemma 5.2 we would conclude $\nabla u = 0$ contradicting $\|\nabla u\|_{q'} > 0$. ■

Next we consider the half-space

(5.8) $H := \{x = (x',x_n) \in \mathbb{R}^n : x' \in \mathbb{R}^{n-1},\ x_n < 0\}$.

For $1 < q < \infty$ and for $i = 0$ let $E_o^q(H)$ be defined by (2.8) and let $E_1^q(H) := E^q(H)$.
Further we put

$$(5.10)\quad \begin{cases} E_o^\infty(H) := \{\nabla p : p \in C_0^\infty(H)\} \\ E_1^\infty(H) := \{\nabla p : p \in C_o^\infty(\bar H)\} = \{\nabla p|_{\bar H} : p \in C_o^\infty(\mathbb{R}^n)\}. \end{cases}$$

Given $\nabla p \in E_i^q(H)$ ($i = 0$ or 1) we put

$$(5.11)\qquad p^i(x) := \begin{cases} p(x) & \text{for } x \in H \\ (-1)^{1+i} p(x',-x_n) & \text{for } x_n \geq 0 \end{cases}, \quad i = 0 \text{ or } 1$$

and for $\phi \in C_o^\infty(\mathbb{R}^n)$ we put for $x \in H$ and $i = 0$ or 1

$$(5.12)\quad (T_i\phi)(x) := \phi(x) + (-1)^{1+i}\, \phi(x',-x_n).$$

Lemma 5.5. Let $1 < q < \infty$.

i) If $\nabla p \in E_i^q(H)$ then $\nabla p^i \in E^q(\mathbb{R}^n)$ and

$$(5.13)\quad \begin{cases} \partial_j p^i(x) = \begin{cases} \partial_j p(x) & \text{for } x \in H \\ (-1)^{1+i}\partial_j p(x',-x_n) & \text{for } x_n \geq 0 \end{cases} & \text{for } i = 0 \text{ or } 1 \text{ and } j = 1,..,n-1 \\ \partial_n p^i(x) = \begin{cases} (\partial_n p)(x) & \text{for } x \in H \\ (-1)^i(\partial_n p)(x',-x_n) & \text{for } x_n \geq 0 \end{cases} & i = 0 \text{ or } 1 \end{cases}$$

ii) For $\phi \in C_o^\infty(\mathbb{R}^n)$ we have

a) $\nabla(T_o\phi) \in E_1^\infty(H)$, $(T_o\phi)(x',0) = 0$ and therefore $\nabla(T_o\phi) \in E_o^q(H)$

b) $\nabla(T_1\phi) \in E_1^\infty(H)$

iii) Let $\nabla p \in E_i^q(H)$ ($i = 0$ or 1). Then for $\phi \in C_o^\infty(\mathbb{R}^n)$

$$(5.14)\quad \langle\nabla p^{(i)},\nabla\phi\rangle_{\mathbb{R}^n} = \langle\nabla p,\nabla(T_i\phi)\rangle_H \quad (i = 0 \text{ or } 1)$$

$$(5.15)\quad \|\nabla p\|_{q,H} \leq \|\nabla p^{(i)}\|_{q,\mathbb{R}^n} \leq 2^{1/q}\|\nabla p\|_{q,H} \quad \text{for } \nabla p \in E_i^q(H)\ (i = 1 \text{ or } 2)$$

$$(5.16)\quad \|\nabla(T_i\phi)\|_{q,H} \leq 2\|\nabla\phi\|_{q,\mathbb{R}^n},\ i = 0,1,\ \phi \in C_o^\infty(\mathbb{R}^n)$$

Proof i) (5.13) and the integrability properties follow by elementary calculations. For $\nabla p \in E_o^q(H)$ observe definition (2.8).
ii) Let $\psi := T_o\phi$ then clearly $\psi(x',0) = 0$. Let $\rho \in C^\infty(\mathbb{R})$, $\rho(t) = 0$ for $|t| \leq 1$, $\rho(t) = 1$ for $|t| \geq 2$, $0 \leq \rho \leq 1$. For $k \in \mathbb{N}$ put $\rho_k(t) := \rho(kt)$. Define $\psi_k(x) := \rho_k(x_n)\psi(x)$. Then $\psi_k \in C_o^\infty(H)$.
If for some $R > 0$ supp $\psi_k \subset Z_R := \{x \in \mathbb{R}^n : |x'| \leq R, -R \leq x_n \leq 0\}$, because of $|\psi(x',x_n) - \psi(x',0)| \leq C(\nabla\psi)|x_n|$ we see $|\psi(x)\ \partial_n\rho_k(x_n)| \leq \text{const}(\nabla\psi)$ and since

$\psi\, \partial_n \rho_k$ is vanishing outside $Z_R \cap \{x = (x',x_n) \in \mathbb{R}^n : -\frac{2}{k} < x_n < -\frac{1}{k}\}$ we get $\|\psi \cdot \partial_n \rho_k\|_q \to 0$ $(k \to \infty)$. Therefore we immediately see $\|\nabla\psi - \nabla\psi_k\|_q \to 0$, therefore $\nabla\psi \in E_o^q(H)$. The remaining statements follow immediately by elementary calculations. ■

We need a further lemma seeming not to be obvious.

Lemma 5.6. Let $1 < q,r < \infty$, let $\nabla p \in E_o^r(H)$ and suppose $\nabla p \in L^q(H)$. Then $\nabla p \in E_o^q(H)$.

Proof. i) Let $p \in C^1(\bar{H})$, $p(x',0) = 0$ for $x' \in \mathbb{R}^{n-1}$ and let $\nabla p \in L^q(H)$. For $R > 0$ define $Z_R := \{x = (x',x_n) \in \mathbb{R}^n : -R < x_n < 0\}$. Then for $x \in Z_R$ we get $p(x',x_n) = -\int_{x_n}^{0} (\partial_n p)(x',t)dt$. Applying Hölder's inequality and integrating with respect to $x \in Z_R$ we get

(5.17) $\|p\|_{q,Z_R} \le R\|\partial_n p\|_{q,Z_R}$.

Since (5.17) holds especially for $p \in C_o^\infty(H)$ we derive from the definition of $E_o^q(H)$ that (5.17) is true for $\nabla p \in E_o^q(H)$ too.

ii) Let now $p \in C_o^1(\bar{H})$, $p(x',0) = 0$. Consider ρ_k like as in part ii) of the proof of Lemma 5.5 and put $p_k(x) = \rho_k(x_n)p(x)$. Then $p_k \in C_o^1(H)$, $\partial_i p_k = \rho_k \partial_i p$ for $i = 1,\dots,n-1$ and $\partial_n p_k = \rho_k \partial_n p + p \cdot \partial_n \rho_k$. Clearly $\|\rho_k \partial_i p - \partial_i p\|_{q,H} \to 0$. Since $|\partial_n \rho_k| \le c \cdot k$ and vanishes outside $Z_{2k^{-1}}$ we get from (5.17)

$\|\partial_n \rho_k \cdot p\|_{q,H} = \|\partial_n \rho_k \cdot p\|_{q,Z_{2k^{-1}}} \le c \cdot k \cdot 2k^{-1} \|\partial_n p\|_{q,Z_{2k^{-1}}} \to 0$ and therefore

$\|\partial_n p_k - \partial_n p\|_{q,H} \to 0$.

For $0 < \epsilon < k^{-1}$ we have $p_{k\epsilon} \in C_o^\infty(H)$, $\|\nabla p_k - \nabla p_{k,\epsilon}\|_q \to 0$.
Therefore $\nabla p \in E_o^q(H)$.

iii) Let now $p \in C^1(\bar{H})$, $p(x',0) = 0$, $\nabla p \in L^q(H)$.
Let $\eta \in C_o^\infty(\mathbb{R}^n)$, $0 \le \eta \le 1$, $\eta(x) = 1$ for $|x| \le 1$, $\eta(x) = 0$ for $|x| \ge 2$ and put $\eta_k(x) := \eta(k^{-1}x)$ for $k \in \mathbb{N}$. By ii) $\nabla(\eta_k \cdot p) \in E_o^q(H)$. Clearly $\|\eta_k \nabla p - \nabla p\|_{q,H} \to 0$. Again by (5.17) because of $|\nabla \eta_k(x)| \le Ck^{-1}$

(5.18) $\|p \cdot \nabla \eta_k\|_{q,H} = \|p \nabla \eta_k\|_{q,Z_{2k}} \le Ck^{-1} \cdot 2k \|\nabla p\|_{q,Z_{2k}} \le C\|\nabla p\|_{q,H}$.

Therefore $\|\nabla(\eta_k p)\|_q \le C$ for all k. Clearly $\nabla p \in E^q(H) \supset E_o^q(H)$. We show now $\nabla(\eta_k p) \to \nabla p$ weakly in $E^q(H)$. Since $E_o^q(H)$ is weakly closed too this implies then $\nabla p \in E_o^q(H)$. Let $F^* \in E^q(H)^*$. We may consider $E^q(H)$ as a closed subspace of $L^q(H)^n$ and we may F^* extend normpreserving to an $\bar{F}^* \in (L^q(H)^n)^*$. Then there is $f = (f_1,\dots,f_n) \in L^{q'}(H)$ such that $\bar{F}^*(g) = \langle f,g \rangle_H = \sum_i \int_H f_i g_i$ for $g \in L^q(H)^n$.

Then $F^*(\nabla p-\nabla(\eta_k p)) = \int_H \sum_i f_i(1-\eta_k)\partial_i p - \int_H \sum_i f_i\partial_i\eta_k p$

Clearly the first integral tends to zero. By the properties of η_k the second integral reduces to an integral taken over $R_k \cap H = \{x \in H : k < |x| < 2k\}$ and therefore by (5.18) $\sum_i |\int_H f_i\partial_i\eta_k \cdot p| \le \sum_i \|f_i\|_{q',R_k\cap H}\cdot C\|\nabla p\|_{q,H} \to 0$.

iv) Let now $\nabla p \in E_o^r(H)$ with $\nabla p \in L^q(H)$. By Lemma 5.5 the extended function satisfies $\nabla p^o \in E_o^r(\mathbb{R}^n)$. Consider the mollified $(p^o)_\epsilon$ with radial depending mollifier kernel j_ϵ. Since $\nabla(p^o)_\epsilon = (\nabla p^o)_\epsilon$ we see moreover $\nabla(p^o)_\epsilon \in L^q(\mathbb{R}^n)$ and $\|\nabla p-\nabla(p^o)_\epsilon\|_{p,H} = \|\nabla p^o|_H-\nabla(p^o)_\epsilon|_H\|_{q,H} \le \|\nabla p^o-\nabla(p^o)_\epsilon\|_{q,\mathbb{R}^n} \to 0$.

Observe by the properties of the mollifier and (5.11) for $i = 0$ that

$\nabla(p^o)_\epsilon(x',0) = 0$ for $x'\in \mathbb{R}^n$. By iii) we conclude $\nabla(p^o)_\epsilon|_H \in E_o^q(H)$ and therefore $\nabla p \in E_o^q(H)$ too. ∎

<u>Remark 5.7</u> The linear space

$$E_v^q(H) := \{\nabla p : p \in C^1(\bar H),\ p(x',0) = 0 \text{ for } x'\in \mathbb{R}^{n-1} \text{ and } \nabla p \in L^q(H)\}$$

satisfies $E_o^\infty(H) \subset E_v^q(H) \subset E_o^q(H)$ and is therefore dense in $E_o^q(H)$.

<u>Lemma 5.8</u> Let $1 < q < \infty$, $1 < r < \infty$.

i) Let $\nabla p \in E_o^r(H)$ and

$$d_o := \sup_{0\neq v\in C_o^\infty(H)} \frac{|\langle\nabla p,\nabla v\rangle|}{\|\nabla v\|_{q',H}} < \infty$$

Then $\nabla p \in E_o^q(H)$ and

$$(5.19)\quad \|\nabla p\|_{q,H} \le C_2 \sup_{0\neq v\in C_o^\infty(H)} \frac{|\langle\nabla p,\nabla v\rangle|}{\|\nabla v\|_{q',H}}$$

Here $C_2 = 2C_1$ with C_1 by Lemma 5.2.

ii) Let $\nabla p \in E^r(H)$ and

$$d_1 := \sup_{0\neq v\in C_o^\infty(\bar H)} \frac{|\langle\nabla p,\nabla v\rangle|}{\|\nabla v\|_{q',H}} < \infty$$

Then $\nabla p \in E^q(H)$ and

$$(5.20)\quad \|\nabla p\|_{q,H} \le C_2 \sup_{0\neq v\in C_o^\infty(\bar H)} \frac{|\langle\nabla p,\nabla v\rangle|}{\|\nabla v\|_{q',H}}.$$

<u>Proof</u>. In both cases we have with $p^{(i)}$ ($i = 0$ or 1) by (5.11)and by Lemma 5.5 that $\nabla p^{(i)} \in E^q(\mathbb{R}^n)$. For $\phi \in C_o^\infty(\mathbb{R}^n)$ we have by (5.14) for $i = 0$ or 1

$$\langle\nabla p^{(i)},\nabla\phi\rangle_{\mathbb{R}^n} = \langle\nabla p,\nabla(T_i\phi)\rangle_H$$

For $i = 0$ by Lemma 5.5 part ii) we get $(T_o\phi)(x',0) = 0$. Then with $v := T_o\phi$ we have $v \in C^1_o(\bar{H})$, $v(x',0) = 0$. Like as in part ii) of the proof of Lemma 5.6 consider $v_k(x) := \rho_k(x_n)\cdot v(x)$. Then $v_k \in C^1_o(H)$ and clearly $v_{k\epsilon} \in C^\infty_o(H)$ for $1 < \epsilon < k^{-1}$. Since $\|\nabla v_k - \nabla v_{k\epsilon}\|_{r'} \to 0$ and $\|\nabla v_k - \nabla v_{k\epsilon}\|_{q'} \to 0$ we see $\langle\nabla p, \nabla v_{k\epsilon}\rangle \cdot \|\nabla v_{k\epsilon}\|^{-1}_{q'} \to \langle\nabla p, \nabla v_k\rangle \|\nabla v_k\|^{-1}_{q'}$ since $\nabla p \in L^r(G)$. Analogously

(compare part ii) of proof of Lemma 5.6) we see $\|\nabla v - \nabla v_k\|_{r'} \to 0$, $\|\nabla v - \nabla v_k\|_{q'} \to 0$ and therefore $\langle\nabla p, \nabla v_k\rangle \|\nabla v_k\|^{-1}_{q'} \to \langle\nabla p, \nabla v\rangle \|\nabla v\|^{-1}_{q'}$. From assumption i) we therefore conclude

$$|\langle\nabla p^{(o)}, \nabla\phi\rangle_{\mathbb{R}^n}| \le |\langle\nabla p, \nabla(T_o\phi)\rangle| \le d_o \|\nabla T_o\phi\|_{q',H} \le 2d_o \|\nabla\phi\|_{q,\mathbb{R}^n}.$$

For $i = 1$ analogously we get

$$|\langle\nabla p^{(1)}, \nabla\phi\rangle_{\mathbb{R}^n}| \le 2d_1 \|\nabla\phi\|_{q',\mathbb{R}^n}$$

By Lemma 5.2 we conclude $\nabla p^i \in L^q(\mathbb{R}^n)$.
Since $\nabla p^{(1)}|_H = \nabla p$ we see for $i = 0$ by Lemma 5.6 that $\nabla p \in E^q_o(H)$. For $i = 1$, from $\nabla p \in E^r(H)$ and $\nabla p \in L^q(H)$ we conclude like at the end of the proof of Lemma 5.2 that $\nabla p \in E^q(H)$. Estimates (5.19) and (5.20) then are trivial. Observe that in (5.19) the sup may be taken for $0 \ne \nabla v \in E^{q'}_o(H)$ and in (5.20) for $0 \ne \nabla v \in E^{q'}(H)$. ■

In the next step we consider a "bended" half-space. Let $\omega \in C^1(\mathbb{R}^{n-1})$ and $x' \in \mathbb{R}^{n-1}$. We suppose that there is some $R = R(\omega) > 0$ such that $\omega(x') = 0$ for $|x'| \ge R$. Then we define

(5.21) $\quad H_\omega := \{x = (x',x_n) \in \mathbb{R}^n : x' \in \mathbb{R}^{n-1},\ x_n < \omega(x')\}$

We want now to extend the results of Lemma 5.8 to H_ω. This will be done by a perturbation argument. For technical reasons we need a density result similar to Corollary 5.4.

Lemma 5.9. Let $1 < q < \infty$ and let Ω denote either H_ω or a bounded domain or an exterior domain G with boundary $\partial G \in C^1$. Then $E^\infty(\Omega) := \{\nabla v : v \in C^\infty(\bar{\Omega})\}$ is dense in $E^q(\Omega)$.

Sketch of proof. By well known techniques (see [5], [6]) given $\nabla v \in E^q(\Omega)$ there exists $\nabla\tilde{v} \in E^q(\mathbb{R}^n)$ such that $\nabla\tilde{v}|_\Omega = \nabla v$ and $\|\nabla\tilde{v}\|_q \le \|\nabla v\|_q$. Apply now Corollary 5.4. ■

Lemma 5.10. Let $1 < q < \infty$, $1 < r < \infty$. Then there exists a constant $K = K(q,r,n) > 0$ with the following property. If $\|\nabla\omega\|_\infty := \sup_{x'\in\mathbb{R}^{n-1}} |\nabla\omega(x')| \le K$ then

i) a) there are constants $C(s) = C(s,K,n)$ such that

$$(5.22) \qquad \|\nabla p\|_{s,H_\omega} \le C(s) \sup_{0\ne v\in C_0^\infty(H_\omega)} \frac{|<\nabla p,\nabla v>|}{\|\nabla v\|_{s',H_\omega}}$$

holds for $\nabla p \in E_0^s(H_\omega)$ and $s = q,q',r,r'$ (here $s' = \frac{s}{s-1}$).

b) If $\nabla p \in E_0^r(H_\omega)$ and $D := \sup_{0\ne v\in C_0^\infty(H_\omega)} \frac{|<\nabla p,\nabla v>|}{\|\nabla v\|_{q',H_\omega}} < \infty$

then $\nabla p \in E_0^q(H_\omega)$ and (5.22) holds for $s = q$.

ii) The assertions of i) hold true if $E_0^s(H_\omega)$ is replaced by $E^s(H_\omega)$ and $C_0^\infty(H_\omega)$ is replaced by $C_0^\infty(\overline{H_\omega})$ $(s = r,q)$.

Proof: i) We define $y : \mathbb{R}^n \to \mathbb{R}^n$ by

$$(5.23) \qquad \begin{cases} y_i(x) := x_i & \text{for } i = 1,\dots,n-1 \\ y_n(x) := x_n - \omega(x') \end{cases}$$

then y maps $\mathbb{R}^n$ one-to-one on $\mathbb{R}^n$, $y \in C^1(\mathbb{R}^n)$.
Further $y\big|_{H_\omega} : H_\omega \to H$ is onto, $y(x',\omega(x')) = (x',0)$ that is $y(\partial H_\omega) = \partial H$.

Further $J[y(x)] = 1$. The inverse map is given by $x_i(y) := y_i$ $(i = 1,..,n-1)$ and $x_n(y) := y_n + \omega(y')$. For $p \in C^1(H_\omega)$ we put $\tilde{p}(y) := p(x(y))$ for $y \in H$.

Then $\tilde{p} \in C^1(H)$, $p(x) = \tilde{p}(y(x))$ and

$$(5.24) \qquad \begin{cases} \partial_i p(x) = (\partial_i\tilde{p})(y(x)) - (\partial_n\tilde{p})(y(x))\partial_i\omega(x') & \text{for } i = 1,\dots,n-1 \\ \partial_n p(x) = (\partial_n\tilde{p})(y(x)) \end{cases}$$

and conversely

$$(\partial_i\tilde{p})(y) = (\partial_i p)(x(y)) + (\partial_n p)(x(y))\partial_i\omega(y') \quad \text{for } i = 1,\dots,n-1$$

$$(\partial_n\tilde{p})(y) = (\partial_n p)(x(y))$$

With the aid of Lemma 5.9 we immediately conclude for s with $1 < s < \infty$: $\nabla p \in E^s(H_\omega)$ if and only if $\nabla\tilde{p} \in E^s(H)$ and $\nabla p \in E_0^s(H_\omega)$ if and only if $\nabla\tilde{p} \in E_0^s(H)$. From (5.24) we derive with a constant $d_1(s) = d_1(s,n) > 0$ for $\nabla p \in E^s(H_\omega)$

(5.25) $\|\nabla p\|_{s,H_\omega} \le d_1(s)(1 + \|\nabla\omega\|_\infty)\ \|\nabla\tilde{p}\|_{s,H}$

Let $\phi \in C_o^\infty(\bar{H}_\omega)$ and define $\tilde{\phi}(y) := \phi(x(y))$ for $y \in H$. Then $\tilde{\phi} \in C_o^1(\bar{H}_\omega)$. If $\nabla p \in E^s(H_\omega)$ then define

$$B_\omega[\nabla\tilde{p},\nabla\tilde{\phi}] := -\sum_{i=1}^{n-1}\int_H(\partial_n\tilde{p}(y)\partial_i\tilde{\phi}(y) + \partial_i\tilde{p}(y)\partial_n\tilde{\phi}(y))\partial_i\omega(y)dy$$

$$+\int_H\sum_{i=1}^{n-1}(\partial_i\omega)^2(y)\partial_n\tilde{p}(y)\partial_n\tilde{\phi}(y)dy$$

and therefore with a constant $d_2(s) = d_2(s,n)$

(5.26) $|B_\omega[\nabla\tilde{p},\nabla\tilde{\phi}]| \le d_2(s)\|\nabla\omega\|_\infty(1+\|\nabla\omega\|_\infty)\|\nabla\tilde{p}\|_{s,H}\|\nabla\tilde{\phi}\|_{s',H}$

From (5.24) via the change of variables formula we immediately derive

(5.27) $\langle\nabla p,\nabla\phi\rangle_{H_\omega} = \langle\nabla\tilde{p},\nabla\tilde{\phi}\rangle_H + B_\omega[\nabla\tilde{p},\nabla\tilde{\phi}]$

and therefore by (5.25) for s' and (5.26) for $\nabla\phi \ne 0$

(5.28)
$$\frac{|\langle\nabla p,\nabla\phi\rangle_{H_\omega}|}{\|\nabla\phi\|_{s',H_\omega}} \ge [d_1(s')(1+\|\nabla\omega\|_\infty)]^{-1}\{\frac{|\langle\nabla\tilde{p},\nabla\tilde{\phi}\rangle|}{\|\nabla\tilde{\phi}\|_{s',H}} -$$
$$- d_2(x)\|\nabla\omega\|_\infty(1+\|\nabla\omega\|_\infty)\|\nabla\tilde{p}\|_{s,H}\}.$$

ii) Choose now $K \le 1$ such that $0 < K \le \min\{(4C_2(s)d_2(s))^{-1} : s = q,q',r,r'\}$ with $C_2(s) > 0$ by Lemma 5.8.
If $\nabla p \in E_o^s(H_\omega)$ we then get from (5.19) and (5.28) if $\|\nabla\omega\|_\infty \le K \le 1$

$$\sup_{0\ne\phi\in C_o^\infty(H_\omega)}\frac{|\langle\nabla p,\nabla\phi\rangle|}{\|\phi\|_{q',H_\omega}} \ge (2d_1(s'))^{-1}\{\sup_{0\ne\tilde{\phi}\in C_o^\infty(H)}\frac{|\langle\nabla\tilde{p},\nabla\tilde{\phi}\rangle|}{\|\nabla\tilde{\phi}\|_{s',H}} - 2d_2(s)K\|\nabla\tilde{p}\|_{s,H}\} \ge$$

$$\ge (2d_1(s')^{-1}\{C_2(s)^{-1}\|\nabla\tilde{p}\|_{s,H} - 2d_2(s)\cdot K\|\nabla\tilde{p}\|_{s,H}\}$$

$$\ge (4d_1(s')C_2(s))^{-1}\|\nabla\tilde{p}\|_{s,H} \ge C(s)^{-1}\|\nabla p\|_{s,H}$$

with $C(s) := (8d_1(s')d_1(s)C_2(s))$.

iii) If $\nabla p \in E^q(H)$ and K is chosen like as in ii) then the analogous calculation using now (5.20) leads to (5.22) in that case too.

iv) In order to prove b) let $\nabla p \in E^r(H_\omega)$. We consider first the case $r \ge q$.

We use a cut-off procedure in order to reduce this case to the half-space. Let $R = R(\omega) > 0$ denotes the constant with $\omega(x') = 0$ for $|x'| \geq R$ and choose $R_1 \geq R$ such that $\max_{|x'|\leq R}|\omega(x')| \leq R_1$. Choose $\varphi \in C^\infty(\mathbb{R}^n)$ such that $\varphi(x) = 0$ for $|x| \leq R_1$, $\varphi(x) = 1$ for $|x| \geq 2R_1$ and $0 \leq \varphi \leq 1$.
Let $L := \{x \in \mathbb{R}^n : x \in H,\ R_1 < |x| < 2R_1\} \subset H_\omega$. Given $h \in C_o^\infty(\bar{H})$ we choose $c(h) \in \mathbb{R}$ such that with $\tilde{h} := h+c(h)$ we have $\int_L \tilde{h}dx = 0$. By the properties of φ we get

$$(5.29)\qquad \begin{aligned} <\nabla(\varphi p),\nabla h> &= <\nabla(\varphi p),\nabla\tilde{h}> = \\ &= <\nabla p,\nabla(\varphi\tilde{h})> - <\nabla p,\nabla\varphi\tilde{h}> + <p\nabla\varphi,\nabla\tilde{h}> \end{aligned}$$

By the Poincaré-inequality (2.2) we get with a constant $c_1 > 0$

$$(5.30)\qquad \|\tilde{h}\|_{q',L} \leq c_1\|\nabla\tilde{h}\|_{q',L} \leq c_1\|\nabla h\|_{q'}$$

Since $r \geq q$ we have $\|\nabla p\|_{q,L} \leq c_3\|\nabla p\|_{r,L}$ and therefore

$$(5.31)\qquad |<\nabla p,\nabla\varphi\tilde{h}>| \leq c_4\|\nabla p\|_{r,L}\cdot\|\nabla h\|_{q'}$$

Since $p \in L^r(L)$ and $r \geq q$ we have $p \in L^q(L)$,

$$(5.32)\qquad \begin{aligned} &|<p\nabla\varphi,\nabla\tilde{h}>| \leq c_5\|p\|_{r,L}\cdot\|\nabla h\|_{q'} \\ &\|\nabla(\varphi\tilde{h})\|_{q'} \leq \|\nabla\varphi\tilde{h}\|_{q',L} + \|\varphi\nabla\tilde{h}\|_{q'} \leq c_6\|\nabla h\|_{q'} \end{aligned}$$

By definition $h \in C_o^\infty(\bar{H})$ if h is the restriction of a $C_o^\infty(\mathbb{R}^n)$-function to $\bar{H}$. Therefore $h \in C_o^\infty(\bar{H}_\omega)$. Clearly $(1-\varphi) \in C_o^\infty(\bar{H}_\omega)$. Then $\varphi_o := \varphi h+c(h)(\varphi-1) \in C_o^\infty(\bar{H}_\omega)$ and since $\varphi\tilde{h} = \varphi_o+c(h)$, $\nabla(\varphi\tilde{h}) = \nabla\varphi_o$ and so

$$(5.33)\qquad |<\nabla p,\nabla(\varphi\tilde{h})>| \leq \sup_{0\neq v\in C_o^\infty(\bar{H}_\omega)} \frac{|<\nabla p,\nabla v>|}{\|\nabla v\|_{q'}}\|\nabla(\varphi\tilde{h})\|_{q'} \leq D\cdot c_6\|\nabla h\|_{q'}$$

Therefore we derive from (5.29) - (5.32)

$$\sup_{0\neq h\in C_o^\infty(\bar{H})} \frac{|<\nabla(\varphi p),\nabla h>|}{\|\nabla h\|_{q'}} \leq c_3\|\nabla p\|_{r,L} + c_4\|p\|_{r,L} + c_5 D < \infty$$

Since $\nabla(\varphi p) \in E^r(H)$ by Lemma 5.8, ii) we conclude $\nabla(\varphi p) \in E^q(H)$ and therefore $\nabla(\varphi p) \in E^q(H_\omega)$. Because of $r \geq q$ clearly $\nabla[(1-\varphi)p] \in E^q(H_\omega)$.

v) In order to prove b) in the case $\nabla p \in E_o^r(H_\omega)$ for $r \geq q$ we proceed similar as in iv). We take any $h \in C_o^\infty(H)$ and put $\tilde{h} = h$ in (5.29). By (5.17) we get

$$(5.34)\qquad \|h\|_{q',L} \leq \|h\|_{q',Z_{2R_1}} \leq 2R_1\|\nabla h\|_{q'}$$

If we replace (5.30) by (5.34) we get again (5.31), (5.32). Clearly

$\varphi h \in C_o^\infty(H_\omega)$ and therefore again $|<\nabla p,\nabla(\varphi h)>| \le Dc_5\|\nabla h\|_{q'}$, and

$$\sup_{0\ne h\in C_o^\infty(H)} \frac{|<\nabla(\varphi p),\nabla h>|}{\|\nabla h\|_{q'}} \le c_3\|\nabla p\|_{r,L} + c_4\|p\|_{r,L} + c_5 D < \infty$$

By Lemma 5.8 i) we see $\nabla(\varphi p) \in E_o^q(H)$.

If $(p_k) \subset C_o^\infty(H)$ is a sequence such that $\|\nabla p_k-\nabla(\varphi p)\|_{q,H} \to 0$. Then clearly (observe (5.17)) $\|\nabla(\varphi p_k)-\nabla(\varphi p)\|_{q,H} \to 0$. Since $\varphi p_k \in C_o^\infty(H_\omega)$ we see $\nabla(\varphi p) \in E_o^q(H_\omega)$. Clearly $\nabla((1-\varphi)p) \in E_o^q(H_\omega) \subset E_o^r(H_\omega)$.

vi) By part ii) and iii) of proof H_ω has property $P_a^i(s)$ for i = 0 and 1 and for $s = r,r',q,q'$. Then by Lemma 5.1 H_ω has property $P_b^i(s)$ too. Again we write $E_i^s(H_\omega)$(i = 0 or 1) and $E_o^\infty(H_\omega) := \{\nabla\phi : \phi \in C_o^\infty(H_\omega)\}$, $E_1^\infty(H_\omega) := \{\nabla\phi : \phi \in C_o^\infty(\overline{H_\omega})\}$. We consider now the case $r < q$ in b). This case can be reduced to the previous one. Let now $\nabla p \in E_i^q(H_\omega)$ and $r < q$. According to the assumption in b) by $F^*(\nabla v) := <\nabla p,\nabla v>$ for $\nabla v \in E_i^\infty(H_\omega)$ a linear functional continuous with respect to $\|\nabla.\|_{q'}$-norm is defined on the dense subspace $E_i^\infty(H_\omega)$ of $E_i^{q'}(H_\omega)$. By property $P_b^i(q)$ the unique extension $\hat{F}^*$ of F^* to the whole space $E_i^{q'}(H_\omega)$ may be represented with a uniquely determined $\nabla\hat{p} \in E_i^q(H_\omega)$ in the form $\hat{F}^*(\nabla v) = <\nabla\hat{p},\nabla v>$ for $\nabla v \in E_i^{q'}(H_\omega)$. For $\nabla v \in E_i^\infty(H_\omega)$ we have $<\nabla p,\nabla v> = F^*(\nabla v) = \hat{F}^*(\nabla v) = <\nabla\hat{p},\nabla v>$. Since $\nabla p \in E^r(H_\omega)$ we see

$$\sup_{0\ne\nabla v\in E_i^\infty(H_\omega)} \frac{|<\nabla\hat{p},\nabla v>|}{\|\nabla v\|_{r'}} \le \|\nabla p\|_r$$

Since now $r < q$ we may apply parts iv) and v) of proof (with interchanged meaning of r and q) and conclude $\nabla\hat{p} \in E_i^r(H_\omega)$. Then $<\nabla p-\nabla\hat{p},\nabla v> = 0$ for $\nabla v \in E_i^\infty(H_\omega)$ and by part a) we conclude $\nabla p = \nabla\hat{p}$. For i = 1 we see immediately because of $\nabla\hat{p} \in E^q(H_\omega)$ that $\nabla p \in E^q(H_\omega)$. For i = 0 consider the function $\tilde{p}$ transformed like as in part i). Then $\nabla\tilde{p} \in E_o^r(H)$ and $\nabla\tilde{p} \in L^q(H)$. By Lemma 5.6 we conclude $\nabla\tilde{p} \in E_o^q(H)$ and transforming back $\nabla p \in E_o^q(H_\omega)$. ■

<u>Lemma 5.11.</u> Let $1 < q < \infty$ and let $G \subset \mathbb{R}^n$ be a domain with boundary $\partial G \subset C^1$. Then for every $x_o \in \partial G$ there exists $R = R(x_o,\partial G,q,n) > 0$ and a constant $C = C(q,n,R) > 0$ with the following properties (write $G_R := G \cap B_R(x_o)$):

a) If $\nabla p \in E^q(G)$ then

$$(5.35)\qquad \|\nabla(\varphi p)\| \quad \le C \sup_{\substack{v\in C_0^\infty(B_R(x_0))\\ 0\ne\nabla v \text{ on } G_R}} \frac{|<\nabla(\varphi p),\nabla v>_{G_R}|}{\|\nabla v\|_{q',G_R}}$$

holds for any $\varphi \in C^\infty(B_{R/2}(x_0))$

b) If $\nabla p \in E_0^q(G)$ then (5.35) holds if the sup is taken over all $0 \ne v \in C_0^\infty(G_R)$

c) If $\nabla p \in E^q(G)$ and $<\nabla p,\nabla\phi> = 0$ for all $\phi \in E^{q'}(G)$, then given $1 < s < \infty$, there is $0 < R' < R$, $R' = R'(s)$, such that $\nabla p \in E^s(G_{R'})$.

d) If $\nabla p \in E_0^q(G)$ and $<\nabla p,\nabla\phi> = 0$ for all $\phi \in E_0^{q'}(G)$, then given $1 < s < \infty$, there is $0 < R' < R$, $R' = R'(s)$, such that $\nabla(\varphi p) \in E_0^s(G_{R'})$ for each $\varphi \in C^\infty(B_R)$.

Proof. i) After a translation we may assume $x_0 = 0$. Since $\partial G \in C^1$ there exists a $\rho > 0$ and a function $\sigma \in C^1(\bar{B}_\rho)$ with $(\nabla\sigma)(0) \ne 0$ such that $G \cap B_\rho = \{x \in B_\rho : \sigma(x) < 0\}$ and $\partial G \cap B_\rho = \{x \in B_\rho : \sigma(x) = 0\}$. A local parametrisation of ∂G most adequate to the problem under consideration is found by projecting $\partial G \cap B_\rho$ on the tangential hyperplane of ∂G at $x_0 = 0$. Essentially this is done in the following. Observe that $|\nabla\sigma(0)|^{-1} \nabla\sigma(0)$ equals the exterior unit normal of ∂G at $x_0 = 0$. This procedure enables us to reduce the situation to that of Lemma 5.10. There exists an orthogonal matrix S such that $S[\nabla\sigma(0)] = |\nabla\sigma(0)|e_n$, where $e_n = (\delta_{1n},\dots,\delta_{nn})$. Define $y(x) := Sx$ and put $\hat\sigma(y) := \sigma(S^{-1}y)$ for $y \in B_\rho$. Let $G_\rho := G \cap B_\rho$, $\hat G = SG$ and $\hat G_\rho := \hat G \cap B_\rho$. For $\nabla v \in E^s(G_\rho)$ we put $\hat v(y) := v(S^{-1}y)$ for $y \in \hat G_\rho$ $(1 < s < \infty)$. Then $\nabla\hat v \in E^s(\hat G_\rho)$ and the norms $\|\nabla\hat v\|_{s,\hat G_s}$ and $\|\nabla v\|_{s,G_\rho}$ are equivalent. Clearly $\nabla v \in E^s(G_\rho)$ if and only if $\nabla\hat v \in E_0^s(\hat G_\rho)$. The most important property (reflecting the invariance of Δ under orthogonal transforms) is that if $\nabla p \in E^s(G_\rho)$, $\nabla v \in E^{s'}(G_\rho)$ then $<\nabla\hat p,\nabla\hat v>_{\hat G_\rho} = <\nabla p,\nabla v>_{G_\rho}$. This is seen by a trivial calulation. Because of these properties we may omit in the sequel the distinction between $\hat v$ and v, $\hat G$ and G, $\hat\sigma$ and σ etc. and assume that the above rotation is performed. Since now $\nabla\sigma(0) = |\nabla\sigma(0)|e_n \ne 0$ by the implicit function theorem we find $0 < \rho' \le \rho$, $h > 0$ and a function $\psi \in C^1(\overline{B'}_{\rho'})$, where $B'_{\rho'} := \{y' \in \mathbb{R}^{n-1}: |y'| < \rho'\}$ with the following properties:

If $Z \equiv Z_{\rho',h} := \{y \in \mathbb{R}^n: |y'| < \rho',\ |y_n| < h\}$ then $Z \subset B_\rho$. For $y' \in B'_{\rho'}$ we have $(y',\psi(y')) \in Z$ and $\sigma(y',\psi(y')) = 0$. Further $\psi(0) = 0$, $(\nabla'\psi)(0) = 0$ (where $\nabla' = (\partial_1,\dots,\partial_{n-1})$) and $\partial G \cap Z = \{y \in Z : y_n = \psi(y')\}$, $G \cap Z = \{y \in Z : y_n < \psi(y')\}$. Let $\eta \in C_0^\infty(\mathbb{R}^{n-1})$ such that $\eta(y') = 1$ for $|y'| \le 1$ and

$\eta(y') = 0$ for $|y'| \geq 2$, $0 \leq \eta \leq 1$ else. For $0 < \lambda < \rho'/2$ put $\eta_\lambda(y') := \eta(\lambda^{-1}y')$ and $\omega_\lambda(y') := \eta_\lambda(y')\psi(y')$ for $|y'| \leq \rho$ and $\omega_\lambda(y') = 0$ otherwise. Since $\psi(0) = |\nabla'\psi(0)| = 0$ we get $\sup\{|\nabla'\omega_\lambda(y')| : y' \in \mathbb{R}^{n-1}\} \to 0$ for $\lambda \to 0$. Let now $1 < q < \infty$ be given. Denote by $K_o := K(q,n) > 0$ the constant according to Lemma 5.10. We choose now $0 < \lambda < \rho'/2$ so small that $\|\nabla\omega_\lambda\|_\infty \leq K$ and define H_{ω_λ} according to (5.21). We choose any $0 < R < \lambda$ such that $B_R \subset\subset Z$.

ii) If $\nabla p \in E^q(G)$ then for $\varphi \in C_o^\infty(B_{R/2})$ clearly $\nabla(\varphi p) \in E^q(G_R)$ where $G_R := G \cap B_R$. By the choice of λ and R we have $G_R \subset H_{\omega_\lambda}$ and we may extend φp by zero to H_{ω_λ}. Denoting the extended function again by φp we have $\nabla(\varphi p) \in E^q(H_{\omega_\lambda})$. By Lemma 5.10 a)

$$(5.36)\quad \|\nabla(\varphi p)\| \quad = \|\nabla(\varphi p)\|_{q,H_{\omega_\lambda}} \leq C_o \sup_{0 \neq v \in C_o^\infty(\bar{H}_{\omega_\lambda})} \frac{|<\nabla(\varphi p),\nabla v>|}{\|\nabla v\|_{q',H_{\omega_\lambda}}}$$

Abbreviate the sup at the right hand side of 5.35) by d. Observe that the Poincaré inequality applies to G_R. Choose now $\psi \in C_o^\infty(B_r)$, $0 \leq \psi \leq 1$ such that $\psi = 1$ on $B_{R/2}$. Let $v \in C_o^\infty(\bar{H}_{\omega_\lambda})$ and let $c := |G_R|^{-1} \int_{G_R} v dy$.

Then because of the Poincaré-inequality there is $C' = C'(R,\psi)$ such that $\|\nabla(\psi(v-c))\|_{q',H_{\omega_\lambda}} \leq C'\|\nabla v\|_{q',H_{\omega_\lambda}}$. By definition of $C_o^\infty(\bar{H}_{\omega_\lambda})$ there is $\tilde{v} \in C_o^\infty(\mathbb{R}^n)$ with $\tilde{v}|_{H_{\omega_\lambda}} = v$. Then $\psi(\tilde{v}-c) \in C_o^\infty(B_R)$. Since $\psi \equiv 1$ on $\operatorname{supp}(\varphi p)$ we get

$$(5.37)\quad |<\nabla(\varphi p),\nabla v>| = |<\nabla(\varphi p),\nabla(v-c)>| = |<\nabla(\varphi p),\nabla(\psi(v-c))>| \leq$$
$$\leq d\|\nabla(\psi(v-c))\|_{q',G_R} \leq d \cdot C'\|\nabla v\|_{q'}$$

and (5.35) follows immediately from (5.36) and (5.37) with $C = C_o \cdot C'$.

iii) Let $Z_R := \{y = (y',y_n) \in \mathbb{R}^n : -R < x_n < \omega_\lambda(x')\}$. By means of the transform (5.23) we immediately see that (5.17) remains true for Z_R and $\nabla v \in E_o^{q'}(H_{\omega_\lambda})$. The proof of part b) is analogous to part ii) with the following changes: We use now (5.22). Given $v \in C_o^\infty(H_\omega)$, then by means of (5.17) we get $\|\nabla(\psi v)\|_{q',G_R} \leq C'\|\nabla v\|_{q'H_{\omega_\lambda}}$, $\psi v \in C_o^\infty(G_R)$ and instead of (5.37) $|<\nabla(\varphi p),\nabla v>| = |<\nabla(\varphi p),\nabla(\psi v)>| \leq dC'\|\nabla v\|_{q'}$.

iv) The proof of c) and d) respectively is performed by induction using the Sobolev embedding theorem. We may assume $1 < q < 2 \le n$ and $\nabla p \in E^q(G_R)$. If $q \ge 2$ then clearly $\nabla p \in E^r(G_R)$ for $1 < r \le q$. Denote by k the biggest integer smaller than $\frac{n}{q}$. Then $k < \frac{n}{q} \le k+1$ and let $q_j := \frac{nq}{n-jq}$ for $j = 0,1,..,k$. Since $k + 1 \ge \frac{n}{q}$ we get $q_k \ge n \ge 2$. Let $K(q_{j-1},q_j) > 0$ for $j = 1,..,k$ denote the constants according to Lemma 5.10 and $K_1 := \min\{K(q_{j-1},q_j) : j = 1,..,k\} > 0$. Choose now $\lambda > 0$ in addition so small that $\|\nabla\omega_\lambda\|_q \le K$ and consider again H_{ω_λ}. Let $0 < R < \lambda$ (R as above). Let $R_j := R2^{-(j+1)}$ for $j = 0,1,..,k+1$. Choose $\varphi_j \in C_o^\infty(B_{R_j})$ such that $0 \le \varphi_j \le 1$ and $\varphi_j = 1$ on $B_{R_{j+1}}$. Let $G_j := G \cap B_{R_j}$ for $j = 0,1,..,k,k+1$. Given $v \in C_o^\infty(\bar{H}_{\omega_\lambda})$ let $v_j := v - c_j(v)$ where $c_j(v) := |G_j|^{-1} \int_{G_j} v\,dx$. Then $\varphi_j v_j \in C_o^\infty(\bar{H}_{\omega_\lambda})$ and $\nabla(\varphi_j v_j) \in E^{q'}(G)$. Therefore

$$0 = \langle\nabla p, \nabla(\varphi_j v_j)\rangle = \langle\nabla(\varphi_j p), \nabla v_j\rangle - \langle p\nabla\varphi_j, \nabla v_j\rangle + \langle\nabla p, v_j\nabla\varphi_j\rangle$$

that is

$$\langle\nabla(\varphi_j p), \nabla v_j\rangle = \langle p\nabla\varphi_j, \nabla v_j\rangle - \langle\nabla p, v_j\nabla\varphi_j\rangle \tag{5.38}$$

We prove now by induction that $\nabla(\varphi_j p) \in E^{q_j}(H_{\omega_\lambda})$ for $j = 0,1,..,k$. The case $j = 0$ is clear. Let now $0 < j \le k$ and suppose $\nabla(\varphi_{j-1} p) \in E^{q_{j-1}}(H_{\omega_\lambda})$. Since $\varphi_{j-1} = 1$ on G_j we conclude $\nabla p|_{G_j} \in L^{q_{j-1}}(G_j)$ and by the Sobolev embedding theorem $p \in L^{q_j}(G_j)$ and

$$\|p\|_{q_j,G_j} \le d_{1j}\|p\|_{H^{1,q_{j-1}}(G_j)} \tag{5.39}$$

Therefore with $M_j := \|\nabla\varphi_j\|_\infty$

$$|\langle p\nabla\varphi_j, \nabla v_j\rangle| \le M_j\|p\|_{q_j,G_j}\|\nabla v_j\|_{q_j',G_j} \le d_{1j}M_j\|p\|_{H^{1,q_{j-1}}(G_j)}\|\nabla v_j\|_{q_j',G_j} \tag{5.40}$$

$$|\langle\nabla p, v_j\nabla\varphi_j\rangle| \le M_j\|\nabla p\|_{q_{j-1},G_j}\|v_j\|_{q_{j-1},G_j} \tag{5.41}$$

Since $\frac{n}{q} - n + 1 < 1 \le j$ we conclude $q_j' = \frac{nq}{nq-n+jq} < n$

and by the Sobolev theorem $v_j \in L^{q_j'^*}(G_j)$ where

$$q_j'^* = \frac{nq_j'}{n-q_j'} = q'_{j-1} \text{ and}$$

$$\|v_j\|_{q'_{j-1},G_j} \le d_{2j}\,\|v_j\|_{H^{1,q_j'}(G_j)} \tag{5.42}$$

Since $\int_{G_j} v_j\,dx = 0$ we get by the Poincaré-inequality

(5.43) $\|v_j\|_{H^{1,q_j'}(G_j)} \le d_{3j}\|\nabla v_j\|_{q_j',G_j} \le d_{3j}\|\nabla v\|_{q_j',H_{\omega_\lambda}}$

By means of (5.40), (5.41) and (5.43) we get from (5.38) for $v \in C_o^\infty(H_{\omega_\lambda})$

$$|<\nabla(\varphi_j p),\nabla v>| = |<\nabla(\varphi_j p),\nabla v_j>| \le d_{4j}\|p\|_{H^{1,q_{j-1}}(G_j)}\|\nabla v\|_{q_j',H_{\omega_\lambda}}.$$

By Lemma 5.10 we conclude $\nabla(\varphi_j p) \in E^{q_j}(H_{\omega_\lambda})$ and therefore $\nabla p|_{G_{k+1}} \in L^{q_j}(G_{j+1})$. At the end follows $\nabla p|_{G_{k+1}} \in L^{q_k}(G_{k+1})$ where $q_k \ge n > 2$. Let now an arbitrary $s > n$ be given. Choose $0 < \epsilon < 1$ such that $s = \frac{n}{\epsilon}$. From the choice of k above we conclude $\frac{n}{n+1} \le q < \frac{n}{k}$. Define $\tilde{q} := \frac{n}{k+\epsilon}$. Then $\frac{n}{k+1} \le \tilde{q} < \frac{n}{k}$ and $k < \frac{n}{\tilde{q}} \le k + 1$ Since we originally assumed $1 < q < 2$ we have $k \ge \frac{n}{q} - 1 > \frac{n}{2} - 1 \ge 0$ we see $\tilde{q} \le \frac{n}{1+\epsilon} \le n \le q_k$ so that by the proof above we have $\nabla p|_{G_{k+1}} \in L^{\tilde{q}}(G_{k+1})$.

We repeat now the induction proof starting with $\tilde{q}_o = \tilde{q}$ and ending with

$\tilde{q}_k = \dfrac{n\tilde{q}}{n-k\tilde{q}} = \dfrac{n}{\epsilon} = s$. But observe that the constant $K_1 = K_1(\tilde{q})$ and therefore λ and especially R have to be taken depending on s.

v) Part d) is proven similiarily: Let $v \in C_o^\infty(H_{\omega_\lambda})$ be given. We no longer need to apply (5.42), instead we apply (5.17) for Z_R and $\nabla v \in E_o^{q'}(H_{\omega_\lambda})$ (see the beginning of part iii) of proof). We put now $v_j \equiv v$ and derive again (5.38). Since $\nabla(\varphi_{j-1}p) \in E_o^{q_{j-1}}(H_{\omega_\lambda}) \subset E_o^{q_{j-1}}(\mathbb{R}^n)$ we immediately get from the Sobolev estimate with a constant $d_{1j} = d_{1j}(q_j,n) > 0$

(5.44) $\|\varphi_{j-1}p\|_{q_j} \le d_{1j}\|\nabla(\varphi_{j-1}p)\|_{q_{j-1}}$

replacing now (5.39). Analogously

(5.46) $\|v\|_{q'_{j-1}} \le d_{2j}\|\nabla v\|_{q_j'}$

(observe $q'_{j-1} = q_j'^{*}$) replacing (5.42). Then observing $p\nabla\varphi_j \equiv (\varphi_{j-1}p)\nabla\varphi_j$ we get

for $v \in C_o^\infty(H_{\omega_\lambda})$

(5.47) $|<p\nabla\varphi_j,\nabla v>| \le M_j d_{1j}\|\nabla(\varphi_{j-1}p)\|_{q_{j-1}}\|\nabla v\|_{q_j'}$

(5.48) $|<\nabla p, v\nabla\varphi_j>| \le M_j d_{2j}\|\nabla(\varphi_{j-1}p)\|_{q_{j-1}}\|\nabla v\|_{q_j'}$

and so for $v \in C_o^\infty(H_{\omega_\lambda})$

$$|<\nabla(\varphi_j p),\nabla v>| \le d_{4j}\|\nabla(\varphi_{j-1}p)\|_{q_{j-1}}\|\nabla v\|_{q_j'}.$$

Again by Lemma 5.10 i) b) $\nabla(\varphi_j p) \subset E_o^{q_j}(H_{\omega_\lambda})$ and since $\mathrm{supp}(\varphi_j p) \subset G_{j+1}$ we have $\nabla(\varphi_j p) \in E_o^{q_j}(G_{j+1})$. By induction we end with $\nabla(\varphi_k p) \in E_o^{q_k}(G_{k+1})$. Choose $R' = R_{k+1}$. If $\varphi \in C_o^\infty(B_{R'})$ then because of $\varphi_k = 1$ on B_{k+1} we have $\varphi\, \varphi_k p = \varphi p$. The remaining considerations are like as in iv). ■

The most difficult hard work is now done. For an easier later application we consider two further lemmas.

Lemma 5.12. Let $1 < q < \infty$ and let $G \subset \mathbb{R}^n$ be a domain. Let $x_o \in G$ and let $R > 0$ be such that $B_R(x_o) \subset\subset G$.

a) Let $\nabla p \in E^q(G)$. Then for $\varphi \in C_o^\infty(B_{R/2}(x_o)), \nabla(\varphi p) \subset E_o^q(B_{R/2}(x_o)) \subset E_o^q(G)$ and with a constant $C = C(R, C_1(q)) > 0$, where C_1 is by Lemma 5.2, we have

$$(5.49) \qquad \|\nabla(\varphi p)\|_q \le C \sup_{0 \ne v \in C_o^\infty(B_R)} \frac{|<\nabla(\varphi p),\nabla v>|}{\|\nabla v\|_{q'}}$$

b) Let $\nabla p \in E^q(G)$ and $<\nabla p,\nabla\phi> = 0$ for all $\phi \in C_o^\infty(G)$. Given $1 < s < \infty$ then there is a $0 < R' \le R$ with $R' = R'(s)$ such that $\nabla(\varphi p) \in E_o^s(B_{R'}(x_o))$ for each $\varphi \in C_o^\infty(B_{R'}(x_o))$.

Proof. The proof is almost identical with that of Lemma 5.12. In the sequel we abbreviate $B_r := B_r(x_o)$ for $r > 0$. i) If $x \in B_{R/2}$ and $0 < \epsilon < R/2$ then for the mollified p_ϵ we have $\nabla p_\epsilon(x) = (\nabla p)_\epsilon(x)$ and therefore $\|\nabla p - \nabla p_\epsilon\|_{q,B_{R/2}} \to 0$. For $\varphi \in C_o^\infty(B_{R/2})$ we have $\varphi p_\epsilon \in C_o^\infty(B_{R/2})$ and clearly $\|\nabla(\varphi p) - \nabla(\varphi p_\epsilon)\|_{q,B_{R/2}} \to 0$ proving $\varphi p \in E_o^q(B_{R/2}) \subset E_o^q(\mathbb{R}^n)$. Now we proceed like as in part ii) of the preceeding lemma applying Lemma 5.2 instead of Lemma 5.10: Choose again $\psi \in C_o^\infty(B_R)$ with $0 \le \psi \le 1$, $\psi = 1$ on $B_{R/2}$. If $v \in C_o^\infty(\mathbb{R}^n)$ put $c := |B_R|^{-1} \int_{G_R} v dy$ and use now the Poincaré-inequality for B_R. Consider again $\psi(v-c) \in C_o^\infty(B_R)$ and proceed analogously.

ii) The proof of b) is literally the same as part iv) of proof of Lemma 5.11, beginning with the 9th line before formula (5.38). Observe that $G_j = B_{R_j}$. ■

Lemma 5.13. Let $1 < q < \infty$ and let $G \subset \mathbb{R}^n$ be an exterior domain, $G = \mathbb{R}^n \setminus \mathring{K}$ with $\emptyset \neq K \subset\subset \mathbb{R}^n$. Let $R > 0$ be such that $K \subset\subset B_R$, let $\varphi \in C^\infty(\mathbb{R}^n)$, $\varphi(x) = 0$ for $|x| \leq 2R$, $\varphi(x) = 1$ for $|x| \geq 3R$.

a) If $\nabla p \in E^q(G)$ then $\nabla(\varphi p) \in E^q(G)$.

b) There is a constant $C = C(q,n,R) > 0$ such that

$$(5.50) \quad \|\nabla(\varphi p)\|_q \leq C \sup_{0 \neq v \in C_o^\infty(\bar{G})} \frac{|<\nabla(\varphi p),\nabla v>|}{\|\nabla v\|_{q',G}}$$

c) If $\nabla p \in E^q(G)$ has the property

$$(5.51) \quad <\nabla p,\nabla v> = 0 \text{ for all } v \in C_o^\infty(\bar{G}).$$

Then $\nabla(\varphi p) \in E^s(G)$ for any $1 < s < \infty$.

Proof. i) Because of the properties of φ clearly $\nabla(\varphi p) \in E^q(G)$.

ii) (φp) may be extended by zero to the whole $\mathbb{R}^n$. Then $\nabla(\varphi p) \in E^q(\mathbb{R}^n)$. Since $\|\nabla v\|_{q',\mathbb{R}^n} \geq \|\nabla v\|_{q',G}$ for $v \in C_o^\infty(\mathbb{R}^n)$ estimate (5.50) follows immediately from (5.5).

iii) We will first show $\nabla(\varphi p) \in E^s(G)$ for $1 < s \leq q$. This is a priori by no means trivial. By definition of $E^q(G)$ we clearly have $\nabla p \in L^s(G_{3R}), p \in L^s(G_{3R})$ for $1 < s \leq q$. If $v \in C_o^\infty(\mathbb{R}^n)$ put $\tilde{v} := v - c$ where $c := |B_{3R}|^{-1} \int_{B_{3R}} v dx$. Then $(\varphi\tilde{v} + c) \in C_o^\infty(\bar{G})$ and by (5.51) we again get

$$0 = <\nabla p,\nabla(\varphi(v-c) + c)> = <\nabla p,\nabla(\varphi(v-c))> =$$

$$= <\nabla(\varphi p)\nabla v> - <p\nabla\varphi,\nabla v> + <\nabla p,\nabla\varphi\tilde{v}>$$

and therefore

$$<\nabla(\varphi p),\nabla v> = <\nabla(\varphi p),\nabla\tilde{v}> = <p\nabla\varphi,\nabla\tilde{v}> - <\nabla p,\tilde{v}\nabla\varphi>$$

Since by (2.2) $\|\tilde{v}\|_{s',B_{3R}} \leq C(s')\|\nabla\tilde{v}\|_{s',B_{3R}} \leq C(s')\|\nabla v\|_{s'}$

we get immediately with $C = C(\varphi,R,s,q) > 0$

$$|<(\varphi p),\nabla v>| \leq C(\|p\|_{q,G_{3R}} + \|\nabla p\|_{q,G_{3R}})\|\nabla v\|_{s'}$$

for $v \in C_o^\infty(\mathbb{R}^n)$ and by Lemma 5.2 $\nabla(\varphi p) \in E^s(\mathbb{R}^n)$.

iv) Because of iii) we may assume that $1 < q < 2 \leq n$. Then we proceed like as in part iv) of the proof of Lemma 5.11: Choose again $k \in \mathbb{N}$, $k < \frac{n}{q} \leq k+1$. Let $R_j := R + j\frac{R}{5k}$ for $j = 0,1,..,k+1$. Let $\varphi_j \in C^\infty(\mathbb{R}^n)$, $0 \leq \varphi_j \leq 1$, $\varphi_j(x) = 0$ for

$|x| \le R_j$, $\varphi_j(x) = 1$ for $|x| \ge R_{j+1}$. Given $v \in C_0^\infty(\mathbb{R}^n)$ let $v_j = v - c_j(v)$ where $c_j(v) := |B_{2R}|^{-1} \int_{B_{2R}} v dx$. Since $(\varphi_j v_j + c_j(v)) \in C_0^\infty(\bar{G})$ we can now go on like as in Lemma 5.11. Once we have shown $\nabla(\varphi p) \in E^q(G)$ and $\nabla(\varphi p) \in E^{q_k}(G)$, we see $\nabla(\varphi p) \in E^s(G)$ for $q \le s \le q_k$: Write $G' = \{x \in G: |\nabla(\varphi p)(x)| \ge 1\}$ and $G'' = \{x \in G: |\nabla(\varphi p)(x)| \le 1\}$
then

$$\int_G |\nabla(\varphi p)|^s dx = \int_{G'} |\nabla(\varphi p)|^s dx + \int_{G''} |\nabla(\varphi p)|^s dx \le$$
$$\le \int_{G'} |\nabla(\varphi p)|^{q_k} dx + \int_{G''} |\nabla(\varphi p)|^q dx. \quad \blacksquare$$

<u>Lemma 5.14.</u> Assume the same hypothesis as in Lemma 5.13.

a) If $\nabla p \in E_0^q(G)$ then $\nabla(\varphi p) \in E_0^q(G)$ and there is a constant $C = C(q,n,\mathbb{R}) > 0$ such that

$$(5.52) \quad \|\nabla(\varphi p)\|_q \le C \sup_{0 \ne v \in C_0^\infty(G)} \frac{|<\nabla(\varphi p), \nabla v>|}{\|\nabla v\|_{q'}},$$

b) Suppose that $\nabla p \in E_0^q(G)$ satisfies

$$<\nabla p, \nabla v> = 0 \text{ for all } v \in C_0^\infty(\bar{G}).$$

Then: i) If $1 < q \le \frac{n}{n-1}$ then $\nabla(\varphi p) \in E^s(G)$ for $1 < s < \infty$.

ii) If $q > \frac{n}{n-1}$ then $\nabla(\varphi p) \in E^s(G)$ for $\frac{n}{n-1} < s < \infty$.

<u>Proof.</u> i) Since $\nabla p \in E_0^q(G)$ there is a sequence $(p_i) \subset C_0^\infty(G)$ such that $\|\nabla p - \nabla p_i\|_q \to 0$. By (2.9) $\|p - p_i\|_{q, G_{3R}} \le C(R) \|\nabla p - \nabla p_i\|_q$. Therefore $\nabla(\varphi p_i) = p_i \nabla\varphi + \varphi \nabla p_i \to \nabla(\varphi p)$ in $L^q(G)$, $\varphi p_i \in C_0^\infty(G)$ and therefore $\nabla(\varphi p) \in E_0^q(G)$.

ii) We first show (5.52) for $1 < q \le \frac{n}{n-1}$. Then $q' \ge n$. Let $\psi \in C^\infty(\mathbb{R}^n)$, $0 \le \psi \le 1$, $\psi(x) = 1$ for $|x| \ge 2R$, $\psi(x) = 0$ for $|x| \le \frac{3}{2} R$. For $v \in C_0^\infty(\mathbb{R}^n)$ let $c_v = |B_{2R}|^{-1} \int_{B_{2R}} v dx$. We will show now that $\tilde{v} := \psi(v - c_v) \in E_0^{q'}(G)$ for $q' \ge n$. For this purpose let $\rho \in C_0^\infty(\mathbb{R}^n)$, $0 \le \rho \le 1$, $\rho(x) = 1$ for $|x| \le 1$, $\rho(x) = 0$ for $|x| \ge 2$ and put $\rho_k(x) := \rho(k^{-1}x)$ for $k \in \mathbb{N}$. Then $\operatorname{supp}(\nabla\rho_k) \subset R_k := \{x \in \mathbb{R}^n : k \le |x| \le 2k\}$. Then $\tilde{v}_k := \rho_k \cdot \tilde{v} \in C_0^\infty(G)$. There is $k_0 \in \mathbb{N}$, $k_0 \ge 2R$, such that $\operatorname{supp} v \subset B_{k_0}$. Then for $k \ge k_0$

$$\begin{aligned} \nabla\tilde{v}_k &= \nabla\rho_k \cdot \psi(v - c_v) + \rho_k (\nabla\psi)(v - c_v) + \rho_k \psi \cdot \nabla v \\ &= -\nabla\rho_k c_v + (v - c_v)\nabla\psi + \psi \cdot \nabla v. \end{aligned}$$

Since $|\nabla\rho_k| \le C \cdot k^{-1}$ and vanishes outside R_k we get $\|\nabla\rho_k\|_{q', R_k} \le c \cdot k^{\frac{n}{q'} - 1} \le$ const for $q' \geq n$. If $F^* \in E_o^{q'}(G)^*$ then it may be extended to a functional $F^* \in (L^{q'}(G)^n)^*$ and therefore represented with $f \in L^q(G)^n$ as $F^*(\nabla\phi) = \langle f,\nabla\phi\rangle$ for all $\nabla\phi \in E_o^{q'}(G)$. Since $|\langle f,\nabla\rho_k\rangle| \leq \|f\|_{q,R_k} \leq$ const $\|f\|_{q,R_k} \to 0$ we conclude $f^*(\nabla\tilde{v}_k) \to F^*(\nabla\tilde{v})$. Since $E_o^{q'}(G)$ is closed it is weakly closed too and therefore $\nabla\tilde{v} \in E_o^{q'}(G)$. Since $\nabla\psi = 0$ on the support of φ we see $\nabla(\varphi p)\cdot((v-c_v)\nabla\psi + \psi\cdot\nabla v) = \nabla(\varphi p)\cdot\nabla v$ and therefore $\langle\nabla(\varphi p),\nabla\tilde{v}\rangle = \langle\nabla(\varphi p),\nabla v\rangle$. By (2.2) (applied to B_{2R}) we see $\|\nabla\tilde{v}\|_{q'} \leq K\|\nabla v\|_{q'}$ and therefore

$$\frac{|\langle\nabla(\varphi p),\nabla v\rangle|}{\|\nabla v\|_{q'}} \leq K\,\frac{|\langle\nabla(\varphi p),\nabla\tilde{v}\rangle|}{\|\nabla\tilde{v}\|_{q'}} \quad \text{if } \nabla\tilde{v} \neq 0.$$

We abbreviate the sup at the right hand side of (5.52) by D. Since $0\neq\nabla\tilde{v} \in E_o^{q'}(G)$ there is a sequence $(v_i) \subset C_o^\infty(G)$ such that $\|\nabla v-\nabla v_i\|_{q'} \to 0$. Then

$$\frac{|\langle\nabla(\varphi p),\nabla\tilde{v}\rangle|}{\|\nabla\tilde{v}\|_{q'}} = \lim_{i\to\infty} \frac{|\langle\nabla(\varphi p),\nabla v_i\rangle|}{\|\nabla v_i\|_{q'}} \leq D.$$

Therefore $\dfrac{|\langle\nabla(\varphi p),\nabla v\rangle|}{\|\nabla v\|_{q'}} \leq KD$ for those $v \in C_o^\infty(\mathbb{R}^n)$ such that $\nabla\tilde{v} \neq 0$. If $\nabla\tilde{v}$ vanishes then $\langle\nabla(\varphi p),\nabla\varphi\rangle$ too. Therefore we derive (5.52) from Lemma 5.2.

iii) Let now $q > \frac{n}{n-1}$ and therefore $1 < q' < n$. Let $\psi \in C^\infty(\mathbb{R}^n)$ be defined as in ii). If $v \in C_o^\infty(\mathbb{R}^n)$ then $\psi\cdot v \in C_o^\infty(G)$ and $\|\nabla(\psi v)\|_{q'} \leq \|v\nabla\psi\|_{q'} + \|\nabla v\|_{q'}$.

Let $q'^* := \frac{nq'}{n-q'}$. Then by the Sobolev theorem and Hölder's inequality

$$\|v\nabla\psi\|_{q'} \leq c(\psi)\|v\|_{q',B_{2R}} \leq c(\psi)\|v\|_{q'^*,B_{2R}}|B_{2R}|^{1/n}$$
$$\leq c(\psi)|B_{2R}|^{1/n}\|\nabla v\|_{q'}$$

and so $\|\nabla(\psi v)\|_{q'} \leq c(R)\|\nabla v\|_{q'}$. Therefore

$$|\langle\nabla(\varphi p),\nabla v\rangle| = |\langle\nabla(\varphi p),\nabla(\psi v)\rangle| \leq D\|\nabla(\psi v)\|_{q'} \leq c(R)\cdot D\,\|\nabla v\|_{q'}$$

where D denotes the sup in (5.52) and again by Lemma 5.2 follows (5.52).

iv) We prove now b). Let $1 < q \leq \frac{n}{n-1}$. We first show

(5.53) $\langle\nabla p,\nabla\varphi\rangle = 0$ for φ by Lemma 5.13.

Choose ρ_k like as in part ii). Then for $k \geq 3R$ we have $\rho_k\varphi \in C_o^\infty(G)$ and by assumption

$$0 = \langle\nabla p,\nabla(\rho_k\varphi)\rangle = \langle\nabla p,\rho_k\nabla\varphi\rangle + \langle\nabla p,\varphi\nabla\rho_k\rangle = \langle\nabla p,\nabla\varphi\rangle + \langle\nabla p,\nabla\rho_k\rangle$$

Remember $\|\nabla\rho_k\|_{q'} \leq$ const. for $q' \geq n$. Therefore

$$|\langle\nabla p,\nabla\rho_k\rangle| = |\langle\nabla p,\nabla\rho_k\rangle_{R_k}| \leq C\,\|\nabla p\|_{q,R_k} \to 0.$$

v) For $v \in C_0^\infty(\mathbb{R}^n)$, $\varphi v \in C_0^\infty(G)$ and therefore admissible in (5.52). We then get

(5.54) $\quad \langle\nabla(\varphi p),\nabla v\rangle = \langle p\nabla\varphi,\nabla v\rangle - \langle\nabla p, v\nabla\varphi\rangle.$

Let $c(v) := |B_{3R}|^{-1}\int_{B_{3R}} v\,dx$ and put $\tilde{v} := v - c(v)$. Then by (5.53) $\langle\nabla p,\tilde{v}\nabla\varphi\rangle = \langle\nabla p, v\nabla\varphi\rangle$.

Let now $1 < s \le q \le \frac{n}{n-1}$. Then $p \in L^s(G_{3R})$ and $\nabla p \in L^s(G_{3R})$. We therefore get

$$|\langle\nabla p,\tilde{v}\nabla\varphi\rangle| \le \|\nabla\varphi\|_\infty \|\nabla p\|_{s,G_{3R}}\|\tilde{v}\|_{s',B_{3R}}$$

and by the Poincaré-inequality

$$\|\tilde{v}\|_{s',B_{3R}} \le C\|\nabla\tilde{v}\|_{s'} = C\|\nabla v\|_{s'}$$

and so we derive from (5.54) for $v \in C_0^\infty(\mathbb{R}^n)$

$$|\langle\nabla(\varphi p),\nabla v\rangle| \le C'(\|p\|_{s,G_{3R}} + \|\nabla p\|_{s,G_{3R}})\,\|\nabla v\|_{s'}$$

and so by Lemma 5.2 we get $\nabla(\varphi p) \in E^s(\mathbb{R}^n)$

vi) If $n = q = 2$ then we are ready. If $n > 2$, then $1 < q \le \frac{n}{n-1} < 2$. Now we proceed similiarily as in parts iv) and v) of the proof of Lemma 5.11: Choose again $k \in \mathbb{N}$, $k < \frac{n}{q} \le k+1$. Let $R_j = R + j\frac{R}{5k}$ for $j = 0,1,\dots,2(k+1)$. Let $\varphi_j \in C^\infty(\mathbb{R}^n)$, $0 \le \varphi_j \le 1$, $\varphi_j(x) = 0$ for $|x| \le R_j$, $\varphi_j(x) = 1$ for $|x| \ge R_{j+1}$. Given $v \in C_0^\infty(\mathbb{R}^n)$, then $\varphi_j v \in C_0^\infty(G)$ and from the assumption we again derive (5.38) now with $v_j := v$. Let now again $0 < j \le k$ and assume that $\nabla(\varphi_{j-1}p) \in E^{q_{j-1}}(G)$. Since $\varphi_{j-1} = 1$ for $|x| \ge R_j$, we conclude $\nabla p \in L^{q_{j-1}}(\{|x| \ge R_j\})$. Since $\operatorname{supp}(\nabla\varphi_j) \subset G_j := \{x \in \mathbb{R}^n\colon R_j \le |x| \le R_{j+1}\}$ again by the Sobolev embedding theorem we get (5.39). Since $q_j' < n$ instead of (5.42) we use the Sobolev theorem for $C_0^\infty(\mathbb{R}^n)$ functions

(5.55) $\quad \|v\|_{q'_{j-1}} \le C\|\nabla v\|_{q_j'}.$

Then we can estimate (5.41) and end with

$$|\langle\nabla(\varphi_j p),\nabla v\rangle| \le C\|p\|_{H^{1,q_{j-1}}(G_j)}\|\nabla v\|_{q_j'}.$$

The remaining arguments are the same as in part iv) of the proof of Lemma 5.11

vii) Assume now $q > \frac{n}{n-1}$. Let $\varphi_o \in C^\infty(\mathbb{R}^n)$, $0 \le \varphi_o \le 1$, $\varphi_o = 0$ on B_R, $\varphi_o(x) = 1$ for $|x| \ge \frac{3}{2}R$ and consider (5.54) with this φ_o. If $n \ge 3$ choose any q_1 with $\frac{n}{n-1} < q_1 < 2$ and $q_1 \le q$. Put $r := \frac{q_1 n}{n+q_1}$. Then $1 < r < \frac{2n}{n+2} < n$ (observe $\frac{2n}{n+2} > 1$ for $n \ge 3$). Further $r \le q$ and $q_1 = \frac{nr}{n-r} = r^*$, $q_1' = \frac{nr}{nr-n+r}$ and $q_1'^* = \frac{nq_1'}{n-q_1'} = r'$. Since $r \le q$ we have $p, \nabla p \in L^r(M_R)$, that is $p \in H^{1,r}(M_R)$, where $M_R := \{x \in \mathbb{R}^n: R < |x| < \frac{3}{2} R\}$. The Sobolev inequality gives

$$\|p\|_{q_1, M_R} = \|p\|_{r^*, M_R} \leq C\|p\|_{H^{1,r}(M_R)}$$

Consider (5.54) for $v \in C_0^\infty(\mathbb{R}^n)$ and φ_o. Then with $C' := \|\nabla\varphi_o\|_\infty$

(5.56) $|\langle p\nabla\varphi, \nabla v\rangle| \leq C'\|p\|_{q_1(M_R)} \|\nabla v\|_{q_1'} \leq C'C \|p\|_{H^{1,r}(M_R)}\|\nabla v\|_{q_1'}$

By the Sobolev inequality for $v \in C_0^\infty(\mathbb{R}^n)$ (see (5.55))

(5.57) $|\langle \nabla p, v\nabla\varphi\rangle| \leq C'\|\nabla p\|_{r, M_R} \|v\|_{r'} \leq C'C\|\nabla p\|_{r, M_R}\|\nabla v\|_{q_1'}$

By Lemma 5.2 we see from (5.54) that $\nabla(\varphi_o p) \in E^{q_1}(\mathbb{R}^n)$ holds. Now we may start the iteration procedure from part iii).
If $n = 2$ and given $s > 2$, we put $\tilde{r} = \frac{2s}{2+s}$. Then $1 < \tilde{r} < 2 = n$, $\tilde{r}^* = \frac{2r}{2-r} = s$ and with estimates analogously to (5.56), (5.57) we conclude again via Lemma 5.2 $\nabla(\varphi_o p) \in E^s(G)$. ∎

<u>Remark 5.15.</u>
If $G \subset \mathbb{R}^n$ is a <u>bounded</u> domain with sufficiently smooth boundary, say e.g. $\partial G \in C^1$ and if $1 < q < s < \infty$, if $\nabla p \in E_0^q(G)$ and $\nabla p \in L^s(G)$, then $\nabla p \in E_0^s(G)$. As we have seen in Lemma 5.6 this conclusion still holds for the half-space. But it is no longer true for an exterior domain: Let $K := B_1$, $G := \mathbb{R}^n \setminus \bar{K} = \{x \in \mathbb{R}^n: |x| > 1\}$ and consider for $x \in G$

$$h(x) := \begin{cases} 1 - |x|^{2-n} & \text{if } n \geq 3 \\ \ln|x| & \text{if } n = 2 \end{cases} \tag{5.58}$$

Then $h \in C^\infty(\bar{G})$, $h|_{\partial G} = 0$, $\Delta h = 0$, $\nabla h \in L^q(G)$ for $q > \frac{n}{n-1}$. Consider again $\rho \in C_0^\infty(\mathbb{R}^n)$, $0 \leq \rho \leq 1$, $\rho(x) = 1$ for $|x| \leq 1$, $\rho(x) = 0$ for $|x| \geq 2$ and $\rho_k(x) := \rho(k^{-1}x)$ for $k \in \mathbb{N}$. Since $h|_{\partial G} = 0$ it is easily seen that $h_k := \rho_k h \in E_0^q(G)$ for $q > \frac{n}{n-1}$. Since $\nabla h_k = h\nabla\rho_k + \rho_k\nabla h$ because of (5.58) and the properties of h one immediately verifies $\|\rho_k h\|_q \to 0$ if $q > n$ for $n \geq 2$. Clearly $\rho_k\nabla h \to \nabla h$ in $L^q(G)$. If $n \geq 3$ and $q = 2$ like as in part iii) of the proof of Lemma 5.6 one verifies $\nabla h_k \to \nabla h$ weakly in $E_0^q(G)$ and therefore $\nabla h \in E_0^q(G)$ in this case too. This rests on the property $\|\nabla\rho_k\|_q \leq$ const for $q \geq n$ and $k \in \mathbb{N}$. If $q = n = 2$, $\nabla \ln|x| \notin L^2(G)$. That these are the optimal q for $\nabla h \in E_0^q(G)$ may be seen as follows: If $p \in C_0^\infty(G) \subset C_0^\infty(\mathbb{R}^n)$ by means of the Sobolev-inequality $\|p\|_{q^*} \leq C\|\nabla p\|_q$ for $\nabla p \in E_0^q(G)$ and $1 \leq q < n$. Clearly $h \notin L^s(G)$ vor $1 \leq s < \infty$. ∎

The best possible result we can expect therefore is

Lemma 5.16. Let $1 < q < \infty$ and let $G \subset \mathbb{R}^n$ be an exterior domain, $G = \mathbb{R}^n \setminus \mathring{K}$, with $\phi \neq K \subset\subset \mathbb{R}^n$. Let $R > 0$ be such that $K \subset\subset B_R$. Suppose there is $1 < q < n$ and $\nabla p \in E^q_o(G)$. Assume in addition $p = 0$ in $G_{2R} := G \cap B_{2R}$. If there is any other $1 < r < \infty$ with $\nabla p \in L^r(G)$. Then $\nabla p \in E^r_o(G)$.

Proof. Without loss of generality we may assume $p \in C^\infty(G)$. Otherwise we consider the mollification p_ϵ. Since p vanishes on G_{2R}, $\nabla(p_\epsilon) = (\nabla p)_\epsilon$ in G for $0 < \epsilon < R$. Clearly $\nabla p_\epsilon \in E^q_o(G)$ too. Consider again our standard cut-off function $\rho \in C^\infty_o(\mathbb{R}^n)$, $0 \le \rho \le 1$, $\rho(x) = 1$ for $|x| \le 1$, $\rho(x) = 0$ for $|x| \ge 2$, $\rho_k(x) := \rho(k^{-1}x)$ and $R_k := \{x \in \mathbb{R}^n : k < |x| \le 2k\}$. Then $\mathrm{supp}(\nabla\rho_k) \subset R_k$. Since by assumption $\nabla p \in E^q_o(G)$ and $1 < q < n$, by the Sobolev embedding theorem $p \in L^{q^*}(G)$ with $q^* = \frac{nq}{n-q}$ and $\|p\|_{q^*} \le C\|\nabla p\|_q$. Define $C_k := |R_k|^{-1} \int_{R_k} p\,dx$

By Hölder's inequality

$$|C_k| \le |R_k|^{-1}\, \|p\|_{q^*,R_k} |R_k|^{\frac{1}{q^{*\prime}}} \le C\|\nabla p\|_q |R_k|^{\frac{1}{q^{*\prime}}-1}.$$

Since $q^{*\prime} = \frac{q^*}{q^*-1} = \frac{nq}{nq-n+q}$ we have $\frac{1}{q^{*\prime}} - 1 = \frac{q-n}{q} < 0$ for $q < n$ and we get $|R_k|^{\frac{1}{q^{*\prime}}-1} \to 0$ $(k \to \infty)$ and therefore $|C_k| \to 0$. Observe the Poincaré-inequality (2.6) giving

$$(5.59) \qquad \|p-C_k\|_{r,R_k} \le k\cdot C_1\, \|\nabla p\|_{r,R_k}$$

(since we assumed $p \in C^1(G)$ clearly $p \in L^r(R_k)$). Define $\varphi(x) = 1-\rho_R(x)$ (with ρ like as above) and $p_k := \varphi(\rho_k(p-C_k)) \in C^\infty_o(G)$.
Then $\nabla p_k = \rho_k(p-C_k)\nabla\varphi + \varphi(p-C_k)\nabla\rho_k + \varphi\rho_k\nabla p$. Since $\varphi \equiv 1$ on $\mathrm{supp}(\nabla p)$ we see $\varphi\rho_k\nabla p \to \nabla p$ in $L^r(G)$. Since p vanishes on $\mathrm{supp}(\nabla\varphi) \subset B_{2R}$, $\rho_k \equiv 1$ for $k > R$, $\|\rho_k(p-C_k)\nabla\varphi\|_r = \|\nabla\varphi\|_\infty\, |C_k|\, |B_{2R}| \to 0$. By (5.59) and $|\nabla\rho_k| \le C\cdot k^{-1}$, $\mathrm{supp}(\nabla\rho_k) \subset R_k$ we see $\|\varphi(p-C_k)\nabla\rho_k\|_r \le Ck^{-1}\|p-C_k\|_{r,R_k} \le C\cdot C'\|\nabla p\|_{r,R_k} \to 0$ $(k \to \infty)$.
Therefore $\|\nabla p - \nabla p_k\|_r \to 0$. ■

Remark 5.17. Let the assumption of Lemma 5.14 be satisfied, expecially (5.52) with $\nabla p \in E^q_o(G)$. If $1 < q \le \frac{n}{n-1}$ and $n \ge 3$ (then $\frac{n}{n-1} < n$) we conclude via Lemma 5.15 that $\nabla(\varphi p) \in E^s_o(G)$ for $1 < s < \infty$. If $n = 2$ and $1 < q < 2$ then $\nabla(\varphi p) \in E^s_o(G)$ for $1 < s < \infty$. If $n \ge 3$ and $\frac{n}{n-1} < q < n$ then $\nabla(\varphi p) \in E^s_o(G)$ for $\frac{n}{n-1} < s < \infty$. These properties perfectly fits togehter with the observations made in Remark 5.15. ■

The proof of estimates (5.35), **(5.49)**, **(5.50) was completely** elementary (e.g. we needed only Sobolev's embedding theorem and **Hölder's** inequality) but demanding lengthy hard work. It was done to prove in addition that solutions $\nabla p \in E^q(G)$ resp. $\nabla p \in E_o^q(G)$ of the homogeneous functional equations have integrability properties with respect to "other" exponents $1 < s < \infty$ and (compare Remark 5.17) belong under certain circumstances to $E_o^s(G)$ too. All this work we need to conclude via the trivial $L^2(G)$-uniqueness of the Dirichlet and Neumann problem the L^q-uniqueness. By means of a partition of unity the desired main theorems are then an easy consequence of the following uniqueness result. Concerning the Dirichlet problem we read off from Remark 5.16 that the uniqueness result is best possible. I'm very much indepted to my colleque Professor Dr. Michael Wiegner, who gave me the example of the "exceptional functions" h in (5.58) and drew my attention in a very early stage of the consideration of exterior problems in the appropriate direction.

Theorem 5.18. (Uniqueness) Let $G \subset \mathbb{R}^n$ be either a bounded or an exterior domain with boundary $\partial G \in C^1$ and let $1 < q < \infty$. Then:

a) Uniqueness of the weak Neumann problem:
If $\nabla p \in E^q(G)$ satisies $<\nabla p, \nabla\phi> = 0$ for all $\nabla\phi \in E^{q'}(G)$, then $\nabla p = 0$.

b) Uniqueness of the weak Dirichlet problem:
If $\nabla p \in E_o^q(G)$ satisfies $<\nabla p, \nabla\phi> = 0$ for all $\nabla\phi \in E_o^{q'}(G)$. Then:

i) If G is bounded, then $\nabla p = 0$ (and therefore $p = 0$)

ii) If G is an exterior domain, and
if $n \geq 3$ and $1 < q < n$, then $\nabla p = 0$
if $n = 2$ and $1 < q \leq 2 = n$, then $\nabla p = 0$
(and $p = 0$ too).

Proof. i) By Lemma 5.11 for each $x_o \in \partial G$ there is $R' = R'(x_o) > 0$ such that for $G \cap B_{R'}(x_o)$ the properties c) respectively d) hold. By compactness of ∂G we find finitely many $x_i \in \partial G$ and $R_i := R'(x_i) > 0$, $i = 1,..,M$, such that $\partial G \subset \bigcup_{i=1}^{M} B_i$, where $B_i := B_{R_i}(x_i)$. If G is bounded, $G_1 := G \cap \bigcap_{i=1}^{M} \mathbb{R}^n \setminus B_i \subset G$ and is compact. By means of Lemma 5.12 we see that G_1 can be covered by finitely many balls $B_i = B_{R_i}(x_i) \subset\subset G$, $i = M+1,...,N$. Then the $B_i, i = 1,...,N$ form an open covering of $\bar{G}$. If G is an exerior domain, $G = \mathbb{R}^n \setminus \bar{K}$, where $\phi \neq K \subset\subset \mathbb{R}^n$, then we choose $R > 0$ such that $K \subset\subset B_R$ and put now $G_2 := G_1 \cap \bar{B}_{3R} \subset G_2$. Again G_2 is compact and may be covered by $B_i \subset\subset G$, $i = M+1,...,N$.

Define $B_o := \{x \in \mathbb{R}^n: |x| > 2R\} \subset G$. Then again the system $B_i, i = 0,1,..,N$ forms an open covering of $\bar{G}$. Construct a partion of unity $\{\varphi_i : i = =,1,..,N\}$ such that $0 \leq \varphi_i$, $\varphi_i \in C_o^\infty(B_i)$ for $i = 1,..,N$ and $\varphi_o \in C^\infty(B_o)$, $\varphi_o = 1$ for $|x| \geq 3R$, φ_o vanishing in a neighborhood of $|x| = 2R$, $\sum_{i=1}^{N} \varphi_i(x) = 1$ for $x \in \bar{G}$.

ii) In case a) we immediately conclude from the hypothesis and Lemmas 5.11 - 5.13 that $\nabla(\varphi_i p) \subset E^2(G)$ (continue $\varphi_i p$ by zero for $x \in G$, $x \notin G \cap B_i$) and therefore $\nabla p = \sum_{i=0}^{N} \nabla(\varphi_i p) \in E^2(G)$. Since $E^\infty(G) = \{\nabla\phi : \phi \in C_o^\infty(\bar{G})\}$ is dense in $E^s(G)$ for $1 < s < \infty$ from $\nabla p \in E^2(G)$ and $0 = <\nabla p, \nabla\phi>$ for $\nabla\phi \in E^s(G)$ we see now $0 = <\nabla p, \nabla p>$ and therefore $\nabla p = 0$.

iii) In case b) and G bounded we similiarly conclude by Lemmas 5.11 and 5.12 d) $\nabla(\varphi_i p) \in E_o^q(G)$ and therefore $\nabla p \in E_o^2(G)$ and again $\nabla p = 0$. Clearly, we conclude $p = 0$ too.

iv) If G is an exterior domain we consider Lemma 5.14 and 5.16: If $n \geq 3$ and $1 \leq q \leq \frac{n}{n-1} < n$, then $\nabla(\varphi_o p) \in E_o^s(G)$ for $1 < s < \infty$. If $\frac{n}{n-1} < q < n$, then again $\nabla(\rho_o p) \in E_o^s(G)$ for $1 < s < \infty$. In any case $\nabla(\varphi_o p) \in E_o^s(G)$. If $n = 2$ and $1 < q < 2$, then by Lemma 5.15 and 5.16 $\nabla(\varphi_o p) \in E_o^s(G)$. The case $q = 2$ is trivial. Since $\nabla(\varphi_i p) \in E_o^s(G)$ for $i = 0,..,N$ at the end $\nabla p = \sum_{i=0}^{N} \nabla(\varphi_i p) \in E_o^2(G)$ and again $0 = <\nabla p, \nabla\phi>$ for all $\phi \in C_o^\infty(G)$ and by density of $\{\nabla\phi : \phi \in C_o^\infty(G)\}$ in $E_o^2(G)$ we again have $\nabla p = 0$ that is $p = 0$. ■

6. Proof of the main theorems.

Proof of Theorem 4.1. First we prove part i). Like as in part i) of the proof of Theorem 5.18 we construct a covering $B_i, i = 0,1,..,N$ of G ($B_o := \phi$ if G is bounded) and a partition of unity (φ_i) with respect to this covering such that (5.35) holds for $i = 1,..,M$, (5.49) for $i = M+1,..,N$ and (5.50) for $i = 0$. Suppose that the statement a) of Theorem 4.1 is not true. Then there exists a sequence $(\nabla p_k) \subset E^q(G)$ such that $\|\nabla p_k\|_{q,G} = 1$ and with

$$(6.1) \quad \epsilon_k := \sup_{0 \neq \nabla v \in E^{q'}(G)} \frac{|<\nabla p_k, \nabla v>|}{\|\nabla v\|_{q'}} \to 0 \ (k \to \infty).$$

Without loss of generality we may assume $\int_G p_k dx = 0$ if G is bounded and $\int_{G_{3R}} p_k dx = 0$ if G is an exterior domain. Then by the Poincaré- inequality we get

$\|p_k\|_{H^{1,q}(G)} \le$ const. if G is bounded, $\|p_k\|_{H^{1,q}(G_{3R})} \le$ const if G is an exterior domain. Since $E^q(G)$ is reflexive there is $\nabla p \in E^q(G)$ and a subsequence (again denoted by p_k) such that ∇p_k converges weakly to ∇p. for $\nabla v \in E^{q'}(G)$ we derive from (6.1) $\langle\nabla p,\nabla v\rangle = \lim_{k\to\infty} \langle\nabla p_k,\nabla v\rangle = 0$. By Lemma 5.18 a) we get $\nabla p = 0$. By the $H^{1,q}$-boundedness we see by means of Rellich's theorem that $p_k \to p$ strongly in $L^q(G)$ resp. $L^q(G_{3R})$. Then $\int_G p dx = 0$ ($\int_{G_{3R}} p dx = 0$) too and therefore $p = 0$, that is $p \to 0$ strongly in L^q on G resp. G_{3R}. Fix now any $i \in \{0,1,..,N\}$. If $i = 0$ let $\Omega := \mathbb{R}^n$, if $i = 1,\dots,N$ let $\Omega := B_{2R_i}(x_i)$. With a constant $C_i' > 0$ we have by (5.35), (5.49) and (5.50)

$$(6.2) \qquad C_i\|\nabla(\varphi_i p_k)\|_q \le \sup_{0\ne v\in C_o^\infty(\Omega)} \frac{\langle\nabla(\varphi_i p_k),\nabla v\rangle}{\|\nabla v\|_{q',G\cap\Omega}} := d_k$$

For each $k \in \mathbb{N}$ there is $v_k \in C_o^\infty(\Omega)$, $\|\nabla v_k\|_{q'} = 1$ and

$$(6.3) \qquad 0 \le d_k - \langle\nabla(\varphi_i p_k),\nabla v_k\rangle \le \frac{1}{k}.$$

Let $\tilde{v}_k := v_k - c_k$, where $c_k := |\Omega|^{-1} \int_\Omega v_k dx$ if $i = 1,..,N$ and $c_k := |G_{3R}|^{-1} \int_{G_{3R}} v_k dx$ for $i = 0$. Then by the Poincaré-inequality we conclude $\|\tilde{v}_k\|_{H^{1,q'}(\Omega)} \le$ const. for $i = 1,\dots,N$ and $\|\tilde{v}_k\|_{H^{1,q'}(G_{3R})} \le$ const for $i = 0$. Again we select a subsequence $(\tilde{v}_k)$ and find $\nabla\tilde{v}_k \to \nabla v$ weakly in $E^{q'}(G)$ and $\tilde{v}_k \to v$ strongly in $L^q(\Omega)$ resp. $L^q(G_{3R})$. By (6.3)

$$\begin{aligned} d_k &\le \frac{1}{k} + \langle\nabla(\varphi_i p_k),\nabla\tilde{v}_k\rangle \\ &= \frac{1}{k} + \langle\nabla p_k,\nabla(\varphi_i\tilde{v}_k)\rangle + \langle p_k\nabla\varphi_i,\nabla\tilde{v}_k\rangle - \langle\nabla p_k,\tilde{v}_k\nabla\varphi_i\rangle \\ &\le \frac{1}{k} + \epsilon_k\|\nabla(\varphi_i\tilde{v}_k)\|_{q'} + |\langle p_k\nabla\varphi_i,\nabla\tilde{v}_k\rangle| + |\langle\nabla p_k,\tilde{v}_k\nabla\varphi_i\rangle| \end{aligned}$$

Since at the support of $\nabla\varphi_i$ we have $p_k \to 0$ strongly in L^q and $\nabla\tilde{v}_k \to \nabla v$ weakly we see $\langle p_k\nabla\varphi_i,\nabla\tilde{v}_k\rangle \to 0$. Analogously $\langle\nabla p_k,\tilde{v}_k\nabla\varphi\rangle \to 0$. Further $\|\nabla(\varphi_i\tilde{v}_k)\|_{q'} \le$ const and $\epsilon_k \to 0$. So we conclude by (6.2) $\|\nabla(\varphi_i p_k)\|_q \to 0$ $(k \to \infty)$ and for $i = 0,1,..,N$. Since $\nabla p_k = \sum_{i=0}^{n} \nabla(\varphi_i p_k)$ we get $\|\nabla p_k\|_q = 1$ forming a contradiction. Part b): By part a) G has property $P_a^1(s)$ for $s = q$ and q', therefore by Lemma 5.1 G has property $P_b^1(s)$ for $s = q$ and q', that is b). ∎

Proof of Theorem 4.2. Part a): Like as in the proof of Theorem 4.1 we construct a covering $B_i = B_{R_i}(x_i), i = 0,1,..,N$ of G and a partition of unity such that for $i = 0,..,N$ e.g. (5.35) holds for $\varphi \in C_0^\infty(B_1)$ and the sup is taken over $v \in C_0^\infty(G \cap B_{2R_1}(x_1))$, analogously for (5.49). For (5.52) the sup is taken over $v \in C_0^\infty(G)$. If G is bounded, let $1 < q < \infty$. If G is an exterior domain and $n \geq 3$, let $1 < q < n$ and if $n = 2$, let $1 < q \leq 2$. Suppose again that (4.4) is not true. Then there is a sequence $(\nabla p_k) \subset E_0^q(G)$ such that $\|\nabla p_k\|_q = 1$ and

$$(6.4) \qquad \epsilon_k := \sup_{0 \neq \nabla v \in E_0^{q'}(G)} \frac{\langle \nabla p_k, \nabla v \rangle}{\|\nabla v\|_{q'}} \to 0.$$

For G bounded, by (2.1)

$$(6.5) \qquad \|p_k\|_q \leq C(G)\, \|\nabla p_k\|_q$$ and for G an exterior domain by (2.9)

$$(6.6) \qquad \|p_k\|_{q,G_{3R}} \leq C(R)\, \|\nabla p_k\|_{q,G}.$$

Again by reflexivity we find a subsequence (again denoted p_k) such that $\nabla p_k \to \nabla p \in E_0^q(G)$ weakly. Since by (6.2) $\langle \nabla p, \nabla v \rangle = 0$ for all $\nabla v \in E_0^{q'}(G)$, by Theorem 5.18 $\nabla p = 0$. By (6.5), (6.6) we conclude $p = 0$. Then by Rellich's theorem $p_k \to 0$ strongly in $L^q(G)$ resp. $L^q(G_{3R})$. Now we proceed like as in the proof of Theorem 4.1: Consider (6.2). Observe that if $x_1 \in \partial G$ then the sup has to be taken over $v \in C_0^\infty(G \cap B_{2R_1}(x_1))$ in (6.2). Again we find v_k with $\|\nabla v_k\|_{q'} = 1$ and (6.3). The use of the Poincaré inequality is replaced by (6.5) and (6.6) for v_k instead of p_k and q' instead of q. The remaining arguments are the same. Part b): If G is bounded, by part a) G has property $P_a^o(s)$ for $s = q$ and q' if $1 < q < \infty$. If G is an exterior domain and $n \geq 3$ G has property $P_a^o(s)$ for $s = q$ and q' if $\frac{n}{n-1} < q < n$. If $n = 2$ the exterior domain G has property $P_a^o(s)$ for $s = q$ and q' if and only if $q = n = 2$. Via Lemma 5.1 part b) follows. ■

7. The exceptional spaces for the Dirichlet problem in exterior domains.

By Theorem 5.18 b) from $\nabla p \in E_0^q(G)$, $G \subset \mathbb{R}^n (n \geq 3)$ exterior domain, and $\langle \nabla p, \nabla \phi \rangle = 0$ for $\nabla \phi \in E_0^{q'}(G)$ we can conclude $\nabla p = 0$ only if $1 < q < n$ and the functional representation (Theorem 4.2) holds only for $\frac{n}{n-1} < q < n$. In the case $G := \{x \in \mathbb{R}^n : |x| > 1\}$ the reason is clear by Remark 5.15. For an arbitrary exterior domain G with $\partial G \in C^1$ we see that the situation in Remark 5.15 is typical:

Theorem 7.1. Let $G \subset \mathbb{R}^n$ $(n \geq 2)$ be an exterior domain with boundary $\partial G \in C^1$, $G = \mathbb{R}^n \setminus \bar{K}$, $\phi \neq K \subset\subset \mathbb{R}^n$. Without loss of generality assume $0 \in K$. Then there exists $h \in C^\infty(G) \cap C^0(\bar{G})$ such that $\Delta h = 0$ in G, $h|_{\partial G} = 0$ and $\nabla h \in E_o^q(G)$ for all q with $n \le q < \infty$ if $n \ge 3$ and $q > 2$ if $n = 2$. Let $0 < r < 1$ such that $B_r \subset\subset G$. Then further there is a harmonic function u in B_r with $u(0) = 0$ and constants $a,b \in \mathbb{R}$, $a \neq 0$, $b \neq 0$ such that for $|x| \ge \frac{1}{r}$

$$(7.1) \qquad h(x) = \begin{cases} a + \frac{b}{|x|^{n-2}} + \frac{1}{|x|^{n-2}} u(\frac{x}{|x|^2}) & \text{for } n \ge 3 \\ a + b \ln|x| + u(\frac{x}{|x|^2}) & \text{for } n = 2 \end{cases}$$

Conversely if $\nabla \tilde{h} \in E_o^q(G)$, where q is subordinate to the restrictions above, and $\langle \nabla \tilde{h}, \nabla \phi \rangle = 0$ for all $\phi \in C_o^\infty(G)$, then there exists $\alpha \in \mathbb{R}$ sucht that $\tilde{h} = \alpha h$.■

The proof (see [9]) is not difficult but somehow lengthy. The main tools are the mean value properties (3.1), (3.2), Weyl's Lemma (= Theorem 3.1), the Kelvin-transform etc.

This theorem has a lot of consequences. To avoid the cumbersome distinction $n \ge 3$ and $n = 2$ we restrict ourselfes in the following considerations to the case $n \ge 3$.

Let $H(G) := \{\alpha \nabla h : \alpha \in \mathbb{R}\}$. For $q \ge n$ let $\nabla h_q := \dfrac{\nabla h}{\|\nabla h\|_q} \in E_o^q(G)$.

Then $H(G) := \{\alpha \nabla h_q : \alpha \in \mathbb{R}\}$. Define $F_q^* : H(G) \to \mathbb{R}$ by

(7.2) $\quad F_q^*(\nabla g) = \alpha \quad$ for $g = \alpha h_q$.

Then $\|F_q^*\| = \sup_{0 \neq g \in H} \dfrac{F_o^*(\nabla g)}{\|\nabla g\|_q} = 1$. Extend F_q^* norm-preserving by the Hahn-Banach theorem to a continuous linear functional defined on the whole space $E_o^q(G)$ and denote if again by F_q^*. Define

(7.3) $\quad P_q(\nabla p) := F_q^*(\nabla p) \nabla h_q \quad$ for $\nabla p \in E_o^q(G)$.

Then $P_q : E_o^q(G) \to H(G)$ is a projection, that is $P_q^2 = P_q$, with the additional property $\|P_q\| = 1$.

Define

(7.4) $\quad P_{c,q} := I - P_q$

and

(7.5) $\quad E_{o,c}^q(G) := P_{c,q}(E_o^q(G))$.

Then we have in the sense of a direct sum

(7.6) $\quad E_o^q(G) = E_{o,c}^q(G) \oplus H(G)$

and there is a constant $K > 0$ such that

(7.7) $$\|\nabla p\|_q \geq K(\|P_{c,q}(\nabla p)\|_q + \|P_q(\nabla p)\|_q)$$

for all $\nabla p \in E_o^q(G)$. Clearly $E_{o,c}^q(G)$ is topologically equivalent to the quotient space $E_o^q(G)/H(G)$.

Concerning the estimates from Theorem 4.2, we have the following extension of Theorem 4.2:

<u>Theorem 7.2.</u> For G assume the same as in Theorem 7.1 and let $n \geq 3$. Then for $1 < q < \infty$ there exists a constant $C = C(G,q) > 0$ such that

a) for $1 < q \leq \frac{n}{n-1}$ (then $q' \geq n$) and all $\nabla p \in E_o^q(G)$

(7.8) $$\|\nabla p\|_q \leq C \sup_{0 \neq \nabla\phi \in E_{o,c}^{q'}(G)} \frac{|\langle\nabla p, \nabla\phi\rangle|}{\|\nabla\phi\|_{q'}} .$$

b) for $\frac{n}{n-1} < q < n$ estimate (4.4) holds

c) for $q \geq n$ (then $q' \leq \frac{n}{n-1}$) and all $\nabla p \in E_{o,c}^q(G)$

(7.9) $$\|\nabla p\|_q \leq C \sup_{0 \neq \nabla\phi \in E^{q'}(G)} \frac{|\langle\nabla p, \nabla\phi\rangle|}{\|\nabla\phi\|_{q'}} . \quad \blacksquare$$

Observe that (7.8) is sharper then (4.4), because the variational class is smaller. (7.9) is an extension of (4.4) to the case $q > n$ but for the narrower class $E_{o,c}^q(G) \subsetneq E_o^q(G)$.

Concerning the functional representation we get:

<u>Theorem 7.3.</u> For $G \subset \mathbb{R}^n$ assume the same as in Theorem 7.1 and let $n \geq 3$. Then for $1 < q < \infty$ with the constants $C = C(G,q) > 0$ by Theorem 7.2 holds:

a) If $1 < q \leq \frac{n}{n-1}$ and given $F^* \in (E_{o,c}^{q'}(G))^*$,

$$\|F^*\|_{(E_{o,c}^{q'})^*} := \sup_{0 \neq \nabla\phi \in E_{o,c}^{q'}} \frac{|F^*(\nabla\phi)|}{\|\nabla\phi\|_q}, \text{ then}$$

there is a unique $\nabla p \in E_o^q(G)$ such that

(7.10) $$F^*(\nabla\phi) = \langle\nabla p, \nabla\phi\rangle \text{ for all } \nabla\phi \in E_{o,c}^{q'}(G)$$

Further

(7.11) $$\|\nabla p\|_q \leq C\|F^*\|_{(E_{o,c}^{q'})^*} \leq C\|\nabla p\|_q$$

b) If $\frac{n}{n-1} < q < n$ then Theorem 4.2 b) holds.

c) If $q \geq n$ and given $F^* \in (E_o^{q'}(G))^*$, then there is a unique $\nabla p \in E_{o,c}^q(G)$ such that

(7.12) $\quad F^*(\nabla\phi) = \langle\nabla p,\nabla\phi\rangle$ for all $\nabla\phi \in E_o^{q'}(G)$

Further $\|\nabla p\|_q \leq C\|F^*\|_{(E_o^{q'})^*} \leq C\|\nabla p\|_q$.

For case a) we have in addition

Theorem 7.4. Let the same assumptions as in Theorem 7.1 hold and let $1 < q \leq \frac{n}{n-1}$ $(n \geq 3)$. Given $F^* \in (E_o^{q'}(G))^*$, then there exists $\nabla p \in E_o^q(G)$ with $F^*(\nabla\phi) = \langle\nabla p,\nabla\phi\rangle$ for all $\nabla\phi \in E_o^q(G)$ if and only if $F^*(\nabla h) = 0$ (h by Theorem 7.1). Then in addition $\|\nabla p\|_q \leq C\|F^*\|_{(E_o^q(G))^*} \leq C\|\nabla p\|_q$.

8. Applications, concluding remarks. We are now able to prove existence of weak solutions for the Neumann- and Dirichlet problem in bounded as well as in exterior domains. E.g. let G be an exterior domain and let $f \in C_o^o(G)$ be given. Let $F(\phi) := \langle f,\phi\rangle$ for $\phi \in C_o^\infty(G)$. Suppose $\operatorname{supp}(f) \subset B_R$ for some $R > 0$. By (2.9) we get $|F(\phi)| \leq \|f\|_{q,G_R}\|\phi\|_{q',G_R} \leq C(R)\|f\|_q \|\nabla\phi\|_{q'}$.
Then for $\frac{n}{n-1} < q < \infty$ there exists $\nabla p \in E_o^q(G)$ with

(7.13) $\quad \langle\nabla p,\nabla\phi\rangle = F(\phi) = \langle f,\phi\rangle$ for $\phi \in E_o^{q'}(G)$.

If in addition $\int_G fh dx = 0$, then for $1 < q \leq \frac{n}{n-1}$ there is again $\nabla p \in E_o^q(G)$ with (7.13). Clearly p is a weak solution of the Dirichlet problem "$-\Delta p = f$ in G and $p|_{\partial G} = 0$". It is not difficult to see that for $|x| \geq R_o$, $R_o > R$ sufficiently big, a representation like as in (7.1) holds (since $\Delta p = 0$ for $|x| > R$). In case $1 < q \leq \frac{n}{n-1}$ follows $a = b = 0$. That means $|p(x)| \leq \frac{c}{|x|^{n-1}}$.
In case $\frac{n}{n-1} < q < n$ follows $a = 0$ and $|p(x)| \leq \frac{c}{|x|^{n-2}}$. Analogous results hold for the Neumann problem.
Most important applications are in connection with the Stokes problem in bounded as well as in exterior domains. With ideas similar to that one used here, Galdi and Simader [3] proved existence, uniqueness and L^q-estimates for the Stokes problem in exterior domains $G \subset \mathbb{R}^3$.
A most convincing application of Theorem 4.1 is given in Simader and Sohr [8] in their proof of the Helmholtz decomposition. Moreover, in turn the Helmholtz

decomposition is equivalent (see [8]) to Theorem 4.1. The results in [8] extend those given by Fujiwara and Morimoto [2] to unbounded domains too.
It is well known (see e.g. [5], p. 337 and p. 341 or [6], p. 99 and p. 103) that for bounded domains G with $\partial G \in C^1$ and $1 < s < \infty$ there is a well defined continuous linear trace operator $V_s : H^{1,s}(G) \to W^{1-1/s,s}(\partial G)$ such that for $p \in C^1(\bar{G})$ we have $V_s p = p|_{\partial G}$. If $W^{-1/s',s'}(\partial G) := (W^{1-1/s,s}(\partial G))^*$ equipped with the "dual space norm", then in [8] is shown that for the subspace $F^{s'}(G) := \{\nabla p \in E^{s'}(G) : \Delta p \in L^{s'}(G)\}$ of $E^{s'}(G)$ equipped with norm $\|\nabla p\|_{F^{s'}} := (\|\nabla p\|_{s'}^{s'} + \|\Delta p\|_{s'}^{s'})^{1/s'}$ there is a continuous linear trace operator $S_{s'} : F^{s'}(G) \to W^{-1/s',s'}(\partial G)$ such that for $p \in C^\infty(\bar{G})$ we have $S_{s'}(\nabla p) = \partial_N p|_{\partial G}$, where $\partial_N p|_{\partial G}(x) = \sum_{i=1}^{n} N_i(x)\partial_i p(x)|_{\partial G}$ and $N(x)$ denotes the outward unit normal vector in x_0 at ∂G.

Via difference quotient methods, using (4.1) and (4.4) respectively, like as in the case $q = 2$ higher differentiability properties of weak solutions of e.g. equation (7.13) can be proved (compare e.g. [7].
The case of arbitrary elliptic operators of second order for G bounded, and under additional asymptotic assumptions for the coefficients for exterior domains, is reduced to Δ by elementary coordinate transform and standard localization procedures.

<u>References:</u>

[1] Alt, H.W.: Lineare Funktionalanalysis. Berlin, Heidelberg, New York: Springer 1985.

[2] Fujiwara, D. and Morimoto, H.: An L_r-theorem of the Helmholtz decomposition of vector fields. J. Fac. Sci. Univ. Tokyo, Sec. I 24, 685-700 (1977).

[3] Galdi, G.P. and Simader, C.G.: Existence, uniqueness and L^q-estimates for the Stokes problem in an exterior domain. To appear in: Arch. Ratl. Mech. Anal.

[4] Gilbarg, D. and Trudinger, N.S.: Elliptic partial differential equations of second order. Berlin, Heidelberg, New York: Springer 1977.

[5] Kufner, A., John, O. and Fucik, S.: Function spaces. Prague: Academia 1977.

[6] Necas, J.: Les méthodes directes en théorie des équations elliptiques. Paris-Prague: Masson, Academia 1967.

[7] Simader, C.G.: On Dirichlet's boundary value problem. Lecture Notes in Math. 268, Springer-Verlag 1972.

[8] Simader, C.G. and Sohr, H.: A new approach to the Helmholtz decomposition and the Neumann problem in L^q-spaces for bounded and exterior domains. To appear.

[9] Simader, C.G. and Sohr, H.: The weak Dirichlet and Neumann problem in L^q for second order elliptic equations in bounded and exterior problems. To appear as a book.

[10] Stein, E.M.: Singular integrals and differentialbility properties of functions. Princeton: Princeton University Press 1970.

BOUNDARY BEHAVIOUR OF DIFFERENTIABLE FUNCTIONS AND RELATED TOPICS

Sergej K. Vodop'yanov
Novosibirsk, USSR

A description of restrictions of differentiable functions to sets $F\subset\mathbb{R}^n$ is a classical topic in the function theory arising from papers by Lebesgue [1] and Whitney [2]. There is a numerous bibliography concerning the Sobolev spaces $W_p^\ell(\mathbb{R}^n)$, $\ell\in\mathbb{N}$, the Nikol'skii spaces $H_p^\ell(\mathbb{R}^n)$, the Besov spaces $B_{p,q}^s(\mathbb{R}^n)$ and the spaces of Bessel potentials or the Liouville spaces $L_p^\ell(\mathbb{R}^n)$, $\ell\in\mathbb{R}_+$. (Note, that $W_p^\ell(\mathbb{R}^n) = L_p^\ell(\mathbb{R}^n)$, $\ell\in\mathbb{N}$, and $H_p^\ell(\mathbb{R}^n) = B_{p,\infty}^\ell(\mathbb{R}^n)$, $\ell\in\mathbb{R}_+$.)

In this paper we study the boundary behaviour of differentiable functions from the Sobolev classes $W_p^\ell(\mathbb{R}^n)$ and the Nikol'skii classes $H_p^\ell(\mathbb{R}^n)$ for $p = \infty$. The domain of definition of the function classes in question is an arbitrary connected open set in the Euclidean space $\mathbb{R}^n$, $n\geq 2$. Without considering regularity properties of the boundary of the domain we need to introduce new concepts and a language for the description of the boundary behaviour. We establish that boundary values exist always but we shall understand them in a special sense.

Our method is based on a new equivalent normalization of the Sobolev and Nikol'skii spaces in domains which include geometrical characteristics of the domain in an explicit way. The geometry of the domain is determined by the modulus of continuity defining the function space. The inner geometry of the domain reflects the substance of the studied problem and represents suitable tools to its resolution.

The first application deals with traces (boundary values) of functions on the boundary of the domain of definition. The boundary can be obtained by means of the completion of the domain with respect to the corresponding metric. The elements of the function space are extended by continuity to the boundary and these traces belong to some function space. It is proved that such a characterization of traces is reversible.

The second application concerns some necessary and sufficient extendibility conditions for differentiable functions across the boundary of the domain of definition. They are formulated in terms of the equivalence of the corresponding metrics in the domain and in the surrounding space for $p = \infty$ and involve some conditions on the measure in spaces with integral metrics.

1. Geometric normalizations in spaces of differentiable functions

1.1. Sobolev and Nikol'skii spaces

For an arbitrary $\alpha\in(0,1]$ we define the inner α-metric $d_{\alpha,G}(x,y)$ on a domain $G\subset\mathbb{R}^n$ as follows

$$d_{\alpha,G}(x,y) = \inf_{\gamma} \sum_{i=1}^{m} |x_i - x_{i-1}|^{\alpha},$$

where the infimum is taken over all broken lines γ consisting of segments $[x_{i-1},x_i]\subset G$, $x_0 = x$, $x_m = y$. It is evident that the 1-metric coincides with the infimum of the lengths of rectifiable curves connecting the points $x,y\in G$, and therefore it is the inner metric of the domain in the commonly accepted sense (see [3]). We denote by G_α the metric space $(G,d_{\alpha,G})$.

The elements of the function space $\mathrm{Lip}(\ell,G_\alpha)$, $\ell\in\mathbb{R}_+ = \{x:\ x>0\}$, $\ell\in(k,k+1]$, $k = 0,1,2,\dots,$. are L_∞-functions $f:G\longrightarrow\mathbb{R}$, whose weak partial derivatives, denoted by $D^j f$, also belong to L_∞ for $|j|\le k$. The norm in $\mathrm{Lip}(\ell,G_\alpha)$ is defined by

$$\|f\,|\mathrm{Lip}(\ell,G_\alpha)\| = \sum_{|j|\le k}\left\{\|D^j f\,|L_\infty(G)\| + \sup \frac{|R_j(x,y)|}{\bar d_{\alpha,G}(x,y)^{\ell-|j|}}\right\},$$

where $\bar d_{\alpha,G}(x,y) = d_{\alpha,G}(x,y)^{1/\alpha}$, $\alpha = \ell-k$ and

$$D^j f(x) = \sum_{|j+s|\le k} \frac{D^{j+s}f(y)}{s!}(x-y)^s + R_j(x,y),\quad 0\le|j|\le k.$$

We use the usual multi-index notation, $j = (j_1,j_2,\dots,j_n)$, $s = (s_1,s_2,\dots,s_n)$, $s! = s_1!s_2!\dots s_n!$, $|j| = j_1+j_2+\dots+j_n$ and $x^s = x_1^{s_1}x_2^{s_2}\dots x_n^{s_n}$.

Two function spaces coincide if the operator of embedding of one of the spaces into the other is a bounded isomorphism.

THEOREM 1.1. Let G be an arbitrary domain in $\mathbb{R}^n$. The following function spaces coincide:

$$\mathrm{Lip}(\ell,G_\alpha) = W^\ell_\infty(G) \text{ for } \ell\in\mathbb{N} \quad\text{and}\quad \mathrm{Lip}(\ell,G_\alpha) = H^\ell_\infty(G) \text{ for } \ell\notin\mathbb{N}.$$

Here the space $W^\ell_\infty(G)$ and $H^\ell_\infty(G)$ is a member of the scale of the Sobolev spaces $W^\ell_p(G)$ [4] and the Nikol'skii spaces $H^\ell_p(G)$ [5, 6], respectively. Let us recall that the elements of $W^\ell_p(G)$, $\ell\in\mathbb{N}$, $p\in[1,\infty]$ and $H^\ell_p(G)$, $\ell\in\mathbb{R}_+$, $\ell\in(k,k+1)$, $k = 0,1,2,\ldots$, $p\in[1,\infty]$ are L_p-functions f whose weak partial derivatives, denoted by D^jf, also belong to L_p for $|j|\le\ell$ and $|j|\le k$, respectively. The norm in $W^\ell_p(G)$ and in $H^\ell_p(G)$ is defined by

$$\|f|W^\ell_p(G)\| = \|f|L_p(G)\| + \|\nabla_\ell f|L_p(G)\|$$

or

$$\|f|H^\ell_p(G)\| = \|f|L_p(G)\| + \sup \frac{\|\Delta_h\nabla_k f|L_p(G)\|}{|h|^\alpha},$$

respectively, where $\nabla_\ell = \{D^jf\colon\ |j| = \ell\}$ and $\alpha = \ell-k$,

$$\Delta_h g(x) = \begin{cases} g(x+h) - g(x) & \text{if the segment } [x,x+h]\subset G, \\ 0 & \text{otherwise.} \end{cases}$$

The Nikol'skii spaces $H^\ell_p(G)$ belong to the scale of the Besov spaces $B^\ell_{p,q}(G)$, $q\in[1,\infty]$, as well, namely,

$$H^\ell_p(G) = B^\ell_{p,\infty}(G) \text{ [see 5, 6].}$$

P r o o f of Theorem 1.1. If $f\in\mathrm{Lip}(\ell,G)$, $\ell\in\mathbb{N}$, then any point $x\in G$ is the center of some ball B in which all the derivatives of the function f of the order $\ell-1$ are bounded and satisfy the Lipschitz condition. This implies that the function f has bounded generalized derivatives of order ℓ. Thus the boundedness of the embedding $\mathrm{Lip}(\ell,G_\alpha)\longrightarrow W^\ell_\infty(G)$, $\ell\in\mathbb{N}$, is proved. The continuity of the embedding

$$\mathrm{Lip}(\ell,G_\alpha)\longrightarrow H^\ell_\infty(G),\quad \ell\in\mathbb{N},$$

is evident.

Now we shall prove the continuity of the converse embedding. Let us consider any function from of the Sobolev or the Nikol'skii space with the norm equal to 1. The estimates $|D^jf(x)|\le c$, $|j|\le\ell$, $x\in G$, are proved in [5].

Therefore it remains to establish the inequalities

$$(1.1)\qquad |R_j(x,y)| \le c_j \bar{d}_{\alpha,G}(x,y)^{\ell-|j|},$$

where $\alpha = \ell-k$, $k\in\mathbb{N}$, $k<\ell\le k+1$, $|j|\le k$ and $c_j = n^{(k-|j|)/2}$. We shall prove the estimate (1.1) by induction. If $|j| = k$, then $R_j(x,y) = D^jf(x) - D^j(y)$. We have the following estimate for the difference of the derivative D^jf, $|j| = k$, along a broken line consisting of segments $[x_{i-1},x_i]\subset G$, $i = 1,\dots,m$, $x_0 = x$, $x_m = y$:

$$(1.2)\qquad |D^jf(x) - D^jf(y)| \le \sum_{i=1}^{m} |D^jf(x_i) - D^jf(x_{i-1})| \le \sum_{i=1}^{m}|x_i - x_{i-1}|^{\alpha},$$

where $\alpha = 1$ if $f\in W_\infty^\ell(G)$, and $\alpha = \ell-k$ if $f\in H_\infty^\ell(G)$. Now minimizing the right hand side of (1.2), we obtain the estimate

$$(1.3)\qquad |R_j(x,y)| = |D^jf(x) - D^jf(y)| \le d_{\alpha,G}(x,y) = d_{\alpha,G}(x,y)^{(\ell-|j|)/\alpha},$$

where $|j| = k$ and α depends on the choice of the function space made in the above mentioned way.

The inequalities (1.3) constitute the basis for the induction. Let us suppose that the estimate (1.1) with $c_j = n^{(k-|j|)/2}$ is proved for all j such that $0\le p<|j|\le k$, $p = 0,1,\dots,k-1$. For any multi-index j, $|j| = p$, we shall estimate $R_j(x,y)$, where the points $x,y\in G$ are fixed. Since the value $R_j(y,y)$ in question is equal to 0, it is

$$(1.4)\qquad |R_j(x,y)| \le \sup\left\{|\nabla R_j(\xi,y)| \sum_i|a_i|:\ \xi\in G,\ d_{\alpha,G}(\xi,y) \le d_{\alpha,G}(x,y) + \varepsilon\right\},$$

where the segments a_j form a broken line with endpoints x and y, for which

$$\sum|a_i|^{\alpha} \le d_{\alpha,G}(x,y) + \varepsilon.$$

Since

$$|\nabla R_j(\xi,y)|^2 = \sum_{|j+k|=p+1} |R_{j+k}(\xi,y)|^2$$

and

$$\sum_i|a_i| \le \left[\sum_i|a_i|^{\alpha}\right]^{1/\alpha} \le \left[d_{\alpha,G}(x,y) + \varepsilon\right]^{1/\alpha}$$

by the reverse Minkowski inequality (see [7]), we obtain from (1.4) that

$$\begin{aligned}|R_j(x,y)| &\le n^{1/2} n^{(k-(p+1))/2}\, \bar{d}_{\alpha,G}(x,y)^{\ell-(p+1)}\, \bar{d}_{\alpha,G}(x,y)\\ &= n^{(k-p)/2}\, \bar{d}_{\alpha,G}(x,y)^{\ell-p}.\end{aligned}$$

This proves the inequality (1.1) for every j, $|j| = p$. □

Another application concerns some necessary and sufficient condition for an extension of differentiable functions beyond the boundary. They are formulated in terms of equivalence of the corresponding metric in the domain and in the surrounding space for $p = \infty$ and they involve certain conditions on the measure in spaces with integral metrics.

1.2. Zygmund spaces

The case of Sobolev and Nikol'skii spaces studied in Section 1.1 differs essentially from the case of the spaces H^{ℓ}_{∞}, $\ell\in\mathbb{N}$, considered here. If $\ell = 1$ then a function f belongs to $H^1_{\infty}(\mathbb{R}^n)$ if and only if it has finite norm

$$\|f\,|H^1_{\infty}(\mathbb{R}^n)\| = \sup_{x\in\mathbb{R}^n} |f(x)| + \sup_{\substack{x\in\mathbb{R}^n \\ 0\neq h\in\mathbb{R}^n}} \frac{|\Delta^2_h f(x)|}{|h|}.$$

As usual, $\Delta^2_h f(x) = f(x-h) - 2f(x) + f(x+h)$. For an arbitrary $\ell\in\mathbb{N}$, we define

$$H^{\ell}_{\infty}(\mathbb{R}^n) = \left\{f:\ f\in H^{\ell-1}_{\infty}(\mathbb{R}^n),\ \|f\,|H^{\ell}_{\infty}(\mathbb{R}^n)\| = \|f\,|H^{\ell-1}_{\infty}(\mathbb{R}^n)\| + \sum_{|j|=\ell-1} \sup_{\substack{x\in\mathbb{R}^n \\ 0\neq h\in\mathbb{R}^n}} \frac{|\Delta^2_h D^j f|}{|h|} < \infty\right\}.$$

Here we discuss the natural definition of the Zygmund spaces in domains. Our definition is in fact a successive realization of the concept that the norm of a function in a domain has to take into account the inner geometry of the domain.

Let g be an open set in $\mathbb{R}^n$ and k be a natural number. We say that a jet $\{f_j:\ |j|\le k-1\}$ of functions defined on G belongs to the class $\Lambda^k(G)$ if there exist collections $\{f_{j,\nu}:\ |j|\le k\}$, $\nu\in\mathbb{N}$, of functions defined on G, and a constant $M>0$ such that for all points $x,y\in G$ the following conditions hold:

$$(1.5)\qquad |f_j(x) - f_{j,\nu}(x)| \le M2^{-\nu(k-|j|)},\quad \nu\ge1,\ |j|\le k-1,$$

$$(1.6)\qquad |f_{j,\nu}(x) - f_{j,\mu}(x)| \le M2^{\mu-\nu},\quad \mu\ge\nu\ge1,\ |j| = k,$$

$$(1.7)\qquad \left|f_{j,\nu}(x) - \sum_{|j+s|\le k} \frac{f_{j+s,\nu}(y)}{s!}(x-y)^s\right| \le M2^{\nu} d_{1,G}(x,y)^{k+1-|j|},\quad |j|\le k,\ \nu\ge1,$$

(1.8) $$|f_{j,1}(x)| \le M, \quad |j|\le k.$$

The norm $\|f|\Lambda^k(G)\|$ of the jet $f = \{f_j: |j|\le k-1\}$ is equal to infimum of all M satisfying conditions (1.5) - (1.8) with some $\{f_{j,\nu}: |j|\le k\}$, $\nu\in\mathbb{N}$.

THEOREM 1.2 [8]. *For all* $k\in\mathbb{N}$ *the spaces* $H^k_\infty(\mathbb{R}^n)$ *and* $\Lambda^k(\mathbb{R}^n)$ *coincide.*

2. Boundary values of differentiable functions

2.1. Trace operator

Recall that $k = 0,1,2,\ldots$ and $\ell\in(k,k+1]$. Let us consider the metric space $G_\alpha = (G, d_{\alpha,G})$, $\alpha\in(0,1]$. Let $\tilde{G}_\alpha$ be the completion of G_α with respect to the metric $d_{\alpha,G}$. It is easy to verify that any function of $Lip(\ell,G_\alpha)$ extends by the continuity to the completion $\tilde{G}_\alpha$. Indeed, if $|j| = k$, then $|f^{(j)}(x) - f^{(j)}(y)| \le R_j(x,y) \le Md_{\alpha,G}(x,y)$, where $f^{(j)}(x) = D^jf(x)$. Hence we have an extension by continuity to $\tilde{G}$ of the functions $f^{(j)}(x)$, $|j| = k$. Let us now suppose that the functions $f^{(j)}(x) = D^jf(x)$, $p\le|j|\le k$, $p = 1,2,\ldots,k$, are already extended to $\tilde{G}_\alpha$. The extension to $\tilde{G}_\alpha$ of the functions $f^{(j)}(x) = D^jf(x)$, $|j| = p-1$, follows from the expansion

$$f^{(j)}(x) - f^{(j)}(y) = \sum_{|j|<|j+s|\le k} \frac{f^{(j+s)}(y)}{s!}(x-y)^s + R_j(x,y)$$

and the inequalities $|R_j(x,y)| \le Md_{\alpha,G}(x,y)^{(\ell-(p-1))/\alpha}$, $|x-y|^\alpha \le d_{\alpha,G}(x,y)$, hold. As there exists a unique extension of the functions from $Lip(\ell,G_\alpha)$ to $\tilde{G}_\alpha$ it is natural to introduce the function class $Lip(\ell,\tilde{G}_\alpha)$, consisting of the jets $\{f^{(j)}(x): |j|\le k\}$ defined on $\tilde{G}_\alpha$ as the extension by continuity of collections from $Lip(\ell,G_\alpha)$ with respect to the metric $d_{\alpha,G}(x,y)$.

In order to describe the class $Lip(\ell,\tilde{G}_\alpha)$ in an invariant way, let us define the mapping $i_\alpha:\tilde{G}_\alpha\longrightarrow G$ as the extension by continuity of the identical mapping on G. This extension exists by the inequality $|i_\alpha(x) - i_\alpha(y)|^\alpha \le d_{\alpha,G}(x,y)$. The elements of the class $Lip(\ell,\tilde{G}_\alpha)$ are the jets $\{f^{(j)}: |j|\le k\}$, consisting of the continuous functions defined on $\tilde{G}_\alpha$. The norm in the space $Lip(\ell,\tilde{G}_\alpha)$ is

$$(2.1) \qquad \|f|\mathrm{Lip}(\ell,\tilde{G}_\alpha)\| = \sum_{|j|\le k}\left\{\sup|f^{(j)}(x)| + \sup\frac{|R_j(x,y)|}{\bar{d}_{\alpha,G}(x,y)^{\ell-|j|}}\right\},$$

where the supremum is taken over all points $x,y\in G_\alpha$, $\alpha = \ell-k$, and

$$f^{(j)}(x) = \sum_{|j+s|\le k}\frac{f^{(j+s)}(y)}{s!}(i_\alpha(x)-i_\alpha(y))^s + R_j(x,y).$$

It is easy to verify that $f^{(j)}(x) = D^j f(x)$ for $x\in G$, $|j|\le k$.

Thus we have proved the following assertion.

PROPOSITION 2.1. *Let* G *be an arbitrary domain in* $\mathbb{R}^n$. *Then there exists a natural isometry of the spaces* $\mathrm{Lip}(\ell,G_\alpha)$ *and* $\mathrm{Lip}(\ell,\tilde{G}_\alpha)$: *any element of* $\mathrm{Lip}(\ell,G_\alpha)$ *possesses a unique extension to* $\tilde{G}_\alpha$ *with respect to the inner* α*-metric* $d_{\alpha,G}(x,y)$ *which is an element of* $\mathrm{Lip}(\ell,\tilde{G}_\alpha)$.

Further, the set $\partial G_\alpha = \tilde{G}_\alpha\setminus G_\alpha$ will be called the *α-boundary* of the domain G. It is natural to consider the restrictions of the elements of $\mathrm{Lip}(\ell,\tilde{G}_\alpha)$ to ∂G_α as *boundary values* or *traces of the functions from* $\mathrm{Lip}(\ell,G_\alpha)$ *to the α-boundary of the domain* G. To give a rigorous definition of the concept of the trace we denote by $\mathrm{Lip}(\ell,\partial G_\alpha)$ the function space whose elements are jets $\{f^{(j)}: |j|\le k\}$, consisting of continuous functions defined on $(\partial G_\alpha, d_{\alpha,G})$ with the finite norm (2.1), where the supremum is taken over all points $x,y\in\partial G_\alpha$. The trace operator $tr_\ell:\mathrm{Lip}(\ell,G_\alpha)\longrightarrow\mathrm{Lip}(\ell,\partial G_\alpha)$ is defined as the superposition of the isometry $i:\mathrm{Lip}(\ell,G_\alpha)\longrightarrow\mathrm{Lip}(\ell,\tilde{G}_\alpha)$ given in Proposition 2.1 and the restriction of the elements of $\mathrm{Lip}(\ell,\tilde{G}_\alpha)$ to ∂G_α, which are contained in the class $\mathrm{Lip}(\ell,\partial G_\alpha)$.

Let us formulate the above as the following theorem.

THEOREM 2.1 [9]. *Let* G *be an arbitrary domain in* $\mathbb{R}^n$. *Then there exist the bounded trace operators*

$$tr_\ell: W^\ell_\infty(G)\longrightarrow\mathrm{Lip}(\ell,\partial G_1),\quad \alpha = 1,\ \ell\in\mathbb{N},$$

$$tr_\ell: H^\ell_\infty(G)\longrightarrow\mathrm{Lip}(\ell,\partial G_\alpha),\quad \alpha = \ell-k,\ \ell\in(k,k+1),\ k = 0,1,\dots,$$

defined by the continuity with respect to the metric $d_{\alpha,G}$.

2.2. Extension operator

The characterization of the trace of the functions from the Sobolev and Nikol'skii classes given by Theorem 2.1, are reversible.

<u>*THEOREM 2.2*</u> [9]. *Let G be an arbitrary domain in* $\mathbb{R}^n$. *Then there exist the linear bounded extension operators*

$$\mathrm{ext}_k : \mathrm{Lip}(\ell, \partial G) \longrightarrow W^{\ell}_{\infty}(G), \quad k = \ell-1, \quad \ell \in \mathbb{N},$$

$$\mathrm{ext}_k : \mathrm{Lip}(\ell, \partial G_\alpha) \longrightarrow H^{\ell}_{\infty}(G), \quad \ell \in (k, k+1), \quad k = 0, 1, \dots, \quad \alpha = \ell - k,$$

such that $\mathrm{tr}_\ell \circ \mathrm{ext}_k$ *is the identical mapping.*

P r o o f. The mapping $i_\alpha : (G_\alpha, d_{\alpha,G}) \longrightarrow (G, |x-y|^\alpha)$ has the following property: If $U \subset G$ is a convex set, then $|i_\alpha(x) - i_\alpha(y)|^\alpha = d_{\alpha,G}(x,y)$ for every $x, y \in U$. Hence, we have the equality

$$d_{\alpha,G}(x, \partial G) = |i_\alpha(x) - \partial G|^\alpha, \quad x \in G, \tag{2.2}$$

where

$$|i_\alpha(x) - \partial G|^\alpha = \inf\left\{|x-y|^\alpha : \ y \in \partial G\right\}$$

and

$$d_{\alpha,G}(x, \partial G_\alpha) = \inf\left\{d_{\alpha,G}(x,y) : \ y \in G_\alpha\right\}.$$

This relation implies that the space $G_\alpha = (G, d_{\alpha,G})$ can be decomposed into the "Whitney cubes" in the same way as it is done for the domain G with respect to the Euclidean metric in the Whitney extension theorem (see [10]).

Let G be an open set in $\mathbb{R}^n$. Then there exists a collection of closed cubes Q_k with sides parallel to the axes and with the following properties:

(a) $G = \cup Q_k$.

(b) The interiors of the cubes Q_k are mutually disjoint.

(c) The distance $d(Q_k, F)$ of Q_k to F satisfies

$$\mathrm{diam}\, Q_k \le d(Q_k, F) \le 4\,\mathrm{diam}\, Q_k, \quad k \in \mathbb{N}.$$

(d) If $Q_k \cap Q_\nu \neq \emptyset$, then

$$\frac{1}{4}\,\mathrm{diam}\, Q_k \le \mathrm{diam}\, Q_\nu \le 4\,\mathrm{diam}\, Q_k.$$

(e) Let ε be a fixed number satisfying $0<\varepsilon<1/4$, and let Q_k^* denote the cube which has the same center as Q_k but is expanded by the factor $1+\varepsilon$. Then each point in G is contained in at most N_0 cubes Q_k^*, where N_0 is a fixed number.

Furthermore, $Q_k^*\cap Q_\nu \neq \emptyset$ if and only if $Q_k\cap Q_\nu \neq \emptyset$.

In connection with the decomposition, we shall use the following notation:

x_k is the center of Q_k,

ℓ_k is the diameter of Q_k,

a_k is the length of the side of Q_k (thus, $\ell_k = \sqrt{n}a_k$).

Now, consider a partition of unity: Let ψ be a C^∞-function satisfying $0\le\psi\le1$, $\psi(x) = 1$ for $x\in Q$ and $\psi(x) = 0$ for $x\notin(1+\varepsilon)Q$, where Q denotes the cube centered at the origin with sides of length 1 parallel to the axes. Define ψ_k by $\psi_k(x) = \psi((x-x_k)/a_k)$, and φ_k by $\varphi_k(x) = \psi_k(x)/\sum_k\psi_k(x)$, $x\in G$. Then $\varphi_k(x) = 0$ for $x\notin Q_k^*$, $\sum_k \varphi_k(x) \equiv 1$ on G, and it is easy to show that for any multi-index j

$$|D^j\varphi_k(x)| \le A_j(\operatorname{diam} Q_k)^{-|j|} \tag{2.3}$$

Using the equality (2.1) we can rewrite the property (c) in the following way:

$$\operatorname{diam} Q_k \le \bar{d}_{\alpha,G}(Q_k,\partial G) \le 4 \operatorname{diam} Q_k. \tag{c'}$$

Note that if $x\in Q_k^*$, then

$$\bar{d}_{\alpha,G}(x,\partial G_\alpha) \sim \operatorname{diam} Q_k \tag{2.4}$$

and

$$\bar{d}_{\alpha,G}(Q_k^*,\partial G_\alpha) \sim \operatorname{diam} Q_k. \tag{2.5}$$

As usual, the symbol $\sim$ denotes equivalent quantities. Furthermore, if $y\in\partial G_\alpha$ and $x\in Q_k^*$, then

$$d_{\alpha,G}(y,p_k) \le d_{\alpha,G}(y,x) + d_{\alpha,G}(x,p_k),$$

where $p_k\in\partial G_\alpha$ is a point for which $d_{\alpha,G}(Q_k,\partial G_\alpha) = d_{\alpha,G}(Q_k,p_k)$. Hence, according to (2.4) and (2.5), for $y\in\partial G_\alpha$ and $x\in Q_k^*$ we have

$$d_{\alpha,G}(y,p_k) \le c d_{\alpha,G}(y,x), \tag{2.6}$$

where c is a constant.

Let $\{f^{(j)}: |j|\le k\}$ be a collection of functions defined on ∂G_α. We define the value of an extension operator ext_k on the collection $\{f^{(j)}\}$ as follows:

$$\operatorname{ext}_k(\{f^{(j)}\}) = \begin{cases} f^{(0)}(x), & x\in\partial G_\alpha, \\ \sum_i' P(x,p_i)\varphi_i(x), & x\in G, \end{cases} \tag{2.7}$$

where $P(x,y)$ is the polynomial giving the "Taylor expansion" of the function f with respect to the point $y\in\partial G_\alpha$, i.e.

$$P(x,y) = \sum_{|s|\le k} \frac{f^{(s)}(y)}{s!}(x-i_\alpha(y))^s, \quad x\in G,\ y\in\partial\tilde{G}_\alpha,$$

and $p_i\in\partial G_\alpha$ is the closest point to the cube Q_i. The symbol $\sum'$ indicates that the summation is taken over all cubes whose distance to ∂G_α does not exceed 1. It is easy to see that the value of the extension operator ext_k on the collection $\{f^{(j)}\}$ is a function $f = \mathrm{ext}_k(\{f^{(j)}\})$ extending the function $f = f^{(0)}$ onto $\tilde{G}_\alpha$.

Assuming

$$P_j(x,y) = \sum_{|j+s|\le k} \frac{f^{(j+s)}(y)}{s!}(i_\alpha(x)-i_\alpha(y))^s, \quad x,y\in\tilde{G}_\alpha,$$

we have $f^{(j)}(x) - P_j(x,y) = R_j(x,y)$, $x,y\in\partial G_\alpha$. We denote $P_0(x,y) = P(x,y)$ and $R_0(x,y) = R(x,y)$.

LEMMA 2.1. Let $a,b\in\partial G_\alpha$ *and* $x\in G$. *Then*

$$P(x,b) - P(x,a) = \sum_{|s|\le k} \frac{R_s(b,a)}{s!}(x-i_\alpha(b))^s$$

or, in the general case,

$$P_j(x,b) - P_j(x,a) = \sum_{|j+s|\le k} \frac{R_{j+s}(b,a)}{s!}(x-i_\alpha(b))^s.$$

Using properties (a) - (e) for the points $x\in Q_k^*$, we have

(2.8) $$\delta(x) = |x-\partial G| = d_{\alpha,G}(x,\partial G_\alpha)^{1/\alpha} \sim d_{\alpha,G}(Q_k,\partial G_\alpha)^{1/\alpha} \sim d_{\alpha,G}(Q_k^*,\partial G_\alpha)^{1/\alpha}.$$

As in the Whitney type extension theorem [10] it is enough to consider points $x\in G$ such that $\delta(x)\le c$ where c is a positive constant. Further, we assume that $\|f\,|\,\mathrm{Lip}(\ell,\partial G_\alpha)\| = 1$.

Now, we shall prove the following inequalities:

(i) $|f(x) - P(x,a)| \le A\tilde{d}_{\alpha,G}(x,a)^\ell$, $x\in\tilde{G}_\alpha$, $a\in\partial G_\alpha$;

(ii) $|f^{(j)}(x) - P_j(x,a)| \le A\tilde{d}_{\alpha,G}(x,a)^{\ell-|j|}$, $x\in\tilde{G}_\alpha$, $a\in\partial G_\alpha$, $|j|\le k$;

(iii) $|f^{(j)}(x)| \le A$, $|j|\le k$;

(iv) $|f^{(j)}(x)| \le A\delta(x)^{\ell-k-1}$, $x\in G$, $|j| = k+1$.

We recall that $\tilde{d}_{\alpha,G}(x,y) = d_{\alpha,G}(x,y)^{1/\alpha}$.

Inequality (i) is valid for points $x\in\partial G_\alpha$ with the constant $A = 1$ because

$f \in Lip(\ell, \partial G_\alpha)$. Supposing $x \in G$, $\delta(x) \le c$, we have

$$f(x) - P(x,a) = \sum_i (P(x,p_i) - P(x,a))\varphi_i(x).$$

Applying Lemma 2.1 we obtain

$$|f(x) - P(x,a)| \le \sum_{|s|\le k} \sum_i \tilde{d}_{\alpha,G}(p_i,a)^{\ell-|s|}\tilde{d}_{\alpha,G}(x,a)^{|s|},$$

where the inner sum is taken over all cubes Q_i such that $x \in Q_i^*$. According to (2.5) we have $\tilde{d}_{\alpha,G}(p_i,a) \le c\tilde{d}_{\alpha,G}(x,a)$. Hence the inequality (i) holds.

To prove (ii) we write

$$f^{(j)}(x) = \sum_i (\partial/\partial x)^j(P(x,p_i))\varphi_i(x) + \text{other summands}.$$

Since $(\partial/\partial x)^j P(x,p_i) = P_j(x,p_i)$, in the above mentioned way we obtain the result for the first summand.

As $\sum_i(\partial/\partial x)^j\varphi_i(x) = 0$, the summands of the type

$$P_{j-s}(x,p_i)(\partial/\partial x)^s\varphi_i(x), \tag{2.9}$$

$|s|>0$ and $s_i \le j_i$, $i \le 1,2,\dots,n$, are equal to

$$\sum_i (P_{j-s}(x,p_i) - P_{j-s}(x,a))(\partial/\partial x)^s\varphi_i(x). \tag{2.10}$$

Using (2.3) and (2.8), we prove the inequality (ii).

The estimate (iii) for points $x \in G$, $\delta(x) \le c$, follows from (ii).

If we differentiate the function $f(x)$ in $\{x \in G: \delta(x) \le c_1\}$, we obtain that $f^{(j)}(x)$ is equal to a sum of expressions of type (2.9). As $|j| = k+1$, then necessarily $|s|>0$ because $P_j(x,p_i) = 0$. Therefore, $f^{(j)}(x)$, $|j| = k+1$, is the sum of type (2.10), where $a \in \partial G_\alpha$ is the closest point to x. By Lemma 2.1, (2.3), (2.6) and (2.8), we obtain an upper bound for the sum of type (2.10):

$$A\tilde{d}_{\alpha,G}(p_i,a)^{\ell-|j|+|s|}\delta(x)^{-|s|} \le A'\delta(x)^{\ell-k-1}.$$

The inequality (iv) follows.

From (i) and (ii) we obtain

$$|R_j(x,y)| \le A\tilde{d}_{\alpha,G}(x,y)^{\ell-|j|}, \quad |j| \le k, \tag{2.11}$$

at points $x \in G$ and $y \in \partial G_\alpha$. Let us prove now that inequalities (2.11) are valid at $x,y \in G$. The main case is when the segment L connecting the points x and y, is contained in the domain G. First we suppose that the length of the segment L is less than the distance from the segment L to the boundary ∂G. By the Taylor formula and the use of (iv) we obtain the bound

$$|R_j(x,y)| \le C \sup_{z\in L} |\nabla_{k+1} f(z)| |x-y|^{k+1-|j|} \le C\, |x-y|^{k+1-|j|} \sup_{z\in L} \delta(z)^{\ell-k-1}.$$

Hence in this case $\delta(z)>|x-y|$, $z\in L$, and the result immediately follows:

$$|R_j(x,y)| \le c|x-y|^{\ell-|j|} \le c\tilde{d}_{\alpha,G}(x,y)^{\ell-|j|}.$$

Now we consider the case when at some point $z\in L$ we have $\delta(x)\le|y-x|$. Then there exist points $z'\in L$ and $y'\in\partial G_\alpha$ such that $d_{\alpha,G}(z',y') \le d_{\alpha,G}(x,y)$. Hence $d_{\alpha,G}(y',x) \le 2d_{\alpha,G}(y,x)$ and $d_{\alpha,G}(y',y) \le 2d_{\alpha,G}(x,y)$. According to Lemma 2.1 and the inequality (2.11) we have

$$|R_j(x,y)| = |f^{(j)}(x) - P_j(x,y) - P_j(x,y') + P_j(x,y')|$$

$$\le \left| R_j(x,y') + \sum_{|j+s|\le k} \frac{R_{j+s}(y',y)}{s!}(x-i_\alpha(y'))^s \right|$$

$$\le A\tilde{d}_{\alpha,G}(x,y')^{\ell-|j|} + \sum_{|j+s|\le k} \tilde{d}_{\alpha,G}(y,y')^{\ell-|j|-|s|}\tilde{d}_{\alpha,G}(x,y')^{|s|}$$

$$\le A\tilde{d}_{\alpha,G}(x,y)^{\ell-|j|}.$$

The last case concerns the situation when the points x and y cannot be connected by a segment contained in the domain G. The proof in this case is fully analogous to the previous one as long as there exist two points $z'\in G$ and $y\in\partial G_\alpha$ for which we have

$$d_{\alpha,G}(z',y') \le d_{\alpha,G}(x,y) \quad \text{and} \quad \max\{d_{\alpha,G}(z',x), d_{\alpha,G}(z',y)\} \le d_{\alpha,G}(x,y).$$

Therefore, $d_{\alpha,G}(y',x) \le 2d_{\alpha,G}(x,y)$ and $d_{\alpha,G}(y',y) \le 2d_{\alpha,G}(x,y)$ what coincides with the above situation. Thus we have proved the theorem. □

2.3. Above mentioned questions in Zygmund spaces

Here we prove analogues of theorems 2.1 and 2.2 for the Zygmund spaces $\Lambda^k(G)$, $k\in\mathbb{N}$, G is an arbitrary domain in $\mathbb{R}^n$. Let $\{f_j: |j|\le k\}$, be an element of the space $\Lambda^k(G)$ and $\{f_{j,\nu}: |j|\le k\}$, $\nu\in\mathbb{N}$, collections of functions corresponding $\{f_j\}$ according to the definition of the space $\Lambda^k(G)$. From (1.5) and (1.8) for $|j|\le k-1$ we have

$$|f_{j,\nu}(x)| \le |f_{j,\nu}(x) - f_j(x)| + |f_j(x) - f_{j,1}(x)| + |f_{j,1}(x)| \le 2M,$$

and for $|j|=k$,

$$|f_{j,\nu}(x)| \le |f_{j,\nu}(x) - f_{j,1}(x)| + |f_{j,1}(x)|$$
$$\le \sum_{i=2}^{\nu} |f_{j,i}(x) - f_{j,i-1}(x)| + |f_{j,1}(x)| \le (\nu-1)2M + M \le 2M\nu.$$

Hence the following relations are valid for all $x \in G$ and $\nu \in \mathbb{N}$:

$$|f_{j,\nu}(x)| \le 2M, \quad |j| \le k-1, \tag{2.12}$$

$$|f_{j,\nu}(x)| \le 2M\nu, \quad |j| = k. \tag{2.13}$$

In the same way as in Section 2.1 we show that elements of the collection $\{f_{j,\nu}: |j| \le k\}$ for every fixed $\nu \in \mathbb{N}$ possess extensions onto $\tilde{G}_1$ by continuity with respect to the metric space $(G_1, d_{1,G})$.

It is natural to consider restrictions of extensions of the jets $\{f_{j,\nu}: |j| \le k\}$, $\nu \in \mathbb{N}$, to the 1-boundary ∂G_1 of the domain G as the boundary values of the element $\{f_j: |j| \le k\}$ from the space $\Lambda^k(G)$ on the 1-boundary of domain G. To give an exact sense to the concept of the trace let us define the function class $\Lambda^k(\partial G_1)$. Jets $\{f_j: |j| \le k-1\}$ of functions defined on the 1-boundary ∂G_1 are called elements of the class $\Lambda^k(\partial G_1)$ if there exist collections $\{f_{j,\nu}: |j| \le k\}$, $\nu \in \mathbb{N}$, of functions defined also on ∂G_1 such that the following conditions hold for $x, y \in \partial G_1$:

$$|f_j(x) - f_{j,\nu}(x)| \le M2^{-\nu(k-|j|)}, \quad \nu \ge 1, \ |j| \le k-1, \tag{2.14}$$

$$|f_{j,\nu}(x) - f_{j,\mu}(x)| \le M2^{\mu-\nu}, \quad \mu \ge \nu \ge 1, \ |j| = k, \tag{2.15}$$

$$|R^k_{j,\nu}(x,y)| \le M2^{\nu} d_{1,G}(x,y)^{k+1-|j|}, \quad \nu \ge 1, \ |j| \le k, \tag{2.16}$$

$$|f_{j,1}(x)| \le M, \quad |j| \le k. \tag{2.17}$$

Here $R^k_{j,\nu}(x,y)$ are defined by

$$f_{j,\nu}(x) = \sum_{|j+s| \le k} \frac{f_{j+s,\nu}(y)}{s!}(i_1(x) - i_1(y))^s + R^k_{j,\nu}(x,y).$$

The norm $\|f|\Lambda^k(\partial G_1)\|$ of the jet $f = \{f_j: |j| \le k-1\}$ is defined as the infimum of all constants M such that the conditions (2.14) - (2.17) hold for some collection $\{f_{j,\nu}: |j| \le k\}$, $\nu \in \mathbb{N}$.

Let G be an arbitrary domain in $\mathbb{R}^n$. We define a trace operator by the following way: Elements of a jet $\{f_{j,\nu}: |j| \le k\}$, $\nu \in \mathbb{N}$, corresponding to a function $f = \{f_j\} \in \Lambda^k(G)$, extend to G by the continuity and then restrict to

∂G_1. As a result we have jets satisfying conditions (2.14) - (2.17) which form an element of the space $\Lambda^k(\partial G_1)$ because, according to (1.5) and (2.14),

$$\lim f_{j,\nu}(x) = f_j(x), \ |j|\le k-1,$$

for any $x\in\partial G_1$. Hence we obtain the following result:

THEOREM 2.3. Let G *be an arbitrary domain in* $\mathbb{R}^n$. *The trace* $T_k f$ *of* $f\in\Lambda^k(G)$ *is an element of the space* $\Lambda^k(\partial G_1)$, *and the trace operator*

$$T_k: \Lambda^k(G) \longrightarrow \Lambda^k(\partial G_1), \ k\in\mathbb{N},$$

is bounded with the norm not exceeding 1.

The characterization of traces of the Zygmund classes given by Theorem 2.3 is reversible.

THEOREM 2.4. Let G *be an arbitrary domain in* $\mathbb{R}^n$. *Then there exists the bounded extension operator*

$$E_k: \Lambda^k(\partial G_1) \longrightarrow \Lambda^k(G), \ k\in\mathbb{N},$$

in the sense that $T_k \circ E_k$ *is the identical mapping.*

Theorems 2.3 and 2.4 have been proved jointly with Yu. Bojarskii and formulated in [11].

P r o o f. Let $\{Q_k\}$ be the above mentioned decomposition of the domain G into cubes and $\{\varphi_k\}$ be a corresponding partition of unity. Here it is more convenient for us to use a double numeration for the Whitney cubes. For $\nu\in\mathbb{N}$, cubes $Q_{\nu,i}$ are the cubes from the collection $\{Q_k\}$ such that $2^{-(\nu+1)} < \operatorname{diam} Q_{\nu,i} \le 2^{-\nu}$. Further, let $\{\varphi_{\nu,i}\}$ be the partition of unity corresponding to the cubes $\{Q_{\nu,i}\}$ and let $p_{\nu,i}\in\partial G_1$ be the closest point to the cube $Q_{\nu,i}$.

Let us consider an element $\{f_j: |j|\le k-1\}$, of the space $\Lambda^k(\partial G_1)$ and, according to the definition, a sequence of jets $\{f_{j,\nu}: |j|\le k\}$, $\nu\in\mathbb{N}$, corresponding to it. We set

$$P_\nu(x,y) = \sum_{|s|\le k} \frac{f_{s,\nu}(y)}{s!}(x-i_1(y))^s \quad \text{for } x\in G, \ y\in\partial G_1.$$

The extension operator E_k is defined on the jet $\{f_j\}\in\Lambda^k(\partial G_1)$ in the following way:

$$E_k(\{f_j\})(x) = \begin{cases} \sum_{\mu=1}^{\infty} \sum_i \varphi_{\mu,i}(x) P_\mu(x, p_{\mu,i}), & x \in G, \\ f(x), & x \in \partial G_1, \end{cases}$$

where for fixed $\mu\in\mathbb{N}$ the summation is taken over all i such that $2^{-(\mu+1)} < \operatorname{diam} Q_{\mu,i} \le 2^{-\mu}$. Let $F(x)$ denote the function $E_k(\{f_j\})$. We define the approximating sequence $\{F_\nu:\ \nu\in\mathbb{N}\}$, corresponding to f, by

$$F_\nu(x) = \begin{cases} \sum_{\mu=\nu+1}^{\infty} \sum_i \varphi_{\mu,i}(x) P_\nu(x, p_{\mu,i}) + \sum_{\mu=1}^{\nu} \sum_i \varphi_{\mu,i}(x) P_\mu(x, p_{\mu,i}), & x \in G, \\ f_\nu(x), & x \in \partial G_1, \end{cases}$$

where the summation over i is the same as above.

Let $P_{j,\nu}(x,y)$, $|j|\le k$, $y\in\partial G_1$, $x\in G$, denote the polynomial $P_\nu(x,y)$ if $j = 0$ and the polynomial $D_x^j P_\nu(x,y)$ if $0<|j|\le k$, respectively, where D_x denotes the differentiation with respect to x. In this notation, we have

$$P_{j,\nu}(x,y) = \sum_{|j+s|\le k} \frac{f_{j+s,\nu}(y)}{s!}(x-i_1(y))^s.$$

Further, we consider a function $\{f_j\}$ whose norm is equal to 1.

LEMMA 2.2. *Let* $a, b\in\partial G_1$, $x\in G$ *and let the conditions* $d_{1,G}(a,b) \le c_1\rho$, $d_{1,G}(x,a) \le c_2\rho$ *hold with some constants* c_1, c_2. *Then there exists a constant* $c>0$ *such that*

$$|P_{j,\nu}(x,a) - P_{j,\nu}(x,b)| \le c2^\nu \rho^{k+1-|j|}, \quad |j|\le k.$$

P r o o f. According to Lemma 2.1 and the inequality (2.16) we have

$$\left|P_{j,\nu}(x,a) - P_{j,\nu}(x,b)\right| = \left|\sum_{|j+s|\le k} \frac{R_{j+s,\nu}(a,b)}{s!}(x-i_1(a))^s\right|$$

$$\le c_1c_2 2^\nu d_{1,G}(a,b)^{k+1-|j+s|} d_{1,G}(x,a)^{|s|} \le c\rho^{k+1-|j|}2^\nu.$$

□

LEMMA 2.3. *Let the points* $a, b\in\partial G_1$, $x\in G$, *satisfy the conditions* $d_{1,G}(a,b) \le 2^{-\nu}$, $d_{1,G}(x,b) \le 2^{-\mu}$ *with* $\mu\ge\nu$. *Then the inequality*

$$|P_{j,\nu}(x,a) - P_{j,\mu}(x,b)| \le c2^{-\nu(k-|j|)}$$

holds for every j, $|j|\le k-1$.

P r o o f. Let us estimate the difference

$$|P_{j,\nu}(x,b) - P_{j,\mu}(x,b)| \le \sum_{|j+s|\le k} \frac{|f_{j+s,\nu}(b) - f_{j+s,\mu}(b)|}{s!}|x - i_1(b)|^{|s|}$$

Using the assumption of the lemma and (2.14) we estimate the summands on the right hand side with $|j|\le k-1$ from above by

$$c2^{-\nu(k+|j+s|)}2^{-\mu|s|} \le c2^{-\nu(k-|j|)}.$$

To estimate summands with $|j+s| = k$ we use (2.15) and

$$c2^{-\mu|s|}2^{\mu-\nu} \le c2^{-\nu|s|} = c2^{-\nu(k-|j|)}.$$

Applying Lemma 2.2 we obtain the result

$$|P_{j,\nu}(x,a) - P_{j,\mu}(x,b)| \le |P_{j,\nu}(x,a) - P_{j,\nu}(x,b)| + |P_{j,\nu}(x,b) - P_{j,\mu}(x,b)|$$

$$\le c2^{-\nu(k-|j|)}. \qquad \square$$

In the folowing lemmas, we shall show that the functions $F_j(x) = D^jF(x)$, $|j|\le k-1$, and the approximating sequence $\{F_{j,\nu}(x)\colon |j|\le k\}$, $\nu\in\mathbb{N}$, of functions $F_{j,\nu}(x) = D^jF_\nu(x)$, possess the properties (1.5) - (1.8).

<u>*LEMMA 2.4.*</u> *If* $x\in\partial G_1$, *then*

$$\lim_{\substack{d_{1,G}(x,y)\to 0\\ y\in G_1}} F_{j,\nu}(y) = f_{j,\nu}(x), \quad |j|\le k, \quad \nu\in\mathbb{N}.$$

If $x\in\tilde{G}_1$, *then*

(2.18) $$|F_j(x) - F_{j,\nu}(x)| \le c2^{-\nu(k-|j|)}, \quad |j|\le k-1, \quad \nu\in\mathbb{N}.$$

P r o o f. If $x\in G$, $d_{1,G}(x,\partial G_1) > 2^{-\nu}$, then $F(x) \equiv F_\nu(x)$ by the construction. Let $x\in G$ be a point such that $2^{-(\tau+1)} < d_{1,G}(x,\partial G_1) \le 2^{-\tau}$, where $\tau\ge\nu$. Then the values $F_j(x)$ and $F_{j,\nu}(x)$ are equal to a finite sum whose summands have the form $D^\ell\varphi_{\mu,i}(x)P_{m,\mu}(x,p_{\mu,i})$ and $D^\ell\varphi_{\mu,i}(x)P_{m,\nu}(x,p_{\mu,i})$, $m+\ell = j$, respectively, and also the cubes participating in the sum satisfy $c_12^{-(\tau+1)} \le \operatorname{diam} Q_{\mu,i} \le c_22^{-\tau}$, where the constants c_1 and c_2 do not depend on x. The number of the summands is bounded by a constant which is independent of x. Now, using Lemma 2.3 and the fact that $\sum_{\mu,i}\varphi_{\mu,i}(x) = 1$ in the domain G, we obtain the estimate

$$|F_j(x) - F_{j,\nu}(x)| \le \sum_{m+\ell=j}\sum_{\mu,i}|D^\ell\varphi_{\mu,i}(x)||P_{m,\mu}(x,p_{\mu,i}) - P_{m,\nu}(x,p_{\mu,i})|$$

$$\le c\sum_{m+\ell=j}2^{\nu|\ell|}2^{-\nu(k-|m|)} \le c2^{-\nu(k-|j|)}.$$

If $x \in \partial G_1$ and $|j| = 0$, then the estimate (2.18) reduces to (2.14). To show (2.18) for $|j|>0$ we prove that $F_{j,\nu}(y) \to f_{j,\nu}(x)$ as $d_{1,G}(y,x) \to 0$. Let us write the difference $F_{j,\nu}(y) - f_{j,\nu}(x)$ in the form

$$\sum_{\mu,i} \varphi_{\mu,i}(y)(f_{j,\nu}(p_{\mu,i}) - f_{j,\nu}(x)) + \text{other summands.}$$

The estimate (2.6) yields $d_{1,G}(x,p_{\mu,i}) \le c d_{1,G}(x,y)$. The function $f_{j,\nu}$ is continuous on ∂G_1 and so the explicitly expressed summands tend to zero as $d_{1,G}(x,y) \to 0$. Since $\sum_{m,i} D^{\alpha}\varphi_{\mu,i}(y) \equiv 0$ for $y \in G$ the other summands can be written in the form

$$\sum_{\mu,i} D^{\alpha}\varphi_{\mu,i}(y)(f_{m,\nu}(p_{\mu,i}) - f_{m,\nu}(x)) + \sum_{0<|m+s|\le k} \frac{f_{m+s,\nu}(p_{\mu,i})}{s!}(y - i_1(p_{\mu,i}))^s,$$

where $\ell+m = j$, $|\ell|>0$. Using the estimates

$$|y - i_1(p_{\mu,i})| \le d_{1,G}(y,p_{\mu,i}) \le c d_{1,G}(x,y)$$

and an analogue of (2.12) for functions of $\Lambda^k(\partial G_1)$ we obtain that in this case the summands tend to 0 as $d_{1,G}(x,y) \to 0$, as well. □

LEMMA 2.5. *Let* $|j| = k$ *and* $x \in \tilde{G}$. *Then*

$$|F_{j,\mu}(x) - F_{j,\nu}(x)| \le c2^{\mu-\nu}, \quad \mu \ge \nu \ge 1. \tag{2.19}$$

The proof is completely analogous to that of Lemma 2.4. The difference $F_{j,\mu}(x) - F_{j,\nu}(x)$ is estimated by a finite sum of terms among which there are expressions of the form $\varphi_{\mu,i}(f_{j,\mu}(p_{\mu,i}) - f_{j,\nu}(p_{\nu,i}))$. We can estimate them by (2.15). In the proof of (2.19) for $x \in \partial G_1$ we use an analogue of (2.13) for functions from $\Lambda^k(\partial G_1)$. □

LEMMA 2.6. *There exists a constant* $A>0$ *such that the inequality*

$$|F_{j,1}(x)| \le A$$

holds for every $x \in \tilde{G}$ *and* $|j| \le k$.

P r o o f. Since the number ν in the expression for $F_1(x)$ is equal to 1 we can write the functions $F_{j,1}(x)$ in the form

$$\begin{aligned} F_{j,1}(x) &= \sum_{\mu,i} \varphi_{\mu,i}(x) P_{j,1}(x,p_{\mu,i}) \\ &+ \sum_{\substack{\ell+m=j \\ |\ell|>0}} D^{\ell}\varphi_{\mu,i}(x)(P_{m,1}(x,p_{\mu,i}) - P_{m,1}(x,b)). \end{aligned} \tag{2.20}$$

The first sum is estimated by

$$|f_{j,1}(p_{\mu,i})| + \sum_{0<|j+s|\le k} \frac{|f_{j+s,1}(p_{\mu,i})|}{s!}|x-i_1(p_{\mu,i})|^s \le c + c2^{-\mu(k-|j|)} \le A$$

where we used 2.17. To evaluate the second sum we use Lemma 2.2. Note that

$$c2^{\mu|\ell|}2^{-\mu(k-|m|)} = c2^{-\mu(k-|j|)} \le c. \qquad \square$$

LEMMA 2.7. Let $x\in G$ *and* $|j| = k+1$. *Then*

$$|F_{j,\nu}(x)| \le c2^{\nu}, \quad \nu\in\mathbb{N}.$$

P r o o f. Since $P_{m,\nu}(x,p_{\mu,i}) \equiv 0$ for $|m|>k$ we can write (cf. (2.20))

$$F_{j,\nu}(x) = \sum_{\mu>\nu}\sum_{i}\sum_{\substack{\ell+m=j\\|\ell|>0}} D^{\ell}\varphi_{\mu,i}(x)(P_{m,\nu}(x,p_{\mu,i}) - P_{m,\nu}(x,b))$$

$$+ \sum_{\mu=1}^{\nu}\sum_{i}\sum_{\substack{\ell+m=j\\|\ell|>0}} D^{\ell}\varphi_{\mu,i}(x)(P_{m,\mu}(x,p_{\mu,i}) - P_{m,\mu}(x,b))$$

where $b\in\partial G_1$ is the closest point to x. Estimating the summands on the right hand side by Lemma 2.2 we obtain the upper bound $c\rho^{-|\ell|}2^{\nu}\rho^{k+1-|m|} = c2^{\nu}$ for the first summand and $c\rho^{-|\ell|}2^{\mu}\rho^{k+1-|m|} \le c2^{\nu}$ for the second one. $\square$

LEMMA 2.8. Let $x\in G$ *and* $a\in\partial G_1$. *Then*

$$|F_{j,\nu}(x) - P_{j,\nu}(x,a)| \le c2^{\nu}d_{1,G}(x,a)^{k+1-|j|}, \quad |j|\le k.$$

P r o o f. Using the definition of functions F_{ν} and Lemma 2.2 we obtain

$$|F_{j,\nu}(x) - P_{j,\nu}(x,a)| = \Big|\sum_{\mu,i}\varphi_{\mu,i}(x)(P_{j,\nu}(x,p_{\mu,i}) - P_{j,\nu}(x,a))$$

$$+ \sum_{\substack{\ell+m=j\\|\ell|>0}}\sum_{\mu,i} D^{\ell}\varphi_{\mu,i}(x)(P_{m,\nu}(x,p_{\mu,i}) - P_{m,\nu}(x,b))\Big|$$

$$\le c2^{\nu}d_{1,G}(x,a)^{k+1-|j|},$$

where $b\in\partial G_1$ is the closest point to x. $\square$

LEMMA 2.9. Let $x,y\in G$ *and let* $R^k_{j,\nu}(x,y)$ *be the remainders in the expansion of the function* $F_{j,\nu}(x) = \sum_{|j+s|\le k}\frac{F_{j+s,\nu}(y)}{s!}(i_1(x)-i_1(y))^s + R^k_{j,\nu}(x,y)$. *Then*

$$|R^k_{j,\nu}(x,y)| \le c2^{\nu}d_{1,G}(x,y)^{k+1-|j|}, \quad |j|\le k.$$

P r o o f. The case when $x\in\partial G$, $y\in\partial G_1$, reduces to (2.16) as $F_{j,\nu}(y) = f_{j,\nu}(y)$ (see Lemma 2.4). The case $x\in G$, $y\in\partial G_1$ is considered in Lemma 2.8.

It remains the case $x,y\in G$. As in Theorem 2.2 we need to study three cases of the mutual disposition of points x,y with respect to the boundary ∂G. The main situation is when the segment L connecting the points x and y is contained in the domain G. First let the length $|x-y|$ of L be less than the Euclidean distance from L to the boundary ∂G. By the Taylor formula and Lemma 2.7 we obtain

$$|R^k_{j,\nu}(x,y)| \le c \sup_{z\in L} |\nabla_{k+1}F_\nu(z)||x-y|^{k+1-|j|} \le c2^\nu d_{1,G}(x,y)^{k+1-|j|}.$$

Now, let us consider the second case when $d_{1,G}(z,\partial G_1) \le |x-y|$ for some point $z\in L$. Then there exist points $z'\in L$ and $y'\in\partial G_1$ such that for any $\nu\in\mathbb{N}$ we have $d_{1,G}(z',y') \le d_{1,G}(x,y)$. According to Lemmas 2.2 and 2.8 we obtain the following bound for the remainders $R^k_{j,\nu}(x,y)$:

$$\begin{aligned}|R^k_{j,\nu}(x,y)| &= |F_{j,\nu}(x) - P_{j,\nu}(x,y) - P_{j,\nu}(x,y') + P_{j,\nu}(x,y')| \\ &\le c2^\nu d_{1,G}(x,y')^{k+1-|j|} + c2^\nu d_{1,G}(x,y')^{k+1-|j|} \\ &\le c2^\nu d_{1,G}(x,y)^{k+1-|j|}.\end{aligned}$$

When it is not possible to connect the points x, y by a segment $L\subset G$ we can find points $z'\in G$ and $y'\in\partial G_1$ such that

$$d_{1,G}(z',y') \le d_{1,G}(x,y), \quad \max\{d_{1,G}(z',x), d_{1,G}(z'y)\} \le d_{1,G}(x,y).$$

Therefore,

$$d_{1,G}(y',x) \le 2d_{1,G}(x,y), \quad d_{1,G}(y',y) \le 2d_{1,G}(x,y),$$

which coincides with the above considered situation. Thus, for the jet $\{F_j: |j|\le k-1\}$ and for the approximating sequence of the jets $\{F_{j,\nu}: |j|\le k\}$, $\nu\in\mathbb{N}$, the conditions (1.5) - (1.8) hold. Note, that (1.5) is Lemma 2.4, (1.6) is Lemma 2.5, (1.7) is Lemma 2.9. It means that the jet $\{F_j: |j|\le k-1\}$ belongs to the space $\Lambda^k(G)$. From Lemma 2.4 we have also that $T_k\circ E_k$ is the identical operator. Theorem 2.4 is proved. □

3. Extension of differentiable functions

An inner α-metric $d_{\alpha,G}(x,y)$, $\alpha\in(0,1]$, in a domain $G\subset\mathbb{R}^n$ is called *locally equivalent* to the α-metric $d_\alpha(x,y) = |x-y|^\alpha$ if there exist numbers $r>0$ and $M>0$

such that for all points $x,y\in G$ with $|x-y|<r$ the inequality

$$\tilde{d}_{\alpha,G}(x,y) \le M|x-y| \tag{3.1}$$

is valid. Let us indicate by

$$d_{\alpha,G}(x,y) \underset{loc}{\sim} d_{\alpha}(x,y)$$

that the metrics $d_{\alpha,G}(x,y)$ and $d_{\alpha}(x,y)$ are locally equivalent in the domain G, and let $M_{\alpha}(r)$ be the least constant for which the inequality (3.1) holds under the condition $|x-y|<r$, $x,y\in G$. Let us also set

$$M_{\alpha} = \lim_{r\to 0} M_{\alpha}(r), \tag{3.2}$$

$$r_{\alpha} = \sup \{r:\ M_{\alpha}(r)<\infty\}. \tag{3.3}$$

It follows at once from the definition that the inner α-metric $d_{\alpha,G}(x,y)$ is locally equivalent to the α metric $d_{\alpha}(x,y)$ in the domain G if and only if the value M_{α} is finite.

Let us consider two seminormed spaces $F(G)$ and $N(\mathbb{R}^n)$ of functions defined on a domain $G\subset\mathbb{R}^n$ and on the Euclidean space $\mathbb{R}^n$, respectively. The mapping $\mathrm{ext}: F(G)\longrightarrow N(\mathbb{R}^n)$ is called *the extension operator* if $\mathrm{ext}(f)|_G = f$ for all functions $f\in F(G)$ and

$$\|\mathrm{ext}\| = \sup_{f\in F(G)} \left\{ \|\mathrm{ext}(f)\,|N(\mathbb{R}^n)\|\ \|f\,|F(G)\|^{-1}:\ f\in F(G)\right\} < \infty.$$

THEOREM 3.1. Let G *be a domain in* $\mathbb{R}^n$ *and let* $\ell\in\mathbb{R}_+$ *and* $k=0,1,\ldots$ *be such that* $k<\ell\le k+1$. *Put* $\alpha = \ell-k$. *If the inner α-metric* $d_{\alpha,G}(x,y)$ *is locally equivalent to the α-metric* $d_{\alpha}(x,y) = |x-y|^{\alpha}$ *in* G, *then there exist linear bounded extension operators*

$$\mathrm{ext}_k: W^{\ell}_{\infty}(G)\longrightarrow W^{\ell}_{\infty}(\mathbb{R}^n) \ \textit{for}\ \ell\in\mathbb{N}$$

and

$$\mathrm{ext}_k: H^{\ell}_{\infty}(G)\longrightarrow H^{\ell}_{\infty}(\mathbb{R}^n) \ \textit{for}\ \ell\notin\mathbb{N}.$$

The norm of the extension operator satisfies the estimate

$$\|\mathrm{ext}_k\| \le \gamma[\max\{M_{\alpha}, 1/r_{\alpha}\}]^{\ell},$$

where the constant γ *does not depend on the domain* G *and the numbers* M_{α}, r_{α} *are defined by* (3.2) - (3.3).

P r o o f. Let ℓ, k and α satisfy the assumptions of the lemma. According to Theorem 1.1 and Proposition 2.1 we have the coincidence of spaces

$$W^{\ell}_{\infty}(G) = Lip(\ell,\tilde{G}_\alpha) \text{ for } \ell\in N \quad \text{and} \quad H^{\ell}_{\infty}(G) = Lip(\ell,\tilde{G}_\alpha) \text{ for } \ell\notin N.$$

Therefore, assuming that $d_{\alpha,G}(x,y) \underset{loc}{\sim} d_\alpha(x,y)$ in the domain G, it suffices to prove the existence of a bounded extension operator

$$ext_k: Lip(\ell,\tilde{G}_\alpha) \longrightarrow Lip(\ell,\mathbb{R}^n)$$

such that $ext_k(f)|_G = f|_G$ for an arbitrary function $f\in Lip(\ell,\tilde{G}_\alpha)$.

To prove this we have to introduce the space $Lip(\ell,F)$, where F is a closed set in $\mathbb{R}^n$ (see [10]). The elements of $Lip(\ell,F)$ are jets $\{f^{(j)}: |j|\le k\}$ consisting of bounded functions such that $f^{(0)} = f$ and

$$(3.4) \qquad f^{(j)}(x) = \sum_{|j+s|\le k} \frac{f^{(j+s)}(y)}{s!}(x-y)^s + R_j(x,y)$$

where

$$(3.5) \qquad |f^{(j)}(x)| \le M, \quad |R_j(x,y)| \le M|x-y|^{\ell-|j|}, \quad x,y\in F, \ |j|\le k.$$

The norm of an element of $Lip(\ell,F)$ is equal to the smallest value M for which the inequalities (3.5) hold. It is easy to verify that $f^{(j)}(x) = D^j f(x)$ in $G = \mathrm{Int}F$.

According to a Whitney type extension theorem (see [10]) there exists a linear bounded extension operator

$$ext_k: Lip(\ell,F) \longrightarrow Lip(\ell,\mathbb{R}^n),$$

whose norm does not depend on the closed set $F\subset\mathbb{R}^n$. To prove Theorem 3.1, it is sufficient to establish the embedding

$$(3.6) \qquad i: Lip(\ell,\tilde{G}_\alpha) \longrightarrow Lip(\ell,\bar{G})$$

such that for each function $f\in Lip(\ell,\tilde{G}_\alpha)$ the relation $(if)(x) = f(x)$, $x\in G$, is valid. The desired extension operator will be the superposition of mappings in the diagram

$$Lip(\ell,\tilde{G}_\alpha) \xrightarrow{\ i\ } Lip(\ell,\bar{G}) \xrightarrow{\ \overline{ext}_k\ } Lip(\ell,\mathbb{R}^n).$$

Thus, $ext_k = (\overline{ext}_k \circ i): Lip(\ell,\tilde{G}_\alpha) \longrightarrow Lip(\ell,\mathbb{R}^n)$.

To find an upper estimate for the norm of the operator ext_k it is sufficient to find the contribution of the geometry of the space G_α to the norm of the operator i, since

$$(3.7) \qquad \|ext_k \circ i\| \le \|i\| \, \|\overline{ext}_k\|$$

and the norm of operator $\overline{ext}_k$ does not depend on the domain G.

Now let us prove the boundedness of the embedding (3.6). Let $\|f|\mathrm{Lip}(\ell,\tilde{G}_\alpha)\|$ be equal to 1. Since $|f^{(j)}(x)| = |D^j f(x)| \le 1$, $x\in G$, $|j|\le k$, it remains to prove the inequalities

$$|R_j(x,y)| \le C|x-y|^{\ell-|j|}, \quad x,y\in G. \tag{3.8}$$

Indeed, it follows from (3.8) that the functions $f^{(j)}$ extend by continuity to $\bar{G}$, the extended functions satisfy the inequalities (3.5) for all points $x,y\in \bar{G}$, and the embedding (3.6) is bounded.

To prove (3.8) we denote by r_m the largest number such that for points $x,y\in G$, $0<|x-y|<r_m$, we have the inequality

$$\tilde{d}_{\alpha,G}(x,y) \le mM_\alpha|x-y|, \quad m\in\mathbb{N}, \tag{3.9}$$

Since $|f^{(j)}(x)|\le 1$, the estimate

$$|R_j(x,y)| \le [C(n,k)/r_m^{(\ell-|j|)}]|x-y|^{\ell-|j|} \tag{3.10}$$

for $|j|\le k$, in $x,y\in G$, $|x-y|\ge r_m$, follows from the expansion (3.4). If $x,y\in G$, $|x-y|<r_m$, then we use (3.9) to obtain the estimate

$$|R_j(x,y)| \le \tilde{d}_{\alpha,G}(x,y)^{(\ell-|j|)} \le (mM_\alpha)^\ell|x-y|^{\ell-|j|}. \tag{3.11}$$

The inequalities (3.10) and (3.11) together yield the estimate (3.8).

It follows that the embedding (3.6) is continuous. If $|j| = 0$, the upper bound of the embedding is obtained from (3.10) and (3.11). As to the norm of the operator $\mathrm{ext}_k = \overline{\mathrm{ext}_k}\circ i$, we have the following estimate:

$$\|\mathrm{ext}_k\| \le \gamma \max\{mM_\alpha, 1/r_m\}^\ell, \tag{3.12}$$

where the constant γ depends only on n and ℓ. To finish the proof of Theorem 3.1 it remains to minimize the inequality (3.12) over all r_m such that

$$\tilde{d}_{\alpha,G}(x,y) \le mM_\alpha|x-y|, \quad m\in\mathbb{N},\ x,y\in G,\ 0<|x-y|^\alpha<r_m.$$

In the case of the Sobolev spaces ($\alpha=1$) Theorem 3.1 was proved by another method in [12].

There exists a relation between the α-metrics resulting from the reverse Minkowski inequality (see [7]): Given any numbers α and β, $0\le\alpha\le\beta\le 1$, the inequalities

$$d_{1,G}(x,y) \le d_{\beta,G}(x,y)^{1/\beta} \le d_{\alpha,G}(x,y)^{1/\alpha}, \tag{3.13}$$

hold for every $x,y\in G$. It follows that the relation $d_{\alpha,G}(x,y) \underset{\mathrm{loc}}{\sim} d_\alpha(x,y)$ is stronger than the relation $d_{\beta,G}(x,y) \underset{\mathrm{loc}}{\sim} d_\beta(x,y)$. For given numbers α and β,

$0<\alpha<\beta\le 1$, we can construct an example of a domain $G\subset\mathbb{R}^2$ such that $d_{\beta,G}(x,y) \underset{loc}{\sim} d_\beta(x,y)$, but $d_{\alpha,G}(x,y) \underset{loc}{\sim} d_\alpha(x,y)$ does not hold.

The method of the extension of differentiable functions explained above can also be applied to other spaces. First, we consider the Zygmund space $\Lambda^k(G)$, $k\in\mathbb{N}$. Before formulation of an exact result we adopt the extension method for the spaces $\Lambda^k(G)$. The first step is to extend the functions $f\in\Lambda^k(G)$ onto $\tilde{G}_1$ so that they belong to the space $\Lambda^k(\tilde{G}_1)$. The class $\Lambda^k(\tilde{G}_1)$ consists of the jets $\{f_j: |j|\le k-1\}$ and their approximating sequences $\{f_{j,\nu}: |j|\le k\}$, $\nu\in\mathbb{N}$, such that the conditions (1.5) - (1.8) hold at all points $x,y\in\tilde{G}_1$. Indeed in Section 2.2 we have proved the following

PROPOSITION 3.1. Let $\{f_j\}$, $|j|\le k-1$ be a collection of functions from the class $\Lambda^k(G)$. Then every f_j can be extended in a unique way onto $\tilde{G}_1$, the extended functions belong to the space $\Lambda^k(\tilde{G}_1)$ and preserve the norms.

Now we shall find conditions for the existence of an extension operator

$$\text{Ext}_k: \Lambda^k(\tilde{G}_1)\longrightarrow\Lambda^k(\mathbb{R}^n) \tag{3.14}$$

such that the equality $\text{Ext}_k(\{f_j\})(x) = f_0(x)$, $x\in G$, holds for each function from the jet $\{f_j: |j|\le k-1\}$.

To prove (3.14) we need the space $\Lambda^k(F)$, $k\in\mathbb{N}$, where F is a closed set in $\mathbb{R}^n$. Jets $\{f_j: |j|\le k-1\}$ of functions defined on the set F are called elements of the class $\Lambda^k(F)$ if there exist collections $\{f_{j,\nu}: |j|\le k\}$, $\nu\in\mathbb{N}$, of functions defined on F, such that the conditions

$$|f_j(x) - f_{j,\nu}(x)| \le M2^{-\nu(k-|j|)}, \quad \nu\ge 1, \quad |j|\le k-1, \tag{3.15}$$

$$|f_{j,\nu}(x) - f_{j,\mu}(x)| \le M2^{\mu-\nu}, \quad \mu\ge\nu\ge 1, \quad |j|=k, \tag{3.16}$$

$$\left|f_{j,\nu}(x) - \sum_{|j+s|\le k} \frac{f_{j+s,\nu}(y)}{s!}(x-y)^s\right| \le M2^{-\tau(k-|j|)}, \quad |x-y|\le a2^{-\tau}, \quad \nu\ge 1, \quad |j|\le k, \quad \tau\ge\nu\ge 1; \tag{3.17}$$

$$|f_{j,1}(x)| \le M, \quad |j|\le k \tag{3.18}$$

hold for every $x,y\in F$. The norm $\|f|\Lambda^k(F)\|$ of the jet $\{f_j: |j|\le k-1\}$ is equal to the smallest constant M such that the conditions (3.15) - (3.18) hold.

THEOREM 3.2. Let F be a closed subset of $\mathbb{R}^n$. Then every element $f = \{f_j : |j| \le k-1\}$ of the space $\Lambda^k(F)$ can be extended to a function $Ext_k(f) \in \Lambda^k(\mathbb{R}^n)$. The function $Ext_k(f)$ is an extension of the jet $f = \{f_j : |j| \le k-1\}$ such that for every $|j| \le k-1$ the restriction of the partial derivative $D^j Ext_k(f)$ to the set F is equal to f_j. Moreover, the extension operator $Ext_k : \Lambda^k(F) \longrightarrow \Lambda^k(\mathbb{R}^n)$ is bounded and it may be chosen in such way that the function $Ext_k(f)$ is smooth outside of F.

P r o o f of the Theorem 3.2 is completely analogous to that of Theorem 2.4. □

REMARK 3.1. In [8] the authors define the space $\Lambda^k(F)$ in a different way setting $\tau = \nu$ in the formula corresponding to (3.17). Nevertheless, both classes, if considered on Euclidean space $\mathbb{R}^n$, coincide with the Zygmund space $H^k_\infty(\mathbb{R}^n)$. Our proof of Theorem 3.2 is based on the Whitney method, as well, but we use another construction of the extension operator $ext : \Lambda^k(F) \longrightarrow H^k_\infty(\mathbb{R}^n)$.

To prove the existence of the operator (3.14) it is sufficient to establish the embedding

$$i : \Lambda^k(\tilde{G}_1) \longrightarrow \Lambda^k(\bar{G}) \tag{3.19}$$

such that the relation $(if)(x) = f(x)$, $x \in G$, holds for every function $f \in \Lambda^k(\tilde{G}_1)$. Then the desired operator is obtained as the superposition of operators in the following diagram

$$\Lambda^k(\tilde{G}_1) \xrightarrow{\ i\ } \Lambda^k(\bar{G}) \xrightarrow{\ \overline{Ext}_k\ } \Lambda^k(\mathbb{R}^n).$$

Indeed, $Ext_k = (\overline{Ext}_k \circ i) : \Lambda^k(\tilde{G}_1) \longrightarrow \Lambda^k(\mathbb{R}^n)$.

To find an upper bound for the norm of the operator Ext_k it suffices to know how the norm of the operator i depends on the geometry of the domain G. The boundedness of the operator (3.19) may be obtained by the geometric restriction to the domain G. To see its character it suffices to compare conditions (1.5) - (1.8) and (3.15) - (3.18), in particular, (1.7) and (3.17). We claim that

$$\lim_{\nu\to\infty} \sup \left\{ d_{1,G}(x,y)/|x-y| : x,y \in G,\ 0<|x-y| \le 2^{-\nu} \right\} = M_0 < \infty. \tag{3.20}$$

Let $\nu_0 \in \mathbb{N}$ be a number such that the condition

$$d_{1,G}(x,y) \le 2M_0 |x-y| \tag{3.21}$$

holds for $|x-y| \le 2^{-\nu}$, $\nu \ge \nu_0$.

Let a jet $\{f_j\colon |j|\le k-1\}$ from the class $\Lambda^k(\tilde{G}_1)$ have the norm equal to 1 and let $\{f_{j,\nu}\colon |j|\le k\}$, $\nu\in\mathbb{N}$, be the corresponding sequence of jets such that the conditions (1.5) - (1.8) hold with the constant $M = 2$. The condition (3.21) implies that

$$d_{1,G}(x,y) \le 2M_0|x-y| \le 2M_0 2^{-\tau}$$

for every $x,y\in G$, $|x-y| \le 2^{-\tau}$, and $\tau\ge\nu\ge\nu_0$. We take the least natural number N such that $2^{N+\nu_0}a \ge 2M_0$ where a is the number from (3.17). Let us consider the sequence of jets $\{g_{j,\nu}\colon |j|\le k\}$, $\nu\in\mathbb{N}$, of functions defined on the domain G such that $g_{j,\nu} = f_{j,\nu-N-\nu_0}$ for $\nu>N+\nu_0$ and $g_{j,\nu} = f_{j,1}$ for $\nu = 1,2,\dots,N+\nu_0$. Then for the jet $\{f_j\colon |j|\le k-1\}$ and the corresponding sequence of jets $\{g_{j,\nu}\colon |j|\le k\}$, $\nu\in\mathbb{N}$, conditions (3.15) - (3.18) hold with a constant M depending on k, n, M_0 and ν_0 only. The validity of the conditions (3.15) - (3.18) at the inner points of the domain G implies that there exist extensions by the continuity to the closure $\bar{G}$ of both the functions from the jet $\{f_j\colon |j|\le k-1\}$ and of the functions from the sequence of the jets $\{f_{j,\nu}\colon |j|\le k\}$, $\nu\in\mathbb{N}$. Therefore, the extended jets belong to the space $\Lambda^k(G)$.

Thus, we proved the following assertion which is due to Yu. Bojarskii and the author.

THEOREM 3.3. *Let* G *be a domain in* $\mathbb{R}^n$ *and let the condition* (3.20) *hold. Then for any* $k\in\mathbb{N}$ *there exists a bounded extension operator*

$$\mathrm{Ext}_k\colon \Lambda^k(G)\longrightarrow H^k_\infty(R^n).$$

The idea of the geometric approach to the spaces of Hölder functions defined on a domain G can be also applied to spaces with the integral metric.

Let ℓ be a real number, $\ell\in(k,k+1)$, $k = 0,1,\dots$, $\alpha = \ell-k$ and $1 \le p,q \le \infty$. We say that a function f belongs to $\Lambda^\ell_{p,q}(G)$ if f has generalized derivatives of the order k and the finite norm

$$\|f|\Lambda^\ell_{p,q}(G)\| = \sum_{|j|\le k} \|D^jf|L_p(G)\| + \sum_{|j|\le k}\left\{\sum_{\nu=0}^{\infty} 2^{\nu(\ell-|j|)q}\left[\iint\limits_{\substack{x,y\in G\\ d_{\alpha,G}(x,y)\le 2^{-\nu}}} \frac{|R_j(x,y)|^p}{d_\alpha(x,y)^n}\,dxdy\right]^{q/p}\right\}^{1/q}, \tag{3.22}$$

where the remainders $R_j(x,y)$ are defined in Section 1.1. For $q = \infty$ the norm is modified in the usual way. We note that $\Lambda^{\ell}_{\infty,\infty}(G) = \mathrm{Lip}(\ell, G_\alpha)$. If G is an (ε,δ)-domain [13] then the space $\Lambda^{\ell}_{p,q}(G)$ coincides with the Besov space $B^{\ell}_{p,q}(G)$ (see the theorem below).

THEOREM 3.4. *Let G be a domain in* $\mathbb{R}^n$ *such that the following conditions hold:*

(i) $\bar{d}_{\alpha,G}(x,y) \underset{\mathrm{loc}}{\sim} |x-y|$;

(ii) *there exist numbers* $c, \rho_0 > 0$ *such that*

$$|B(x,\rho) \cap G| \geq c\ |B(x,\rho)|,\ x \in G,\ 0<\rho<\rho_0.$$

Then there exists a bounded linear extension operator

$$\mathrm{ext}_\ell : \Lambda^{\ell}_{p,q}(G) \longrightarrow B^{\ell}_{p,q}(R^n). \tag{3.23}$$

In particular, if G is an (ε,δ)*-domain, then the spaces* $\Lambda^{\ell}_{p,q}(G)$ *and* $B^{\ell}_{p,q}(G)$ *coincide.*

P r o o f. We indicate the main steps of the proof. The condition (i) of the theorem permits to pass from the norm (3.22) to a similar one with $|x-y|$ in place of $\bar{d}_{\alpha,G}(x,y)$. The condition (ii) implies that the measure of the boundary ∂G is equal to 0 and so at the end we pass to the space $\Lambda^{\ell}_{p,q}(\bar{G})$. In [14] the Whitney type extension theorem is proved establishing the existence of the extension operator $\mathrm{ext}_\ell : \Lambda^{\ell}_{p,q}(G) \longrightarrow B^{\ell}_{p,q}(\mathbb{R}^n)$. The corresponding diagram has the form

$$\Lambda^{\ell}_{p,q}(G) \longrightarrow \Lambda^{\ell}_{p,q}(\bar{G}) \xrightarrow{\overline{\mathrm{ext}_\ell}} B^{\ell}_{p,q}(\mathbb{R}^n).$$

The superposition of the operators in the diagram is the desired extension operator.

To prove the last assertion we note that an (ε,δ)-domain satisfies both conditions (i) [15] and (ii) [16]. Therefore there exists not only the extension operator (3.23) but also the extension operator $\mathrm{ext}: B^{\ell}_{p,q}(G) \longrightarrow B^{\ell}_{p,q}(\mathbb{R}^n)$ [16]. Thus, if both the restriction operators $\mathrm{rest}: B^{\ell}_{p,q}(\mathbb{R}^n) \longrightarrow \Lambda^{\ell}_{p,q}(G)$ and $\mathrm{rest}: B^{\ell}_{p,q}(\mathbb{R}^n) \longrightarrow B^{\ell}_{p,q}(G)$ are bounded, then the spaces $\Lambda^{\ell}_{p,q}(G)$ and $B^{\ell}_{p,q}(G)$ coincide. □

4. Necessary extension conditions for differentiable functions

4.1. Metric extension conditions

THEOREM 4.1 [9]. *If there exists an extension operator*

$$ext_1: W_p^1(G) \longrightarrow W_p^1(\mathbb{R}^n), \quad p>n \ (\alpha = 1)$$

or

$$ext_\ell: H_p^\ell(G) \longrightarrow H_p^\ell(\mathbb{R}^n), \quad \ell p>n \ (\alpha = \ell < 1),$$

then the inner α-metric $d_{\alpha,G}(x,y)$ *is locally equivalent to the α-metric* $d_\alpha(x,y) = |x-y|^\alpha$ *in the domain* G *and*

$$\|ext_\alpha\| \geq K M_\alpha^{\ell-n/p}$$

where K *does not depend on the domain* G.

In the case $p = \infty$, the related topics can be found in [12, 15].

In the same way we prove the following

PROPOSITION 4.1. Let G *be a domain in* $\mathbb{R}^n$. *If there exists an extension operator*

$$ext: H_p^1(G) \longrightarrow H_p^1(\mathbb{R}^n), \quad n<p\leq\infty,$$

then the inner metric $d_{1,G}(x,y)$ *is locally equivalent to the Euclidean metric* $|x-y|$ *in* G.

4.2. Regularity condition

It is well known that the space $W_p^\ell(\mathbb{R}^n)$ is embedded not only into the space of continuous functions but also into the Hölder space. In what follows we need only two embeddings:

$$(4.1) \qquad i: W_p^\ell(\mathbb{R}^n) \longrightarrow H_\infty^{\ell-n/p}(\mathbb{R}^n), \qquad i: B_{p,q}^\ell \longrightarrow H_\infty^{\ell-n/p}(\mathbb{R}^n).$$

With the help of the embeddings (4.1) we prove

PROPOSITION 4.2. Let $x,y\in\mathbb{R}^n$ *and let* $\ell p>n$. *Let* $f\in W_p^\ell(\mathbb{R}^n)$ *be a function such that* $f(x) = 1$, $f(y) = 0$. *Moreover, if* $\ell-n/p > 1$, *let* f *be equal to* 1 *in some neighbourhood of the point* x. *Then the estimate*

$$\frac{1}{|x-y|^{\ell-n/p}} \leq \gamma \|f \,|\, W_p^\ell(\mathbb{R}^n)\|$$

is valid, where the constant γ is independent of x *and* f. *The analogous result holds for functions* $f\in B^{\ell}_{p,q}(\mathbb{R}^n)$.

P r o o f. The boundedness of the embeddings (4.1) yields the estimate

$$\|f|H_{\infty}^{\ell-n/p}(\mathbb{R}^n)\| \le \|1\|\ \|f|W_p^{\ell}(\mathbb{R}^n)\|. \tag{4.2}$$

If $\ell-n/p$ is a fractional number, then

$$\frac{1}{|x-y|^{\ell-n/p}} \le \frac{|R_0(x,y)|}{|x-y|^{\ell-n/p}} \le \|f|H_{\infty}^{\ell-n/p}(\mathbb{R}^n)\|,$$

since $|R_0(x,y)| = |f(y) - f(x)| = 1$ by the assumption.

If $\ell-n/p = m+1$ is an integer then the left hand side of (4.2) is estimated from below by

$$\frac{1}{|x-y|^{m+1}} \le \sup_{\substack{h,z\in\mathbb{R}^n\\ h\ne 0}} \frac{|\Delta_h^2\nabla_m g(z)|}{|h|} \le \|f|H_{\infty}^{m+1}(\mathbb{R}^n)\|,$$

where $g(z) = f((x-z)/|x-z|)$. Let us note that

$$\inf_{g} \sup_{\substack{h,z\in\mathbb{R}^n\\ h\ne 0}} \frac{|\Delta_h^2\nabla_m g(z)|}{|h|} \ge \beta > 0,$$

where the infimum is taken over all functions equal to 1 in some neighbourhood of the origin and vanishing at some point of the unit sphere $S(0,1)$. The proof is finished. □

A domain G in $\mathbb{R}^n$ is said to be *regular at a point* $x\in G$ if there exist positive constants $\delta = \delta(x)$ and $\gamma = \gamma(x)$ such that for any ball $B(x,r)$, $0<r<\delta(x)$, the inequality

$$|B(x,r)\cap G| \ge \gamma|B(x,r)| \tag{4.3}$$

holds. The domain G is said to be *regular* if it is regular at every point $x\in G$.

<u>*THEOREM 4.2.*</u> *Let* G *be a domain in* $\mathbb{R}^n$ *and let* $p>n$. *If the inequality*

$$\sup_{\substack{u,v\in G,\\ u\ne v}} \frac{|f(u)-f(v)|}{|u-v|^{1-n/p}} \le c\|f|W_p^1(G)\| \tag{4.4}$$

holds for every function $f\in W_p^1(G)$, *then the domain* G *is regular.*

P r o o f. Let $\varphi:\mathbb{R}\longrightarrow\mathbb{R}^n$ be a smooth function supported in the unit ball, and such that $0\le\varphi(x)\le 1$ for every $x\in\mathbb{R}^n$ and φ is equal to 1 in some neighbourhood of the origin. Let us fix points $x\in\overline{G}$, $y\in G$ and put $r = |x-y|$ and

$f(z) = \varphi((z-x)/r)$. Then $f(z) = 1$ in a neighbourhood of x, $f(y) = 0$ and $|\nabla f| \le a/r$. The inequality (4.4) yields

$$\frac{|B(x,r)|^{1/p}}{r} = \frac{\gamma}{|x-y|^{1-n/p}} \le c(\|f|L_p(G)\| + ar^{-1}\|\chi_{B(x,r)}|L_p(G)\|)$$

$$\le c(|B(x,r)\cap G|^{1/p} + ar^{-1}|B(x,r)\cap G|^{1/p}).$$

Comparing the first and the last term we obtain

$$|B(x,r)\cap G| \ge c^{-p}(a+r)^{-p}|B(x,r)|.$$

Thus for the domain G the inequality (4.3) is satisfied with the constants $\gamma = c^{-p}(a+1)^{-p}$ and $\delta=1$ independent of x. □

The last theorem has a generalization for the space W_p^{ℓ}. The formulation and the method of the proof is obvious from the following assertion.

THEOREM 4.3. Let G be a domain in $\mathbb{R}^n$ and let $n/\ell < p < \infty$. If there exists the bounded extension operator $ext: W_p^{\ell}(G) \longrightarrow W_p^{\ell}(\mathbb{R}^n)$, then the domain G is regular.

P r o o f. Let $x\in\bar{G}$, $y\in G$ and let f be the function from the previous proof corresponding to points x and y. Then $f(z) = 1$ in some neighbourhood of x, $f(y) = 0$ and $|\nabla_{\ell} f| \le b/r^{\ell}$. Using the inequality

$$\|ext(f)|W_p^{\ell}(\mathbb{R}^n)\| \le \|ext\| \; \|f|W_p^{\ell}(G)\|$$

and Proposition 4.2, we obtain

$$\frac{|B(x,r)|^{1/p}}{r^{\ell}} = \frac{\gamma}{|x-y|^{\ell-n/p}} \le c\|ext(f)|W_p^{\ell}(\mathbb{R}^n)\|$$

$$\le c\|ext\|[|B(x,r)\cap G|^{1/p} + br^{-\ell}|B(x,r)\cap G|^{1/p}].$$

It follows that

$$|B(x,r)\cap G| \ge c^{-p}(b+r^{\ell})^{-p}|B(x,r)|.$$

Thus, we have proved the regularity of the domain G, namely (4.3) holds with the constants $\gamma = c^{-p}(b+1)^{-p}$ and $\delta = 1$ independent of the point $x\in\bar{G}$. □

An analogous result is valid for the Nikol'skii-Besov spaces.

THEOREM 4.4. Let G be a domain in $\mathbb{R}^n$. Suppose that $\ell p>n$ if $1\le p<\infty$, $1\le\theta\le\infty$, and $\ell p\ge n$ if $1\le p<\infty$, $\theta = 1$. If there exists a bounded extension operator $ext: B_{p,\theta}^{\ell}(G) \longrightarrow B_{p,\theta}^{\ell}(\mathbb{R}^n)$, then the domain G is regular.

REFERENCES

[1] I. M. Tumakov: *Henri Leon Lebesgue* (1875-1941) (Russian). Nauka, Moscow, 1975.

[2] H. Whitney: *Analytic extension of differentiable functions defined in closed sets*. Trans. Amer. Math. Soc. **36** (1934), 63-89.

[3] A. D. Aleksandrov: *Inner geometry of convex surfaces* (Russian). Gostekhizdat, Moscow, 1948. (German translation of the enlarged version: Die innere Geometrie der konvexen Flächen. Akademie-Verlag, Berlin, 1955.)

[4] S. L. Sobolev: *Some applications of functional analysis in mathematical physics*. (Russian) Izdat. Leningrad. Gos. Univ., Leningrad, 1950. (English translation: Amer. Math. Soc., Providence, R. I., 1963.)

[5] S. M. Nikol'skii: *Approximation of functions of several variables and imbedding theorems* (Russian). Nauka, Moscow, 1975. (English translation: Springer Verlag, Berlin - Heidelberg - New York, 1975.)

[6] O. V. Besov, V. P. Il'in, S. M. Nikol'skii: *Integral representation of functions and embedding theorems* (Russian). Nauka, Moscow, 1975. (English translation: Vol. 1-2, John Wiley & Sons, New York, 1979.)

[7] G. H. Hardy, J. E. Littlewood, G. Polya: *Inequalities*. Cambridge University Press, Cambridge, 1934.

[8] A. Jonsson, H. Wallin: *The traces to closed sets of functions in* $\mathbb{R}^n$ *with second difference of order* $O(h)$. J. Approx. Theory **26** (1979), 159-184.

[9] S. K. Vodop'yanov: *Equivalent normalizations of Sobolev and Nikol'skii spaces in domains. Boundary values and extension*. In: Function spaces and applications. Proc. US-Swedish Seminar, Lund 1986. Lecture Notes in Mathematics **1302**, Springer, Berlin - New York, 1988, 397-409.

[10] E. M. Stein: *Singular integrals and differentiability properties of functions*. Princeton University Press, Princeton, 1970.

[11] Yu. I. Bojarskii, S. K. Vodop'yanov: *Boundary values of functions of Zygmund classes* (Russian). Proc. School on Approx. Theory. Lutsk, September 1989. Kiev, 1989.

[12] V. N. Konovalov: *Descriptions of traces of certain classes of functions of several variables* (Russian). Akad. Nauk Ukrain. SSR Inst. Mat. Preprint 1984, no. 21, 64 pp.

[13] P. W. Jones: *Quasiconformal mappings and extendibility of functions in Sobolev spaces*. Acta Math. **147** (1981), 71-88.

[14] A. Jonsson, H. Wallin: *Function spaces on subsets of* $\mathbb{R}^n$. Math. Rep. **2** (1984), no. 1, Harwood Academic Publ., Chur, 1984.

[15] F. W. Gehring, O. Martio: *Quasiexternal distance domains and extension of quasiconformal mappings*. J. Analyse Math. **45** (1985), 181-206.

[16] P. A. Shvartsman: *Local approximations of functions and embedding theorems* (Russian). Preprint, Yaroslavl', 1983, Manuscript No. 2025-83, deponed VINITI, 1983.

L I S T O F P A R T I C I P A N T S

Abbreviations:

FSU Jena	= Sektion Mathematik, Friedrich-Schiller-Universität, UHH, 6900 Jena
KWI Berlin	= Karl-Weierstrass-Institut für Mathematik, AdW der DDR, Mohrenstr. 39, 1086 Berlin
MFF UK Praha	= Faculty of Mathematics and Physics, Charles University, Sokolovská 83, 186 00 Praha 8
MI Praha	= Mathematical Institute, Czechoslovak Academy of Sciences, Žitná 25, 115 67 Praha 1
MI Tbilisi	= Mathematical Institute, Academy of Sciences of Georgian SSR, Rukhadze 1, 380093 Tbilisi
TU Plzeň	= Department of Mathematics, Technical University of Plzeň, Nejedlého sady 14, 306 14 Plzeň
Univ. Malaga	= Departamento de Matemáticas, Facultad de Ciencias, Universidad de Malaga, Campus de Teatinos, 29071 Malaga
Univ. Luleå	= Deparment of Mathematics, Luleå University, S-95187 Luleå
Univ. Oulu	= Department of Mathematics, Faculty of Science, University of Oulu, Linnanmaa, SF-90570 Oulu

Argentina

Macías Roberto	INTEC, Güemes 3450, 3000 Santa Fe

Australia

Love Eric Russell	Department of Mathematics, University of Melbourne, Parkville, Vic 3052
Love Elvie	

Belgium

Carton-Lebrun Christiane	Department of Mathematics, University of Mons, B-7000 Mons

Canada

Gabardo Jean-Pierre	Department of Mathematics and Statistics, McMaster University, 1280 Main Street West, Hamilton, Ontario L8S 4K1
Heinig Hans Paul	Department of Mathematics and Statistics, McMaster University, 1280 Main Street West, Hamilton, Ontario L8S 4K1
Sinnamon Gordon	Department of Mathematics, University of Western Ontario London, Ontario N6A 5B7

Czechoslovakia

Bock Igor	Slovak Technical University, Faculty of Electro-engineering, Department of Mathematics, Mlynská dolina, 812 19 Bratislava
Daněček Josef	Technical University of Brno, Barvičova 85, 662 37 Brno
Doktor Pavel	MFF UK Praha
Dolanský Petr	TU Plzeň
Drábek Pavel	TU Plzeň
Franců Jan	Březová 32, 637 00 Brno
Gurka Petr	Štětínská 341, 181 00 Praha 8
Jarušek Jiří	ÚTIA ČSAV, Pod vodárenskou věží 4, 182 08 Praha 8
John Oldřich	MFF UK Praha
Kodnár Rudolf	Institute of Applied Mathematics, Komenský University, Mlynská dolina, 842 15 Bratislava
Konečný Milan	739 44 Brušperk 500
Kottas Jiří	MFF UK Praha
Kováčik Ondrej	Zvolenská 15, 010 08 Žilina
Krbec Miroslav	MI Praha

Kufner Alois	MI Praha
Lovíšek Ján	Sokolská 4, 811 04 Bratislava
Marko Ľubomír	Institute of Applied Mathematics, Komenský University, Mlynská dolina, 842 15 Bratislava
Mošová Vratislava	Marxova 14, 682 01 Vyškov
Nečas Jindřich	Mathematical Institute of the Charles University, Sokolovská 83, 186 00 Praha 8
Opic Bohumír	MI Praha
Pick Luboš	MI Praha
Rákosník Jiří	MI Praha
Rudolf Boris	Slovak Technical University, Faculty of Electro-engineering, Department of Mathematics, Mlynská dolina, 812 19 Bratislava
Stará Jana	MFF UK Praha
Tomiczek Petr	TU Plzeň
Finland	
Berkovits Juha	Univ. Oulu
Kittila Arto	Univ. Oulu
Mustonen Vesa	Univ. Oulu
Nieminen Esko	Department of Mathematics, University of Jyväskylä, Seminaarinkatu 15, SF-40100 Jyväskylä
Federal Republic of Germany	
Farwig Reinhard	Fachbereich 17 Mathematik, Universität-GH Padeborn, Warburger Str. 100, 4790 Padeborn
Simader Christian G.	Universität Bayreuth, Postfach 101251, 8580 Bayreuth
German Democratic Republic	
Albinus Günter	KWI Berlin
Geisler Michael	FSU Jena
Gröger Konrad	KWI Berlin
Leopold Hans-Gerd	FSU Jena
Malarski Mircea	FSU Jena
Rempel Stephan	KWI Berlin
Schmeisser Hans-Jürgen	FSU Jena
Sickel Winfried	FSU Jena
Zacharias Klaus	KWI Berlin
Hungary	
Bognár Gabriella	Mathematical Institute, Mechanical University of Heavy Industries, 3515 Miskolc-Egyetemváros
Ladányi Márta	Department of Mathematics and Computer Science, University of Horticulture & Feed Industry, Villányi út. 29, H-1114 Budapest
Weisz Ferenc	Department of Numerical Analysis, Eötvös Loránd University, Bogdánfy n. 10/B, H-1117 Budapest
Japan	
Sato Shuichi	Department of Mathematics, Faculty of Education, Kanazawa University, Kanazawa, 920
Poland	
Banasiak Jacek	Instytut Matematyki Politechniki Lódzkiej, ul. Zwirki 36, Lódz
Gawinecki Jerzy	Department of Mathematics, Military Technical Academy, ul. S. Kaliskiego, 00-908 Warszawa
Jaroszewska Magdalena	Institute of Mathematics, Adam Mickiewicz University, Matejki 48/49, 60-769 Poznań
Liskowski Marian	Institute of Mathematics, Technical University of Poznań, 60-965 Poznań
Mikosz Miroslawa	Institute of Mathematics, Adam Mickiewicz University, Matejki 48/49, 60-769 Poznań

Soviet Union

Ayrian Edik	Yerevan Physics Institute, Alikhanian Brothers St. 2, 375036 Yerevan
Gabidzashvili Merab	Gldanis Chevi, corp. IV, kv. 80, 380053 Tbilisi
Genebashvili Josef	MI Tbilisi
Gogatishvili Amiran	MI Tbilisi
Gol'dman Michail L.	Krupskoj 8, corp. 1, kv. 187, 117311 Moscow
Hanin Leonid G.	Institutskij prospect 29, kv. 45, 194021 Leningrad
Kazarian Kazaros S.	Mathematical Institute, Academy od Sciences of Armenian SSR, 375019 Yerevan
Kokilashvili Vachtang	MI Tbilisi
Ktitarev Dmitrij V.	Joint Institute for Nuclear Research, Head Post Office, P. O. Box 79, 101000 Moscow
Puzyrev Vladimir	Institute of Applied Mathematics and Mechanics, Academy of Sciences of Ukrainian SSR, R. Luxemburg 74, 340114 Donetsk
Rabinovich Ilya	Yerevan Physics Institute, Alikhanian Brothers St. 2, 375036 Yerevan
Tararykova Tamara V.	Yuzhno-Moravskaya St. 40-43, 394086 Voronezh
Vodop'yanov Sergej K.	Novosibirsk State University, 630 090 Novosibirsk

Spain

Atencia Enrique	Departamento de Matemáticas, Facultad de Ciencias, Campus de Teatinos, 29080 Malaga
Blasco Oscar	Departamento de Matemáticas, Universidad de Zaragoza, 50009 Zaragoza
Gallardo Diego	Univ. Malaga
García-Cuerva José	Departamento de Matemáticas, C-XVI-Universidad Autonoma de Madrid, 28049 Madrid
Martin Reyes Francisco J.	Univ. Malaga
Ortega Salvador Pedro	Departamento de Estadística y Econometría, Facultad de Ciencias Económicas, Universidad de Malaga, 29013 Malaga
de la Torre Alberto	Univ. Malaga

Sweden

Koskela Pekka	Mittag-Leffler Institute, Auravägen 17, S-18262 Djursholm
Lindström Torsten	Univ. Luleå
Nilsson Lena	
Persson Lars	Univ. Luleå
Strömberg Thomas	Univ. Luleå
Zhang Genkai	Department of Mathematics, University of Stockholm, Box 6701, S-11385 Stockholm

United Kingdom

Roach Gary F.	Department of Mathematics, University of Strathclyde, Livingstone Tower, 26 Richmond Street, Glasgow G1 1XH

USA

Brown Phyllis Humphrey	
Brown Richard Clark	Department of Mathematics, University of Alabama, Tuscaloosa, Ala. 35487-0350

Vietnam

Yen Nguyen Dong	Institute of Mathematics, P. O. Box 631 BO HO, Hanoi

Yugoslavia

Ostrogorski Tatjana	Matematički Institut, Knez Mihailova 35, 11000 Beograd

Diese Reihe wurde geschaffen, um eine schnellere Veröffentlichung mathematischer Forschungsergebnisse und eine weitere Verbreitung von mathematischen Spezialvorlesungen zu erreichen. TEUBNER-TEXTE werden in deutsch, englisch, russisch oder französisch erscheinen. Um Aktualität der Reihe zu erhalten, werden die TEUBNER-TEXTE im Manuskriptdruck hergestellt, da so die geringeren drucktechnischen Ansprüche eine raschere Herstellung ermöglichen. Autoren von TEUBNER-TEXTEN liefern an den Verlag ein reproduktionsfähiges Manuskript. Nähere Auskünfte darüber erhalten die Autoren vom Verlag.

This series has been initiated with a view to quicker publication of the results of mathematical research-work and a widespread circulation of special lectures on mathematics. TEUBNER-TEXTE will be published in German, English, Russian or French. In order to keep this series constantly up to date and to assure a quick distribution, the copies of these texts are produced by a photographic process (small-offset printing) because its technical simplicity is ideally suited to this type of publication. Authors supply the publishers with a manuscript ready for reproduction in accordance with the latter's instructions.

Cette série des textes a été créée pour obtenir une publication plus rapide de résultats de recherches mathématiques et de conférences sur des problèmes mathématiques spéciaux. Les TEUBNER-TEXTE seront publiés en langues allemande, anglaise, russe ou française. L'actualité des TEUBNER-TEXTE est assurée par un procédé photographique (impression offset). Les auteurs des TEUBNER-TEXTE sont priés de fournir à notre maison d'édition un manuscrit prêt à être reproduit. Des renseignements plus précis sur la form du manuscrit leur sont donnés par notre maison.

Эта серия была создана для обеспечения более быстрого публикования результатов математических исследований и более широкого распространения математических лекций на специальные темы. Издания серии ТОЙБНЕР-ТЕКСТЕ будут публиковаться на немецком, английском, русском или французском языках. Для обеспечения актуальности серии, ее издания будут изготовляться фотомеханическом способом. Таким образом, более скромные требования к полиграфическому оформлению обеспечат более быстрое появление в свет. Авторы изданий серии ТОЙБНЕР-ТЕКСТЕ будут предоставлять издательству рукописи, удовлетворящие требованиям фотомеханического печатания. Более подробные сведения авторы получат от издательства.

B. G. Teubner Verlagsgesellschaft KG, Leipzig
O - 7010 Leipzig, Postfach 930